METHODS IN MOLECULAR BIOLOGY

Series Editor
John M. Walker
School of Life and Medical Sciences
University of Hertfordshire
Hatfield, Hertfordshire, AL10 9AB, UK

For further volumes:
http://www.springer.com/series/7651

Yeast Cytokinesis

Methods and Protocols

Edited by

Alberto Sanchez-Diaz

Instituto de Biomedicina y Biotecnología de Cantabria, CSIC, Universidad de Cantabria, Santander, Spain; Departamento de Biología Molecular, Facultad de Medicina, Universidad de Cantabria, Santander, Spain

Pilar Perez

Instituto de Biología Funcional y Genómica, CSIC, Universidad de Salamanca, Salamanca, Spain

Humana Press

Editors
Alberto Sanchez-Diaz
Instituto de Biomedicina y Biotecnología de Cantabria
CSIC, Universidad de Cantabria
Santander, Spain

Departamento de Biología Molecular
Facultad de Medicina, Universidad de Cantabria
Santander, Spain

Pilar Perez
Instituto de Biología Funcional y Genómica
CSIC, Universidad de Salamanca
Salamanca, Spain

ISSN 1064-3745 ISSN 1940-6029 (electronic)
Methods in Molecular Biology
ISBN 978-1-4939-3144-6 ISBN 978-1-4939-3145-3 (eBook)
DOI 10.1007/978-1-4939-3145-3

Library of Congress Control Number: 2015947262

Springer New York Heidelberg Dordrecht London

Printed on acid-free paper

Humana Press is a brand of Springer
Springer Science+Business Media LLC New York is part of Springer Science+Business Media (www.springer.com)

Preface

The final stage of the eukaryotic cell cycle is cytokinesis, when a mother cell is separated into two cells by a contractile actomyosin ring. This contractile ring is normally assembled in the middle of the cell and is linked to the plasma membrane such that, when it constricts, it creates a cleavage furrow that partitions the cell in two. The process is conserved between fungi and animals. Defects in the placement, timing, assembly, and contraction of the actomyosin ring can influence processes such as ploidy, cell differentiation, establishment of tissue architecture, and cell morphogenesis. Understanding the mechanisms underlying cytokinesis is therefore crucial for our understanding of cell proliferation, differentiation, and disease.

Besides actin and myosin, a large number of actomyosin ring components have been identified, most of which are conserved between yeasts and human. For that reason, budding and fission yeasts have been along the years ideal model systems to study cell division, allowing a detailed molecular understanding of the spatial mechanisms regulating assembly and dynamics of the cytokinetic ring. In this context, we present this book of protocols on yeast cytokinesis studies that will help Molecular and Cellular Biology researchers in the use of these microorganisms to approach the study of general or specific key questions in cytokinesis. In this thematic issue we have collected a number of methods which cover different aspect of yeast cytokinesis. The aim of the book is to provide practical and step-by-step detailed protocols useful for a wide audience ranging from experienced researchers to beginners in the use of yeasts.

The first six chapters provide an up-to-date description of the fluorescence microscopy, which provides a powerful tool to visualize cellular and subcellular processes in live yeast cells. Each chapter describes a specific technique or protocol, including the novel high-speed fluorescence photoactivation localization microscopy (FPALM). These protocols will enable researchers to study in great detail distinct cytokinesis steps and dynamic behavior of proteins in time and space.

The following three chapters address the use of Transmission Electron microscopy, an invaluable tool for ultrastructural examination of yeast cells during cytokinesis. The last of these chapters focuses on electron tomography of freeze-substituted section and cryo-electron tomography of frozen sections to determine the organization of the septin cytoskeleton.

Next chapters concentrate on essential methodologies that will allow the characterization of the main players in the cytokinesis process such as methods for the isolation and partial purification of the actomyosin ring (Chapter 10); measurements of myosin-II motor activity (Chapter 11); characterization of the biochemical and biophysical properties of purified actin binding proteins (Chapter 12); of F-BAR proteins (Chapter 13); and of exocyst components (Chapter 14). Following chapters describe protocols necessary for the mechanistic understanding of the cytokinesis process such as analysis of Rho-GTPase activity (Chapter 15); the phosphorylation profile of cytokinetic proteins (Chapter 16); the architecture of cytokinetic protein complexes (Chapter 17); and the generation of degron conditional mutants for screening of suppressor genes (Chapter 18).

The book ends with a series of general methods such as the efficient synchronization of yeast cells (Chapters 19 and 20), the selection of fluorescence proteins and image analysis used in yeast (Chapters 21 and 22), and the tools for protein tertiary structure prediction and structure-function studies (Chapters 23 and 24). This set of methods and techniques is not only useful in the study of cytokinesis but should be useful in a wide range of molecular and cellular biology studies using yeast as model organisms.

The editors wish to thank the authors of all the chapters whose commitment and contributions made this volume possible. We also thank John Walker for providing expert editorial advice and professional guidance during all the process of editing this volume.

Santander, Spain ***Alberto Sanchez-Diaz***
Salamanca, Spain ***Pilar Perez***

Contents

Contributors

IRENE ARCONES • *Department of Microbiology and Genetics, Instituto de Biología Funcional y Genómica, CSIC/USAL, Salamanca, Spain*

STEVE BAGLEY • *Imaging and Cytometry, Cancer Research UK Manchester Institute, University of Manchester, Manchester, UK*

MOHAN K. BALASUBRAMANIAN • *Division of Biomedical Cell Biology, Warwick Medical School, University of Warwick, Coventry, UK*

YVES BARRAL • *ETH Zürich, Institute of Biochemistry, Zürich, Switzerland*

AURÉLIE BERTIN • *Biochemistry, Biophysics and Structural Biology Division, Department of Molecular and Cell Biology, University of California, Berkeley, CA, USA; Institut Curie, CNRS UMR 168, Paris, France*

JOERG BEWERSDORF • *Department of Molecular Biophysics and Biochemistry, Yale University, New Haven, CT, USA; Department of Biomedical Engineering, Yale University, New Haven, CT, USA*

MAJA BIALECKA-FORNAL • *Department of Developmental and Cell Biology, Center for Complex Biological Systems, University of California, Irvine, CA, USA*

ALESSIO BOLOGNESI • *ETH Zürich, Institute of Biochemistry, Zürich, Switzerland*

TING GANG CHEW • *Division of Biomedical Cell Biology, Warwick Medical School, University of Warwick, Coventry, UK*

JUAN CARLOS G. CORTÉS • *Instituto de Biología Funcional y Genómica, Consejo Superior de Investigaciones Científicas/Universidad de Salamanca, Salamanca, Spain*

RESHMA DAVIDSON • *Department of Molecular Genetics, The Ohio State University, Columbus, OH, USA*

ASLI DEVREKANLI • *Department of Molecular Biology and Genetics, Canik Basari University, Samsun, Turkey*

GANG DONG • *Max F. Perutz Laboratories (MFPL), Department of Medical Biochemistry, Vienna Biocenter (VBC), Medical University of Vienna, Vienna, Austria*

MAGDALENA FOLTMAN • *Instituto de Biomedicina y Biotecnología de Cantabria, Consejo Superior de Investigaciones Científicas/Universidad de Cantabria, Santander, Spain; Departamento de Biología Molecular, Facultad de Medicina, Universidad de Cantabria, Facultad de Medicina, Santander, Spain*

KENNETH S. GERIEN • *Department of Molecular Genetics, The Ohio State University, Columbus, OH, USA*

KATHLEEN L. GOULD • *Department of Cell and Developmental Biology, Vanderbilt University School of Medicine, Nashville, TN, USA*

FANG HUANG • *Department of Molecular Biophysics and Biochemistry, Yale University, New Haven, CT, USA*

JUNQI HUANG • *Division of Biomedical Cell Biology, Warwick Medical School, University of Warwick, Coventry, UK*

MASATO T. KANEMAKI • *Center of Frontier Research, National Institute of Genetics, Research Organization of Information and Systems, Mishima, Shizuoka, Japan;*

Department of Genetics, SOKENDAI, Mishima, Shizuoka, Japan; JST, PRESTO, Kawaguchi, Saitama, Japan
DAVID R. KOVAR • *Department of Molecular Genetics and Cell Biology, The University of Chicago, Chicago, IL, USA; Department of Biochemistry and Molecular Biology, The University of Chicago, Chicago, IL, USA*
ARUN KUMAR • *Centre for Genomic Regulation (CRG), Barcelona, Spain; Universitat Pompeu Fabra (UPF), Barcelona, Spain*
CAROLINE LAPLANTE • *Department of Molecular Cellular and Developmental Biology, Yale University, New Haven, CT, USA*
JOHANNES LESIGANG • *Max F. Perutz Laboratories (MFPL), Department of Medical Biochemistry, Vienna Biocenter (VBC), Medical University of Vienna, Vienna, Austria*
YAJUN LIU • *Department of Molecular Genetics, The Ohio State University, Columbus, OH, USA*
MATTHEW LORD • *Department of Molecular Physiology and Biophysics, University of Vermont, Burlington, VT, USA*
TATYANA MAKUSHOK • *Department of Biochemistry and Biophysics, University of California, San Francisco, San Francisco, CA, USA*
NATHAN A. MCDONALD • *Department of Cell and Developmental Biology, Vanderbilt University School of Medicine, Nashville, TN, USA*
LIAM JAMES MCGUFFIN • *School of Biological Sciences, University of Reading, Reading, UK*
FRANZ MEITINGER • *DKFZ-ZMBH Alliance, German Cancer Research Center (DKFZ), University of Heidelberg, Heidelberg, Germany; Ludwig Institute for Cancer Research, La Jolla, CA, USA*
MANUEL MENDOZA • *Centre for Genomic Regulation (CRG), Barcelona, Spain; Universitat Pompeu Fabra (UPF), Barcelona, Spain*
MITHILESH MISHRA • *Department of Biological Sciences, Tata Institute of Fundamental Research (TIFR), Mumbai, India*
IAGO MOLIST • *Instituto de Biomedicina y Biotecnología de Cantabria, Consejo Superior de Investigaciones Científicas/Universidad de Cantabria, Santander, Spain; Departamento de Biología Molecular, Facultad de Medicina, Universidad de Cantabria, Santander, Spain*
SERGIO MORENO • *Instituto de Biología Funcional y Genómica, Consejo Superior de Investigaciones Científicas/Universidad de Salamanca, Salamanca, Spain; Instituto de Investigación Biomédica de Salamanca (IBSAL), Salamanca, Spain*
ALISHA N. MORGANTHALER • *Department of Molecular Genetics and Cell Biology, The University of Chicago, Chicago, IL, USA*
EVA NOGALES • *Biochemistry, Biophysics and Structural Biology Division, Department of Molecular and Cell Biology, University of California, Berkeley, CA, USA; Howard Hughes Medical Institute, University of California, Berkeley, CA, USA; Life Science Division, Lawrence Berkeley National Laboratory, Berkeley, CA, USA*
MASAYUKI ONISHI • *Department of Genetics, Stanford University School of Medicine, Stanford, CA, USA*
SARAVANAN PALANI • *Division of Biomedical Cell Biology, Warwick Medical School, University of Warwick, Coventry, UK*
ANNE PAOLETTI • *Centre de Recherche, Institut Curie, Paris, France; CNRS UMR144, Paris, France*

GISLENE PEREIRA • *DKFZ-ZMBH Alliance, German Cancer Research Center (DKFZ), University of Heidelberg, Heidelberg, Germany*
LIVIA PÉREZ-HIDALGO • *Instituto de Biología Funcional y Genómica, Consejo Superior de Investigaciones Científicas/Universidad de Salamanca, Salamanca, Spain; Instituto de Investigación Biomédica de Salamanca (IBSAL), Salamanca, Spain*
LUTHER W. POLLARD • *Department of Molecular Physiology and Biophysics, University of Vermont, Burlington, VT, USA*
THOMAS D. POLLARD • *Department of Molecular Cellular and Developmental Biology, Yale University, New Haven, CT, USA; Department of Molecular Biophysics and Biochemistry, Yale University, New Haven, CT, USA; Department of Cell Biology, Yale University, New Haven, CT, USA*
RUPALI PRASAD • *ETH Zürich, Institute of Biochemistry, Zürich, Switzerland*
JOHN R. PRINGLE • *Department of Genetics, Stanford University School of Medicine, Stanford, CA, USA*
SUSANNE M. RAFELSKI • *Department of Developmental and Cell Biology, Center for Complex Biological Systems, University of California, Irvine, CA, USA*
JUAN CARLOS RIBAS • *Instituto de Biología Funcional y Genómica, Consejo Superior de Investigaciones Científicas/Universidad de Salamanca, Salamanca, Spain*
SERGIO A. RINCON • *Centre de Recherche, Institut Curie, Paris, France; CNRS UMR144, Paris, France*
DANIEL BARRY ROCHE • *Institut de Biologie Computationnelle, LIRMM, CNRS, Université de Montpellier, Montpellier, France; CEA, DSV, IG, Genoscope, Évry, France; CNRS-UMR8030, Évry, France; Université d'Évry Val d'Essonne, Évry, France; PRES UniverSud Paris, Saint-Aubin, France*
CESAR RONCERO • *Department of Microbiology and Genetics, Instituto de Biología Funcional y Genómica, CSIC/USAL, Salamanca, Spain*
ALBERTO SANCHEZ-DIAZ • *Instituto de Biomedicina y Biotecnología de Cantabria, Universidad de Cantabria, CSIC, Santander, Spain; Departamento de Biología Molecular, Universidad de Cantabria, Facultad de Medicina, Santander, Spain*
MATTHIAS SIPICZKI • *Department of Genetics and Applied Microbiology, University of Debrecen, Debrecen, Hungary*
ANDRZEJ SLIWA-GONZALEZ • *ETH Zürich, Institute of Biochemistry, Zürich, Switzerland*
CRISTIAN SUAREZ • *Department of Molecular Genetics and Cell Biology, The University of Chicago, Chicago, IL, USA*
QING TANG • *Department of Molecular Physiology and Biophysics, University of Vermont, Burlington, VT, USA*
MARTA TORMOS-PÉREZ • *Instituto de Biología Funcional y Genómica, Consejo Superior de Investigaciones Científicas/Universidad de Salamanca, Salamanca, Spain; Instituto de Investigación Biomédica de Salamanca (IBSAL), Salamanca, Spain*
JIAN-QIU WU • *Department of Molecular Genetics, The Ohio State University, Columbus, OH, USA; Department of Molecular and Cellular Biochemistry, The Ohio State University, Columbus, OH, USA*
DENNIS ZIMMERMANN • *Department of Molecular Genetics and Cell Biology, The University of Chicago, Chicago, IL, USA*

Chapter 1

Time-Lapse Fluorescence Microscopy of Budding Yeast Cells

Arun Kumar and Manuel Mendoza

Abstract

The discovery of green fluorescent protein (GFP) allowed visualization of a wide variety of processes within living cells. Thanks to the development of differently colored fluorophores, it is now possible to simultaneously follow distinct subcellular events at the single cell level. Here, we describe a basic method to visualize multiple events during cytokinesis by time-lapse fluorescence microscopy in the budding yeast *Saccharomyces cerevisiae*. In this organism, contraction of an actomyosin-based ring drives ingression of the plasma membrane at the mother–bud division site to partition the cytoplasm of the dividing cell. Simultaneous visualization of distinct cytokinesis steps in living cells, such as ring contraction and membrane ingression, will facilitate a complete understanding of the mechanisms of eukaryotic cell division.

Key words Time-lapse, Fluorescence microscopy, Budding yeast, Cytokinesis

1 Introduction

Imaging of fluorescently labeled proteins in living cells provides spatial and temporal information on their subcellular localization. This is essential to understand their role in dynamic processes such as cell division. In addition, time-lapse microscopy allows direct analysis of the correlation (or lack thereof) between distinct events at the single cell level, which could be masked in bulk population studies such as time-courses of synchronized cultures. Such correlations might reveal functional relationships, coordination of events in space and time, and feedback regulatory loops (*see* [1] for a detailed discussion of these topics).

In budding yeast cells, the site of cytokinesis (termed the bud neck) is marked by the presence of a membrane-associated, septin-based collar [2]. Cytokinesis begins at the end of mitosis when the septin collar splits into two rings, defining a compartment where actomyosin ring contraction takes place [3]. Closure of this contractile ring and septum deposition drive the overlying plasma membrane inward, in a process akin to furrow ingression in animal

Alberto Sanchez-Diaz and Pilar Perez (eds.), *Yeast Cytokinesis: Methods and Protocols*, Methods in Molecular Biology, vol. 1369, DOI 10.1007/978-1-4939-3145-3_1, © Springer Science+Business Media New York 2016

cells (*see* [4, 5] for reviews). The ring then disassembles, allowing fission of the contracted membrane, in a process termed abscission [3]. After completion of cytokinesis, cell wall degradation at the septation site allows separation of mother and daughter cells [6]. Here, we describe optimized methods to image multiple cytokinesis events in living yeast cells, namely septin ring dynamics, actomyosin ring contraction, and plasma membrane ingression at the mother–bud neck.

2 Materials

2.1 Strains, Culture Media, and Chemicals

1. Yeast strains: gene tagging with fluorescent reporters can be achieved by standard PCR-based methods [7]. A list of strains used in this chapter is provided in Table 1 (*see* **Note 1**).
2. Yeast extract peptone dextrose (YPD) plates: 2 % bacto-agar, 1 % bacto-yeast extract, 2 % bacto-peptone, 2 % dextrose, and 0.004 % adenine. Dissolve the components in distilled water by mixing with the help of a magnetic stirrer and sterilize by autoclaving. Pour in Petri plates and allow drying at room temperature for 2–3 days. Plates can be stored in sealed plastic bags at 4 °C for several months.
3. Synthetic Complete (SC) medium: 0.067 % yeast nitrogen base (YNB) without amino acids (Becton Dickson), 0.004 % adenine, 0.002 % uracil, 0.002 % tryptophan, 0.002 % histidine, 0.003 % lysine, 0.003 % leucine, 0.002 % methionine (Sigma), 2 % glucose. 10× Stock solutions of each component should be prepared separately using distilled water. Amino acids and glucose stocks should be sterilized with a 0.22 μm filter; YNB can be autoclaved. Prepare fresh 1× SC media a few days before each experiment.
4. Concanavalin A (ConA): from *Canavalia ensiformis* (jack bean) Type IV, lyophilized powder (Sigma). A stock solution (1 mg/ml) can be prepared by dissolving the lyophilized powder in PBS, pH 7.4. Filter-sterilize with a 0.22 μm filter and aliquot for storage at −20 °C for several months.
5. β-Estradiol (Sigma): A 10 mM stock solution should be prepared in 100 % ethanol and kept at −20 °C for long-term storage.

Table 1
Yeast strains used in this chapter

Myo1-mCherry, GFP-CAAX	*Mat* a *ADEGV::URA3GFP-CAAX::HIS3 MYO1-mCherry:: natNT2 leu2Δ lys2-801 ade2-101 trp1Δ63*
Cdc3-mCherry, GFP-CAAX	*Mat* a *ADEGV::URA3GFP-CAAX::HIS3 CDC3-mCherry::hphNT1 leu2Δ lys2-801 ade2-101 trp1Δ63*

A working stock solution (36 μM) diluted with ethanol can be kept for a couple of weeks and used for daily experiments.

6. Microscopy chambers: 8-well Lab-Tek (Nunc) (*see* **Note 2**).

2.2 Setup for Time-Lapse Video Microscopy

1. In principle, any good epifluorescence or fast-acquisition (spinning disk) confocal microscope can be used. One critical aspect is to strike a good balance between high sensitivity and low photo-toxicity (*see* **Note 3**). Ideally the microscope should be equipped with a z-stepper (such as a piezo motor), an automated stage for the parallel acquisition of multiple fields of view, a temperature controller (such as a closed chamber box to provide temperature-controlled environment and protection from ambient light) and a vibration-free table.

3 Methods

3.1 Sample Preparation

1. Inoculate a freshly grown yeast colony from a YPD plate in 5 ml SC medium and incubate at the desired temperature (preferably 30 °C for wild-type strains) overnight in an incubator shaker at 200 rpm (*see* **Note 4**).
2. Next day, dilute the overnight culture into 10 ml fresh SC minimal media in a 50 ml flask to an O.D. of 0.1 (or about 0.3×10^7 cells/ml) and grow for 3–4 h until mid-log phase (O.D. 0.6–0.8, or 1.8×10^7 cells/ml). If visualizing GFP-CAAX (*see* Subheading 3.2, **step 3**), add β-estradiol to 90 nM (2.5 μl/ml from a 36 μM β-estradiol working stock solution) 2 h before cell imaging.
3. Cells can be synchronized if a larger number of cells in a specific cell cycle stage is required (*see* **Note 5**). In this case, cells can be arrested in a given cell cycle stage by treatment with α-factor, hydroxyurea or nocodazole. Subsequently, cells can be released from the block and imaged at the desired cell cycle stage.
4. While cells are growing, apply 250 μl of 1 mg/ml concanavalin A in PBS to the glass surface of a microscopy chamber (Nunc Lab-Tek) for 10–20 min. Chambers are then washed two times with synthetic minimal medium (*see* **Note 6**).
5. Pellet a small volume (0.5ml) of yeast log phase culture at low speed in a bench top centrifuge ($3000 \times g$ for 2–3 min) and remove the media by aspiration. Resuspend cells into fresh SC medium to diminish cell crowding in the microscopy chamber. Generally, resuspending in 250–300 μl of fresh SC minimal media achieves the desired density ($0.3–0.9 \times 10^7$ cells/ml). Cells should be resuspended gently with the help of micropipette

and be applied directly to the Lab-Tek cover glass chamber for sample preparation.

6. Place a 50–100 μl drop of yeast culture on one concanavalin A-coated chamber. Wait for 10–15 min to allow yeast cells to settle down and attach to the glass.
7. Remove the excess of the yeast culture by gentle aspiration, and remove any unbound cells by washing two times with approximately 0.5 ml of fresh SC medium with the help of a micropipette. Finally, place 100–250 μl of fresh SC medium in the chamber (*see* **Note 7**). The cells are now ready to be imaged.

3.2 Image Acquisition

1. Ensure that the microscope room (and/or temperature chamber, if available) is pre-equilibrated at the appropriate temperature for the experiment. It is advised to adjust the temperature 20–30 min prior to imaging, as temperature fluctuations can affect the microscope stage and lead to shift of the selected field and/or focal plane during time-lapse imaging.
2. Place the chamber on the slide chamber of the microscope stage and fix it with the help of holder pins. Using bright-field illumination, find a region where yeast cells are not densely packed. This is especially important for long-term imaging (more than 2 h). If an automated stage is available, select several fields of view for parallel imaging not too far from each other, to facilitate rapid switching between the fields while maintaining the focus during the imaging session.
3. Image acquisition settings will depend on the proteins being imaged (see below and **Note 8**). To visualize the plasma membrane, we use a *GAL,10* promoter-driven GFP fused to the CAAX motif from Ras2. Upon posttranslational modification (palmitoylation), the CAAX motif directs targeting of GFP to the inner leaflet of the plasma membrane. As high levels of GFP-CAAX can result in labeling of inner cell membranes like the endoplasmic reticulum, we use a constitutively expressed chimeric transcription factor (*ADH1pr*-Gal4-ER-VP16) [8] to drive GFP-CAAX expression. Optimal levels of GFP-CAAX are obtained by addition of 90 nM β-estradiol 2 h before imaging. This system has the additional advantage that glucose can be used as the carbon source throughout the experiment. In addition, we use Myo1-mCherry to monitor ring contraction and Cdc3-mCherry to visualize septin ring morphology, both expressed from their native promoters at their endogenous loci.

Figure 1 shows two cells imaged in a spinning disk microscope (*see* **Note 3**) with the following settings: three different fields of view and two fluorescent channels were defined (GFP and mCherry), with 150 ms exposure time for each channel using argon laser illumination. We use 20 *Z*-stacks with 200 nm step size resolution to cover the whole yeast cell surface. Cells were imaged

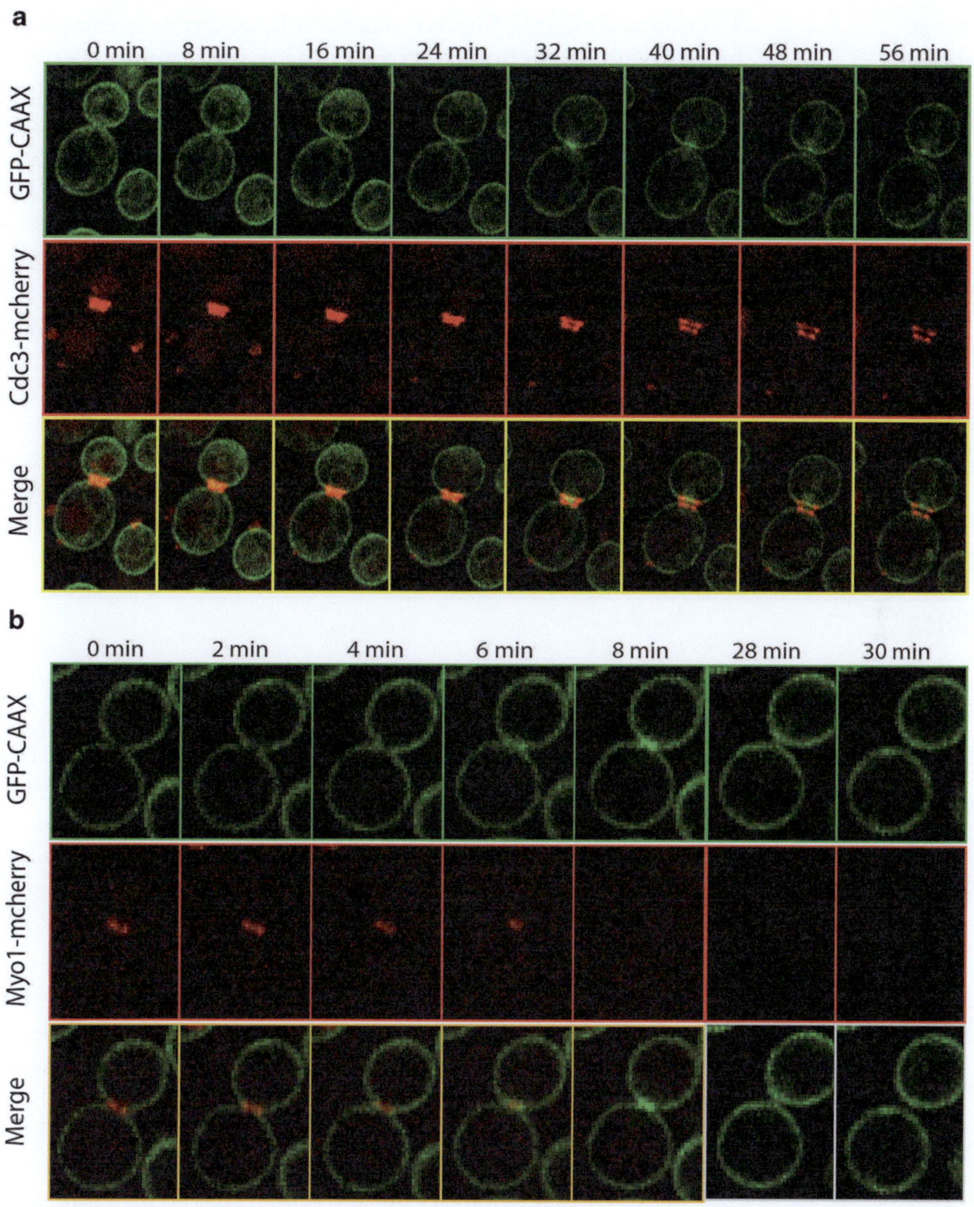

Fig. 1 Time-lapse of cells expressing GFP-CAAX and Cdc3-mCherry (**a**) or Myo1-mCherry (**b**). Maximal projections are shown for the mCherry channel but a single medial confocal slice is shown for GFP. The time is indicated in minutes

for 2 h with a 2-min time interval. Contraction of the myosin ring (Myo1-mCherry) closes the bud neck plasma membrane (GFP-CAAX) and divides the cytoplasm of mother and daughter cells (Fig. 1b). Similarly, the septin collar (Cdc3-mCherry) splits into two thinner rings during bud neck closure (Fig. 1a). Inspection of single confocal slices allows pinpointing of the precise times of plasma membrane closure and its subsequent resolution into two distinct layers.

3.3 Image Visualization

Time-lapse imaging files are relatively large and it is therefore useful to ensure that enough memory space and computer processing power are available to process them (each of the image fields in the above example represent 1200 [20 × 60] frames for each channel, between 0.5 and 1.5 GB in size). Open-source image analysis software such as ImageJ (NIH) (http://imagej.nih.gov) and Fiji [9] greatly facilitate handling of multi-dimensional time-lapse data. For instance, image hyperstacks (containing x, y, z, and time dimensions) can be easily converted to two-dimensional movies after projection of multiple Z slices into a single plane. ImageJ/Fiji also supports export of images into a variety of formats (TIFF, Jpeg, .avi, etc.). In Fig. 1, Fiji was used to generate maximal projections for the Myo1 and Cdc3 channels, whereas the GFP-CAAX channel shown represents a single optical plane through the cell middle to better visualize membrane closure and resolution at the bud neck.

4 Notes

1. A variety of plasmids can be used to amplify fluorescent protein tags by PCR, which are then introduced into the yeast genome by homologous recombination. Modified versions of fluorescent proteins are continuously being developed with increasing brightness, photostability, and faster maturation times [10], and can be obtained from a variety of non-profit sources such as Euroscarf (www.euroscarf.de) or Addgene (www.addgene.org).
2. When using 8-well Lab-Tek chambers (Nunc), the cell suspension is applied to a single well (0.8 cm^2, 0.4 ml capacity) so that cells attach to the ConA-coated borosilicate glass bottom. A lid avoids media evaporation during long-term imaging (8–10 h). Multiple wells can be used at once, and unused wells can be saved for later experiments. Good alternatives to Lab-Tek chambers include 96-well glass bottom plates (MoBiTec); FCS2 Closed Chamber System (Bioptechs) with a temperature-controlled system to heat the sample directly; glass-bottom Petri dishes (MatTek); and Ludin Chambers (Life Imaging Services) that can directly hold glass coverslips.
3. We find optimal results with a spinning disk confocal microscope (Andor Revolution XD) equipped with diode lasers (488 and 561 nm). A high-resolution objective with higher numerical aperture (NA) is required to visualize yeast subcellular structures. An oil immersion 100× lens with NA 1.4 or 63× NA 1.3 objective are both appropriate. Finally, a camera with capabilities for recording large number of frame rates per sec-

ond with fast speed, sensitivity, and low background noise is essential for a good time-lapse resolution imaging. Cooled CCD (such as Hamamatsu, ORCA-R2) and high-speed EMCCD (Hamamatsu) are both widely used.

4. Yeast cells grown in rich media such as YPD can show poor attachment to ConA-coated glass surfaces (*see* Subheading 3.1, **step 4**). To avoid this, it is advisable to grow yeast cultures in SC medium prior to imaging. However, when using α-factor to synchronize cells in G_1 (*see* **Note 5**) we prefer to use YPD medium during this step as it improves the efficiency of the arrest.
5. Sometimes it is convenient to have a population of yeast cells that are traversing the cycle synchronously. This can be achieved by treating a log phase culture of *MAT* **a** cells with α-factor (Sigma) at 2 mg/ml for 2 h at 25 °C to arrest cells in G_1. Alternatively, HU (200 mM) or nocodazole (15 μg/ml) can be used to arrest cells in early S phase or M phase, respectively. To release the cells from the arrest, centrifuge gently and wash twice with water, then resuspend in fresh medium. The kinetics of cell cycle re-entry needs to be established for each strain separately, as they might differ depending on genetic background and growth conditions.
6. ConA binds to α-D-mannosyl and α-D-glucosyl residues in various sugars, glycoproteins, and glycolipids present at the yeast surface and thus anchors cells to the glass, restricting their movement during imaging. ConA solution can be stored in small aliquots at −20 °C to avoid repeated freeze–thaw cycles that can lead to formation of aggregates. A milky ConA solution is generally indicative of degradation and should not be used. Precoated glass chambers can be stored at 4 °C for 2–3 days before use.
7. To reduce background fluorescence use SC medium lacking tryptophan (except of course if the imaged cells are *trp-auxotrophs*) and filter-sterilized rather than autoclaved glucose solution.
8. A good imaging protocol is a fine balance of number of fields chosen, number of Z-slices, exposure time, interval between frames and illumination intensity. Avoid using excessive illumination and long exposure times that can cause bleaching of fluorophores and phototoxic stress. Thus it is advisable that the illumination periods do not exceed the duration of the darkness period. A good indication that cells are not suffering stress during imaging is if the doubling time of cells under the microscope is similar to that of cells in liquid cultures.

Acknowledgments

We thank all lab members for helpful suggestions. Research in our laboratory is supported by the Spanish ministry of Economy and Competitiveness (BFU09-08213).The research leading to these results has received funding from the European Research Council under the European Union's Seventh Framework Programme (FP7/2007-2013) / ERC grant agreement n° 260965. It reflects only the authors' views and the Union is not liable for any use that may be made of the information contained therein.

We acknowledge support of the Spanish Ministry of Economy and Competitiveness, 'Centro de Excelencia Severo Ochoa 2013-2017', SEV-2012-0208.

References

1. Muzzey D, van Oudenaarden A (2009) Quantitative time-lapse fluorescence microscopy in single cells. Annu Rev Cell Dev Biol 25:301–327
2. Gladfelter AS, Pringle JR, Lew DJ (2001) The septin cortex at the yeast mother-bud neck. Curr Opin Microbiol 4:681–689
3. Dobbelaere J, Barral Y (2004) Spatial coordination of cytokinetic events by compartmentalization of the cell cortex. Science 305:393–396
4. Balasubramanian MK, Bi E, Glotzer M (2004) Comparative analysis of cytokinesis in budding yeast, fission yeast and animal cells. Curr Biol 14:R806–R818
5. Roncero C, Sánchez Y (2010) Cell separation and the maintenance of cell integrity during cytokinesis in yeast: the assembly of a septum. Yeast 27:521–530
6. Colman-Lerner A, Chin TE, Brent R (2001) Yeast Cbk1 and Mob2 activate daughter-specific genetic programs to induce asymmetric cell fates. Cell 107:739–750
7. Janke C, Magiera MM, Rathfelder N et al (2004) A versatile toolbox for PCR-based tagging of yeast genes: new fluorescent proteins, more markers and promoter substitution cassettes. Yeast 21:947–962
8. Louvion JF, Havaux-Copf B, Picard D (1993) Fusion of GAL4-VP16 to a steroid-binding domain provides a tool for gratuitous induction of galactose-responsive genes in yeast. Gene 131:129–134
9. Schindelin J, Arganda-Carreras I, Frise E et al (2012) Fiji: an open-source platform for biological-image analysis. Nat Methods 9:676–682
10. Lee S, Lim WA, Thorn KS (2013) Improved blue, green, and red fluorescent protein tagging vectors for S. cerevisiae. PLoS One 8:e67902

Chapter 2

Real-Time Visualization and Quantification of Contractile Ring Proteins in Single Living Cells

Reshma Davidson, Yajun Liu, Kenneth S. Gerien, and Jian-Qiu Wu

Abstract

Single-cell microscopy provides a powerful tool to visualize cellular and subcellular processes in wild-type and mutant cells by observing fluorescently tagged proteins. Here, we describe three simple methods to visualize fission yeast cells: gelatin slides, coverslip-bottom dishes, and tetrad fluorescence microscopy. These imaging methods and data analysis using free software make it possible to quantify protein localization, dynamics, and concentration with high spatial and temporal resolution. In fission yeast, the actomyosin contractile ring is essential for cytokinesis. We use the visualization and quantification of contractile ring proteins as an example to demonstrate how to use these methods.

Key words Contractile ring, Cytokinesis, Fimbrin, Fission yeast, Gelatin, GFP, ImageJ, Microscopy, Quantification, Septin, Tetrad fluorescence microscopy

1 Introduction

The discovery and modification of the GFP (Green Fluorescent Protein) have revolutionized biomedical research and triggered the development of highly automated live cell fluorescence microscopy [1–6]. Analysis of fluorescence images and time-lapse movies has redefined the understanding of many biological processes. The unicellular fission yeast *Schizosaccharomyces pombe* is a particularly useful organism to study cytokinesis in live cells. The actomyosin contractile ring is essential for cytokinesis in fission yeast. Characteristics such as less genetic redundancy, regular cell size and shape, and efficient homologous recombination make *S. pombe* ideal for observing and studying contractile ring proteins during cytokinesis [7–9].

Visualization of a protein in *S. pombe* involves gene targeting to tag the protein of interest with a fluorescent protein at one of its termini, ideally expressed at its endogenous level [10]. A variety of options are available for choosing fluorescent proteins and techniques for imaging. Here, we demonstrate the need to choose the

Alberto Sanchez-Diaz and Pilar Perez (eds.), *Yeast Cytokinesis: Methods and Protocols*, Methods in Molecular Biology, vol. 1369, DOI 10.1007/978-1-4939-3145-3_2,

best monomeric fluorescent proteins with two examples. Depending on the duration and temperature of imaging, we describe how to visualize fission yeast cells using gelatin slides [11–14] and coverslip-bottom dishes [15, 16]. Visualization using gelatin slides is ideal for counting protein molecules and making short time-lapse movies (1–2 h) at lower temperatures. Using coverslip-bottom dishes allows imaging for longer durations (>2 h) and at higher temperatures.

We also describe a novel microscopy technique called "tetrad fluorescence microscopy" that we have developed to unequivocally determine the phenotypes of deletions of essential genes and synthetic lethal interactions [16, 17]. The technique is a promising new approach to determine the cause of cell death in synthetic lethal double mutants or in deletions of essential genes by following the progenies of a meiotic event. Using both a tetrad dissection microscope and a sensitive fluorescence microscope, this fusion microscopy technique can distinguish the possible causes of cell death in cells expressing certain fluorescent markers with a much higher certainty and resolution than the traditional random spore assays. Because spore germination and growth are often highly variable, imaging cells with known genotypes shortly after spore germination simplifies the process necessary to obtain images of cell polarization and the first few mitotic divisions [16].

Then we discuss some basic consideration on microscopy settings to obtain the best images. Finally, we briefly explain how to quantify the timing of contractile ring proteins using established methods and free software ImageJ [9, 11, 18]. We hope that these detailed methods are useful for studies of cytokinesis and other cellular processes.

1.1 How to Choose the Best Fluorescent Proteins for Imaging

The published guides for choosing suitable fluorescent proteins should be followed [19–21]. Here, we emphasize two important considerations. Besides maturing rapidly and having a bright signal, ideal fluorescent proteins should be thermal stable and functional when fused with the protein of interest. A good example to illustrate this point is by comparing the tags GFP(S65T) and EGFP(F64L, S65T). The plasmid pFA6a-GFP(S65T)-kanMX6 for gene targeting was described in [10, 22, 23] and widely distributed/used in yeast community. At 36 °C, cells expressing fimbrin Fim1 tagged with enhanced GFP (EGFP) were normal (Fig. 1a). However, cells expressing fimbrin Fim1-GFP(S65T) had severe cytokinesis defects such as accumulating multiple and/or misplaced/aberrant septa (Fig. 1b). Thus, GFP(S65T) should not be used except for some special experiments.

Another important consideration is the difference between regular (often forming a weak dimer) and monomeric fluorescent proteins. For example, YFP (Yellow fluorescent protein) is a weak dimer and mYFP (monomeric YFP) has a dimer interface breaking

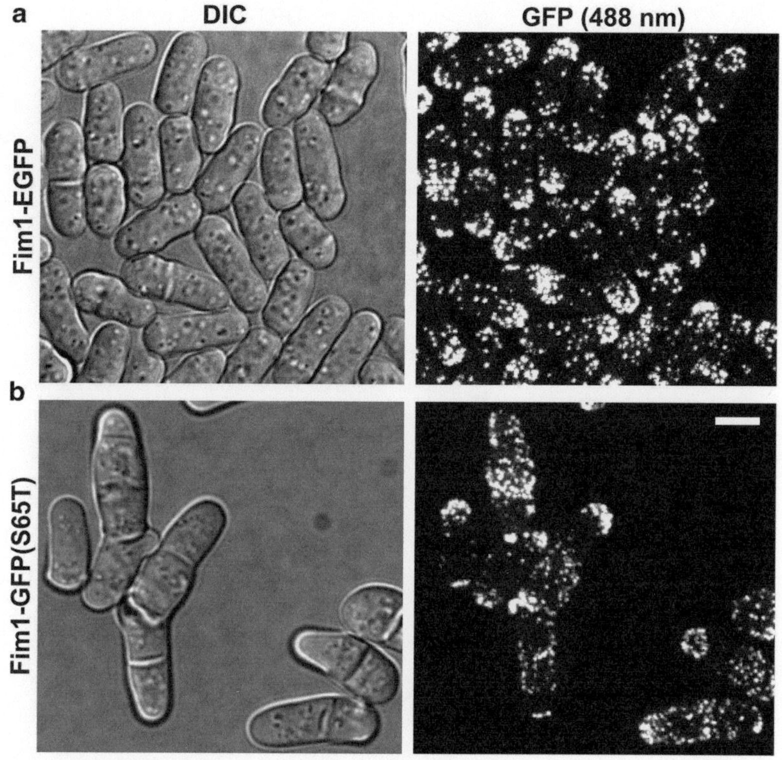

Fig. 1 Confocal microscopy of (**a**) fimbrin Fim1-EGFP and (**b**) Fim1-GFP(S65T). DIC (Differential Interference Contrast) images are shown to the *left* and the corresponding fluorescence images to the *right*. The fluorescence images are a maximum intensity projection of 21 slices with 0.3 μm spacing. Cells were grown in YE5S liquid medium for ~36 h at 25 °C, then shifted to 36 °C for 10 h before imaging at 36 °C. Cells were imaged using coverslip-bottom dish at 36 °C. Bar, 5 μm

mutation A206K [20, 24]. Septin Spn1 tagged with mYFP showed normal septin localization to the double ring at the division site (Fig. 2a). In contrast, Spn1-YFP formed atypical structures such as bars or spirals in the cytoplasm in addition to its normal division site localization (Fig. 2b). Thus, monomeric fluorescent proteins should be used whenever possible to avoid artificial dimerization of the tagged proteins.

2 Materials

The recipes for the standard fission yeast media can be found at PombeNet of the Forsburg lab at http://www-bcf.usc.edu/~forsburg/media.html. The *S. pombe* strains used in this study are listed in Table 1.

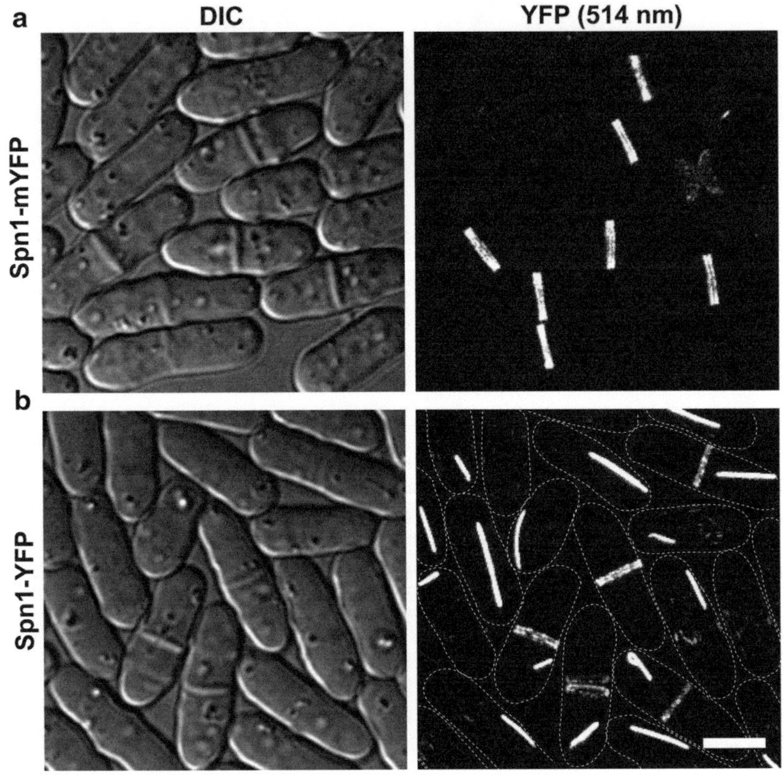

Fig. 2 Confocal microscopy of (**a**) septin Spn1-mYFP and (**b**) Spn1-YFP. DIC images are shown to the *left* and the corresponding fluorescence images to the *right*. The fluorescence images are a maximum intensity projection of 31 slices with 0.2 μm spacing. The *dash lines* mark the cell boundary in (**b**). Cells were cultured in YE5S liquid medium for ~48 h at 25 °C and imaged on slides with YE5S medium at 24 °C. Bar, 5 μm

Table 1
***S. pombe* strains used in this study**

Strain	Genotype	Source
JW811	*h-spn1-YFP-kanMX6 ade6-M210 leu1-32 ura4-D18*	This study
JW1092	*h-spn1-mYFP-kanMX6 ade6-M210 leu1-32 ura4-D18*	This study
JW1143	*h-fim1-GFP(S65T)-kanMX6 ade6-M210 leu1-32 ura4-D18*	This study
JW1211	*h+fim1-EGFP-kanMX6 ade6-M210 leu1-32 ura4-D18*	This study
JW6716	*kanMX6-Pmyo2-mEGFP-myo2 sad1-mCherry-natMX6 ade6-M210 leu1-32 ura4-D18*	This study

2.1 Materials for Microscopy Using Gelatin Slides

1. Edinburgh minimal medium + Supplements (EMM5S) liquid medium (minimal medium with low autofluorescence).
2. Yeast extract medium + Supplements (YE5S) agar plates and YE5S liquid medium (rich medium with slightly higher autofluorescence).
3. 10× n-propyl-gallate (n-PG) stock solution: 50 μM n-propyl-gallate in EMM5S or YE5S. Keep at −20 °C in the dark.
4. Gelatin from porcine skin.
5. Valap: 1:1:1 of Parafin, Vaseline, and Lanoline. Melt equal weight of each component completely at ~80 °C, mix thoroughly, and aliquot into 1.5 ml microcentrifuge tubes before solidification. Keep at room temperature.
6. Regular glass slides and 22 mm No. 1.5 square coverslip.
7. Fluorescence microscope.

2.2 Materials for Microscopy Using Coverslip-Bottom Dishes

1. YE5S agar plates and YE5S liquid media.
2. Delta TPG 35-mm culture dish with No. 1.5 coverslip bottom (0.17 mm thick, round).
3. 18 mm × 18 mm No. 1.5 square coverslip.
4. Fluorescence microscope.

2.3 Materials for Tetrad Fluorescence Microscopy

1. Sporulation medium + Supplements (SPA5S) plates for crosses and sporulation.
2. YE5S agar plates with flat surface for tetrad dissection: 25 ml melted YE5S agar in each 100 × 15 mm petri dish. Lay out singly on a flat surface before agar solidification. Dry for 2 days at room temperature and then keep at 4 °C for storage.
3. Tetrad dissection microscope.
4. No.1.5 coverslips: 22 mm × 50 mm and 24 mm × 60 mm
5. Rectangular plastic frame to prevent agar medium from drying: the dimensions of the homemade frame are 50 mm × 23 mm × 3 mm (outer edge) and 40 mm × 17 mm × 3 mm (inner edge).
6. Fluorescence microscope with an automated stage.

2.4 Software for Quantification and Data Analysis

1. Image acquisition: Volocity Beta 6.3.1 or UltraVIEWERS or others.
2. Quantification and data analyses: ImageJ (NIH) at http://imagej.nih.gov/ij/.

3 Methods

3.1 Preparing Cells for Microscopy

1. Wake up cells from −80 °C stocks and grow on a YE5S plate for 2 days at 25 °C (or other plates and temperatures dependent on the strains). Re-streak cells on the plate twice.

2. Inoculate a strain in YE5S liquid medium (or other medium) in a baffled 50 ml flask and grow cells in 25 °C water bath with shaking. Prior to microscopy, cells are usually grown in YE5S liquid for ~48 h (*see* **Note 1**). The optical density (OD) of cells is measured at 595 nm. The OD_{595} should always be maintained at less than 0.5 (1×10^7 cells/ml) to assure that cells are growing in the log phase. Thus, the culture needs to be diluted using fresh YE5S medium every 9–15 h.
3. Spin down 1 ml of the cells at 3000 rpm (approximately $850 \times g$) for 30 to 60 s. Alternatively, spin cells for 30 s, rotate the microcentrifuge tube 180°, and then spin for another 30 s. Remove the supernatant.
4. Wash the cells in 1 ml fresh EMM5S to reduce background fluorescence (*see* **Note 2**), and spin down the cells as described in **step 3**. Remove the supernatant.
5. Add 900 μl of fresh EMM5S and 100 μl of 10× n-PG stock solution made in EMM5S (thawed and warmed up to room temperature) to resuspend the cells and keep the cells in the dark for several minutes (*see* **Note 3**). Spin down the cells as described in **step 3** just before the slide is ready.
6. Remove extra supernatant and resuspend the cells in the remaining 20–100 μl (the volume dependent on cell concentration) of medium.

3.2 Microscopy Using a Gelatin Slide

1. Add 0.9 ml of EMM5S (or YE5S if the autofluorescence from the medium is not a concern) liquid medium to a microcentrifuge tube.
2. Add 0.25 g of gelatin to the 0.9 ml of EMM5S. Mix immediately by shaking vigorously so that gelatin will not clog at the top of the tube.
3. Place the tube at 65 °C hot plate (dry bath) for ~5 min to melt gelatin.
4. Add 100 μl of 10× n-PG stock solution in EMM5S (or in YE5S if YE5S is used in **step 1**) to the tube (*see* **Note 4**). Invert the tube a few times to mix thoroughly. Keep the tube at 65 °C and cover with foil. Wait for approximately 20 min until all air bubbles disappear.
5. Wipe slides clean. Then slowly take 100 to 150 μl of the melted EMM5S + gelatin + n-PG using a 1 ml pipette and add to the center of the slide. Avoid introducing air bubbles on the slide (*see* **Note 5**). Place another clean slide on top of the gelatin medium and clip one small paper binder on each end of the slides (*see* **Note 6**). The two slides should stagger a little bit to aid the eventual slide separation. Keep the clipped slides in the dark for 15–20 min to solidify the gelatin medium (*see* **Note 7**). Make two or three slides for each strain to ensure at least one is good.

6. Keep the microcentrifuge tube with extra gelatin medium at room temperature in the dark (*see* **Note 8**).
7. When the cell sample is ready to be loaded onto the slide (*see* **Note 9**), slowly separate two slides from one side by removing one binder. Separate the slides with a razor blade if necessary. Then remove the other binder. If no cracks appear in the middle of the solidified gelatin then the slide is usable since the entire gelatin pad will stick to one of the slides.
8. Pipet 5 μl of the concentrated cells from **step 6** of **'Preparing cells for microscopy'** to the center of the gelatin pad. Cover the cells with a 22 mm No. 1.5 square coverslip. Gently tap the coverslip with the blunt end of a 1 ml pipet tip to spread the cells evenly and avoid cell overlaps.
9. Seal the edges of the coverslip properly using Valap melted at 65 °C to prevent the slide from drying out.
10. Observe the slide on a microscope.

3.3 Microscopy Using a Coverslip-Bottom Dish

1. If cells are grown and being observed at temperatures other than the room temperature, all used materials (dish, agar plate, liquid YE5S, n-PG, tips, microcentrifuge tubes, etc.) should be pre-warmed to the desired temperature (*see* **Note 10**).
2. Cut YE5S agar medium in a circle shape with a diameter of ~20 mm using a sterile razor blade from a YE5S plate (*see* **Note 11**). To be consistent from experiments to experiments, the YE5S agar plate for tetrad dissection is recommended. It is better to add n-PG to a final concentration of 5 μM when making the plate.
3. Place 2–10 μl of concentrated cells washed with 900 μl YE5S plus 100 μl 5 μM n-PG in YE5S onto the center of the coverslip of the dish.
4. Place pre-cut agar on top of the cells.
5. Place an 18 mm × 18 mm No. 1.5 coverslip on top of the agar to slow down the evaporation.
6. Gently tap on the top of the 18 mm × 18 mm coverslip to spread cells evenly with the blunt end of a 1 ml pipet tip.
7. Cover the dish with its lid and observe the dish on a microscope.

3.4 Tetrad Fluorescence Microscopy

1. Tetrad dissection is carried out from a 2- or 3-day-old cross as usual with one modification. Place the spores 2.5 mm apart from each other in both X and Y axes using the micromanipulator and glass-fiber needle (*see* **Note 12**).
2. Grow the dissected tetrads at 25 °C or other desired temperatures for 12–24 h or longer if necessary (*see* **Note 13**).

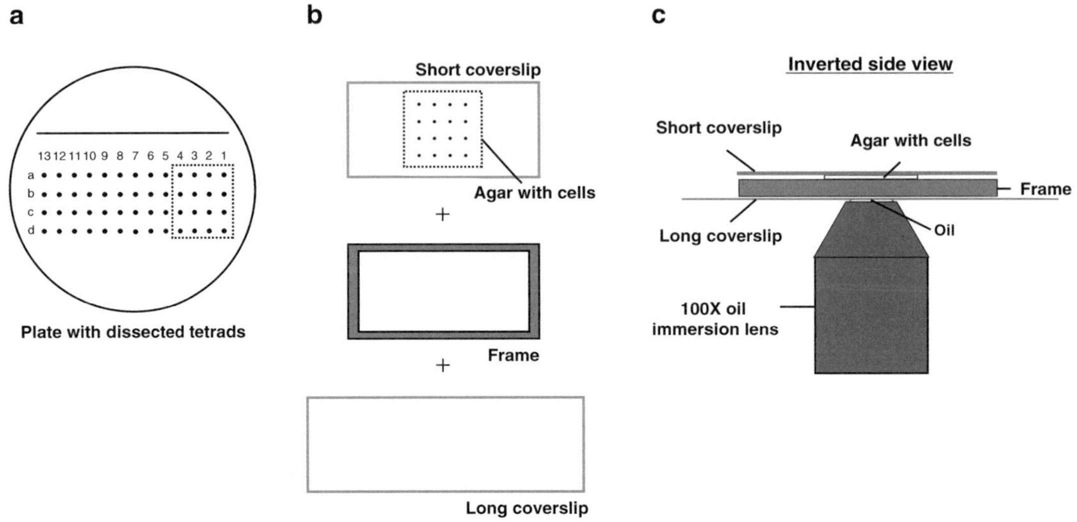

Fig. 3 Illustration of tetrad fluorescence microscopy. (**a**) Tetrads were dissected onto a YE5S plate made for tetrad dissection. (**b**) Four tetrads were cut out from the plate and placed (cells facing up) on a 22 mm × 50 mm glass coverslip. A plastic frame is then placed around the agar piece followed by a longer glass coverslip (24 mm × 60 mm) over the agar. (**c**) The entire setup from (**b**) was inverted and imaged using a 100× oil immersion lens

3. Preparing the agar plates with dissected tetrads for fluorescence microscopy: Since locating single or a few cells derived from germinating spores under a high magnification microscope can be a challenge, we mark the tetrads for imaging by punching holes in the agar near the cells using the dissection needle under the tetrad dissection microscope (*see* **Note 14**).
4. Preparing slides for the tetrad fluorescence microscopy: Typically, imaging 3 or 4 tetrads on a fluorescence microscope with an automated stage would yield images with enough temporal and spatial resolution (Fig. 3a). We cut an agar piece with 3–4 tetrads using the 22 mm × 50 mm coverslip. Then insert the coverslip underneath the agar and scoop up the agar piece that has been cut (tetrads facing up) (Fig. 3b, top). Center the agar on the coverslip and place the plastic frame around it (Fig. 3b). Then cover the tetrads with a slightly bigger 24 mm × 60 mm coverslip. Make sure the bigger coverslip is resting evenly on the agar without any air bubbles (*see* **Note 15**).
5. Invert the entire setup (tetrads facing down), clamp and rest it on slide holder over an objective lens (immersion oil added) for 5–10 min before imaging (Fig. 3c).
6. Use the marks in the agar to focus and to assist in finding colonies. Alternatively, the automated stage can be used to find the colonies based on the regular spacing between them.
7. Save each colony coordinates using the imaging software so they can be revisited for imaging at each time point.

3.5 Image Acquisition

Settings of image acquisition depend on the imaging system, camera, acquisition software, and signals of the samples. Using appropriate microscopy settings is important to obtain a good fluorescence image. Laser power or other excitation lights must be adjusted so that they do not cause excessive photobleaching and phototoxicity. Whenever possible, use lower excitation power and longer exposure times without sacrificing the temporal resolution to obtain decent images. One method to boost the signal artificially is to adjust gain/sensitivity setting that amplifies the signals at the camera/detector. Another way to improve image brightness while slightly compromising spatial resolution is to adjust the binning. For regular CCD camera (like Hamamastu ORCA AG), we use 2×2 binning due to its small pixel size. However, binning is not recommended for EMCCD camera due to its bigger pixel size.

3.6 Quantification of Microscopy Data

A tremendous amount of informative and quantitative data can be extracted from microscopy images, like absolute global and local protein concentrations and timing of cytokinesis events. We and others have discussed how to count protein molecules extensively elsewhere [14, 17, 19, 25–28]. Here, we use quantification of the timing of contractile-ring proteins during cytokinesis as an example to explain how to analyze microscopy data using free software ImageJ (*see* **Note 16**).

1. Download and install ImageJ from http://imagej.nih.gov/ij/. Choose the version for your operating system. You can also download many useful Plugins from the website.
2. Open the microscopy data file in ImageJ as a TIFF file by importing it as "Image sequence" or "Bio-Formats" (*see* **Note 17**). The images are usually grouped in separate channels, like DIC and fluorescence channels.
3. Combine different Z-slices at the same time point using "Z projection" function in the Stacks option under Image tab (*see* **Note 18**). Typically this action gives a better signal for analysis. However, depending on the purpose, sometimes single slice is needed for analysis as well.
4. Open the "Brightness/Contrast" function in the Adjust option under Image tab. This allows the user to adjust the image. "Auto" option will usually give a good contrast of the image. If needed, "Set" option allows the user to manually enter the minimum and maximum displayed values (*see* **Note 19**).
5. Quantify the timing of contractile ring proteins by comparing it to different markers. The contractile ring protein, myosin II heavy chain Myo2, is a commonly used marker because the distinct stages of cytokinesis are easily distinguishable (Fig. 4). The timings of other proteins can be compared to the stages defined by Myo2. We usually use the spindle pole body protein

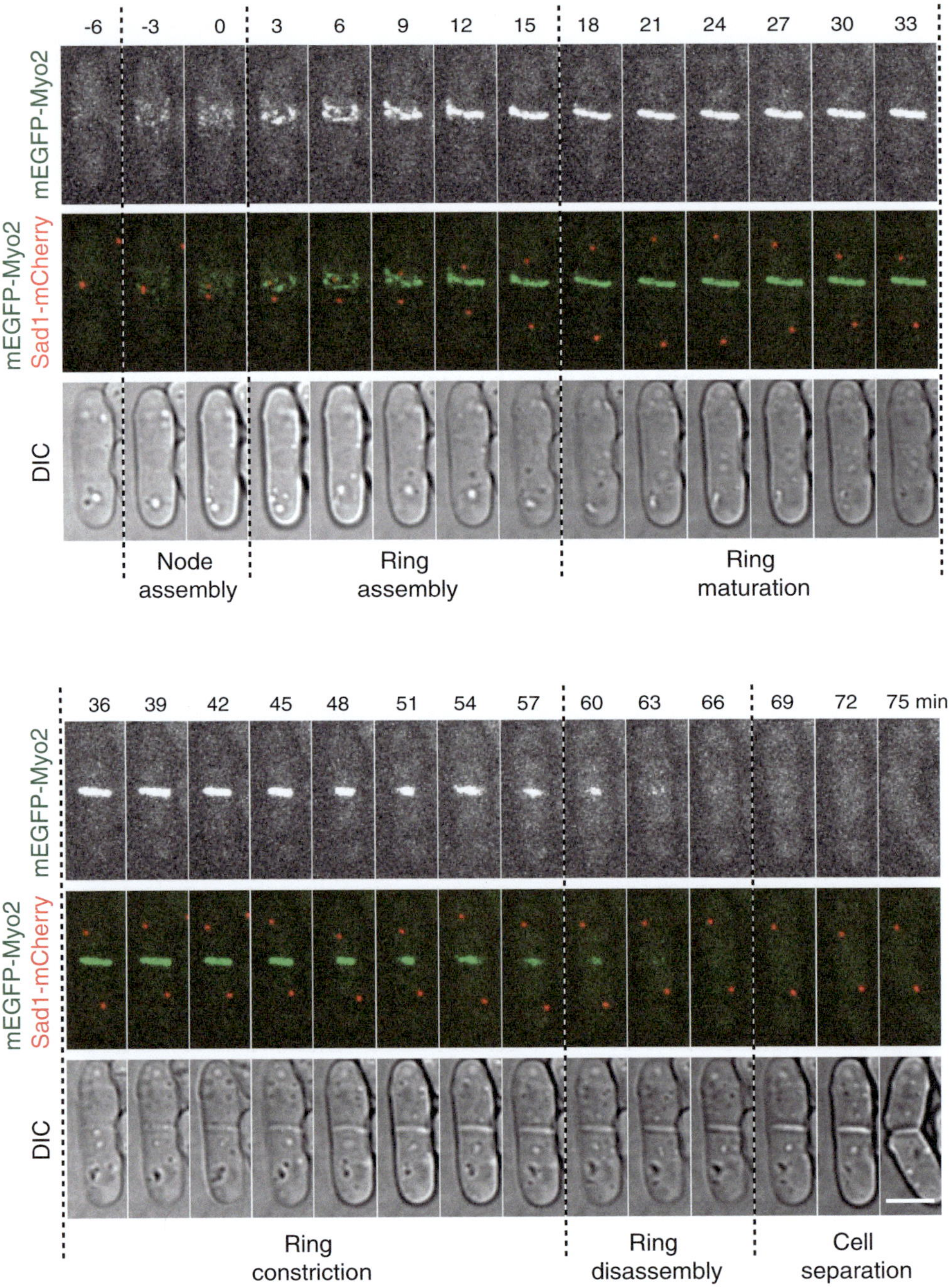

Fig. 4 Stages of cytokinesis in fission yeast. Wild-type cells expressing mEGFP-Myo2 and Sad1-mCherry at their endogenous levels were used. The time course of a representative cell from a movie with 90 s interval is shown. The fluorescence images are a maximum intensity projection of 13 slices with 0.5 μm spacing. Time zero marks SPB separation. The dotted lines separate the various stages of cytokinesis. Bar, 5 μm

Sad1 as an accurate marker for the cell-cycle stages (Fig. 4). Spindle pole body separation marks the entry of mitosis and is generally considered as time zero (*see* **Note 20**).

(a) Cytokinesis node assembly: This stage starts from the appearance of cytokinesis nodes at the division site to the time point just before nodes begin to condense around time zero (*see* **Note 21**).

(b) Contractile ring assembly: This stage starts from the beginning of node condensation to the formation of a compact ring without lagging nodes.

(c) Contractile ring maturation: This stage begins when a compact ring is formed until the ring starts to constrict.

(d) Contractile ring constriction: This stage begins when the contractile ring starts to constrict and ends when the ring has constricted to a dot (with the highest Myo2 pixel intensity). The Myo2 fluorescence intensity at the division site drops rapidly right after the end of ring constriction. The septum forms during ring constriction.

(e) Contractile ring disassembly: This stage is from the completion of ring constriction to the disappearance of Myo2 signal at the division site. This is a narrow definition of ring disassembly since ring components also disassemble during ring constriction.

4 Notes

1. The long culture time is necessary to ensure data consistency. For mutants that pick up a suppressor quickly, cells are usually grown in liquid medium for a shorter time.
2. This wash step is most useful for epifluorescence microscopy or for proteins with low abundance. It is not absolutely required for microscopy with spinning disk confocal microscope. When imaging in a dish with YE5S agar, the cells need to be washed only once with 900 μl of YE5S + 100 μl of 10× n-PG made in YE5S. Then directly proceed to **step 6**.
3. n-PG (Sigma P-3130) protects cells from free radicals during microscopy. It is light sensitive. An aliquot can be kept at room temperature in the dark for 1 day. Stocks should be kept at −20 °C for long-term storage.
4. Medium with gelatin is viscous. Do not use a pipette to mix.
5. A trick to pipet the liquid onto the slide without introducing bubbles is to avoid dispensing the last tiny drop of liquid in the pipette tip. If the bubbles are still introduced, they can be sucked up using the pipette tip. Because the gelatin medium is viscous, do not use a 200 μl pipette to dispense it.

6. When putting on the paper clips on both ends of the slides, it is better to hold one end while clipping the other. This ensures that the gelatin pad is even and not torn apart.
7. If the slide is kept longer than 30 min before use, the slides can be put in an airtight container with wet paper towels to prevent the gelatin pad from drying out.
8. The leftover EMM5S + gelatin + n-PG should be removed from the 65 °C hot plate after making the slides. Prolonged incubation at 65 °C may damage the vitamins and other components in the medium. In addition, it takes only a few minutes to re-melt the medium.
9. Begin preparing cells for imaging ~10 min after preparing the slides and keeping them in the dark.
10. When imaging cells at a temperature other than the room temperature, all the steps should be done on a surface warmed to that temperature. We use a small portable incubator that has been preheated for this purpose. The microcentrifuge tube with cells is kept in a beaker with water at the desired temperature during the collection and wash.
11. Before cutting, you can draw a circle on the bottom of the petri dish using a coin (e.g. a five-cent coin struck by the United States Mint with a diameter ~21 mm) as the guide.
12. Typically spores from tetrads are placed 5 mm apart. However, it takes too long for the automated stage to travel that distance over ten times at each time point during time-lapse imaging so 2.5 mm spacing is used.
13. The growth time depends on when the phenotype shows up in the mutant. For example, if the mutant dies during the first cell division or cell cycle, shorter incubation times are appropriate.
14. These holes in the medium are easily visible to the eye and under the microscope. Different patterns may be used to mark the location of the cells/colonies/spores.
15. Place the coverslip on the agar from one end and then release it slowly to prevent the formation of air bubbles. At this point the punched holes on the agar medium become less visible to eyes. Thus, it is helpful to make a mark on the coverslip using a permanent marker. Do not shift the coverslip once it has been placed on the agar. The frame is designed to be slightly thinner than the agar. Alternatively, one can use an 18 × 18 mm coverslip to scoop up the agar with tetrads and then invert it onto a coverslip-bottom dish for imaging.
16. Fiji is an image processing package that is akin to an advanced version of ImageJ and can also process super-resolution microscopy data. It is a free software and can be downloaded from www.Fiji.sc/Fiji.

17. "Image sequence" import is used for TIFF files. "Bio-Formats" import can be used for library files from Volocity multi-file library.
18. Typically, the Z-slices are combined using "Max Intensity" projection. If the signal intensity needs to be quantified or compared, "Sum Slices" projection should be used.
19. The "Brightness/Contrast" window shows a peak for each image. Typically, the minimum intensity value is displayed at the left bottom of the peak.
20. The time of arrival of various contractile ring proteins to the division site can be measured relative to the distance of SPB separation. The maximum SPB separation marks the end of anaphase B.
21. For some node proteins, the signal at the division site may start with one dot or the dot signal may appear and then disappear. It is hard to define the signal appearance with just one dot, so typically the appearance of signal is defined as the appearance of ~1/3 of the total number of nodes that exist at the beginning of node condensation for this certain protein. Due to higher autofluorescence at the GFP channel than YFP channel, GFP-tagged proteins may "appear" later than YFP-tagged same proteins (*see* Fig. 4 and [11, 14]).

Acknowledgement

We would like to thank Valerie Coffman, Damien Laporte, I-Ju Lee, Ning Wang, Yanfang Ye, and Yihua Zhu for developing and improving the various microscopy techniques currently used in the lab. We thank Tom Pollard for yeast strains. The work in the Wu lab was supported by the National Institute of General Medical Sciences of the NIH grant GM086546 and American Cancer Society grant RSG-13-005-01-CCG to J.-Q.W.

References

1. Prasher DC, Eckenrode VK, Ward WW, Prendergast FG, Cormier MJ (1992) Primary structure of the *Aequorea victoria* green-fluorescent protein. Gene 111:229–233.
2. Chalfie M, Tu Y, Euskirchen G, Ward WW, Prasher DC (1994) Green fluorescent protein as a marker for gene expression. Science 263:802–805.
3. Shimomura O (1979) Structure of the chromophore of *Aequorea* green fluorescent protein. FEBS Lett 104:220–222. doi:10.1016/0014-5793(79)80818-2
4. Deerinck TJ, Martone ME, Lev-Ram V, Green DP, Tsien RY, Spector DL, Huang S, Ellisman MH (1994) Fluorescence photooxidation with eosin: a method for high resolution immunolo-

calization and in situ hybridization detection for light and electron microscopy. J Cell Biol 126:901–910.

5. Dean KM, Palmer AE (2014) Advances in fluorescence labeling strategies for dynamic cellular imaging. Nat Chem Biol 10:512–523. doi:10.1038/nchembio.1556
6. Gao L, Shao L, Higgins CD, Poulton JS, Peifer M, Davidson MW, Wu X, Goldstein B, Betzig E (2012) Noninvasive imaging beyond the diffraction limit of 3D dynamics in thickly fluorescent specimens. Cell 151:1370–1385. doi:10.1016/j.cell.2012.10.008
7. Roberts-Galbraith RH, Gould KL (2008) Stepping into the ring: the SIN takes on contractile ring assembly. Genes Dev 22:3082–3088. doi:10.1101/gad.1748908
8. Laporte D, Zhao R, Wu J-Q (2010) Mechanisms of contractile-ring assembly in fission yeast and beyond. Semin Cell Dev Biol 21:892–898. doi:10.1016/j.semcdb.2010.08.004
9. Lee I-J, Coffman VC, Wu J-Q (2012) Contractile-ring assembly in fission yeast cytokinesis: recent advances and new perspectives. Cytoskeleton 69:751–763. doi:10.1002/cm.21052
10. Bähler J, Wu J-Q, Longtine MS, Shah NG, McKenzie A III, Steever AB, Wach A, Philippsen P, Pringle JR (1998) Heterologous modules for efficient and versatile PCR-based gene targeting in *Schizosaccharomyces pombe*. Yeast 14:943–951. doi:10.1002/(SICI)1097-0061(199807)14:10<943::AID-YEA292>3.0.CO;2-Y
11. Wu J-Q, Kuhn JR, Kovar DR, Pollard TD (2003) Spatial and temporal pathway for assembly and constriction of the contractile ring in fission yeast cytokinesis. Dev Cell 5:723–734.
12. Wu J-Q, Sirotkin V, Kovar DR, Lord M, Beltzner CC, Kuhn JR, Pollard TD (2006) Assembly of the cytokinetic contractile ring from a broad band of nodes in fission yeast. J Cell Biol 174:391–402. doi:10.1083/jcb.200602032
13. Coffman VC, Nile AH, Lee I-J, Liu H, Wu J-Q (2009) Roles of formin nodes and myosin motor activity in Mid1p-dependent contractile-ring assembly during fission yeast cytokinesis. Mol Biol Cell 20:5195–5210. doi:10.1091/mbc.E09-05-0428
14. Laporte D, Coffman VC, Lee I-J, Wu J-Q (2011) Assembly and architecture of precursor nodes during fission yeast cytokinesis. J Cell Biol 192:1005–1021. doi:10.1083/jcb.201008171
15. Zhu Y-H, Ye Y, Wu Z, Wu J-Q (2013) Cooperation between Rho-GEF Gef2 and its binding partner Nod1 in the regulation of fission yeast cytokinesis. Mol Biol Cell 24:3187–3204. doi:10.1091/mbc.E13-06-0301
16. Lee I-J, Wang N, Hu W, Schott K, Bähler J, Giddings TH Jr, Pringle JR, Du L-L, Wu J-Q (2014) Regulation of spindle pole body assembly and cytokinesis by the centrin-binding protein Sfi1 in fission yeast. Mol Biol Cell 25:2735–2749. doi:10.1091/mbc.E13-11-0699
17. Coffman VC, Sees JA, Kovar DR, Wu J-Q (2013) The formins Cdc12 and For3 cooperate during contractile ring assembly in cytokinesis. J Cell Biol 203:101–114. doi:10.1083/jcb.201305022
18. Wang N, Lo Presti L, Zhu Y-H, Kang M, Wu Z, Martin SG, Wu J-Q (2014) The novel proteins Rng8 and Rng9 regulate the myosin-V Myo51 during fission yeast cytokinesis. J Cell Biol 205:357–375. doi:10.1083/jcb.201308146
19. Coffman VC, Wu J-Q (2012) Counting protein molecules using quantitative fluorescence microscopy. Trends Biochem Sci 37:499–506. doi:10.1016/j.tibs.2012.08.002
20. Shaner NC, Steinbach PA, Tsien RY (2005) A guide to choosing fluorescent proteins. Nat Methods 2:905–909. doi:10.1038/nmeth819
21. Remington SJ (2006) Fluorescent proteins: maturation, photochemistry and photophysics. Curr Opin Struct Biol 16:714–721. doi:10.1016/j.sbi.2006.10.001
22. Wach A, Brachat A, Alberti-Segui C, Rebischung C, Philippsen P (1997) Heterologous HIS3 marker and GFP reporter modules for PCR-targeting in *Saccharomyces cerevisiae*. Yeast 13:1065–1075. doi:10.1002/(SICI)1097-0061(19970915)13:11<1065::AID-YEA159>3.0.CO;2-K
23. Longtine MS, McKenzie A III, Demarini DJ, Shah NG, Wach A, Brachat A, Philippsen P, Pringle JR (1998) Additional modules for versatile and economical PCR-based gene deletion and modification in *Saccharomyces cerevisiae*. Yeast 14:953–961. doi:10.1002/(SICI)1097-0061(199807)14:10<953::AID-YEA293>3.0.CO;2-U
24. Zacharias DA, Violin JD, Newton AC, Tsien RY (2002) Partitioning of lipid-modified monomeric GFPs into membrane microdomains of live cells. Science 296:913–916. doi: 10.1126/science.1068539
25. Wu J-Q, Pollard TD (2005) Counting cytokinesis proteins globally and locally in fission

yeast. Science 310:310–314. doi:10.1126/science.1113230

26. Coffman VC, Wu J-Q (2014) Every laboratory with a fluorescence microscope should consider counting molecules. Mol Biol Cell 25:1545–1548. doi:10.1091/mbc.E13-05-0249
27. Joglekar AP, Salmon ED, Bloom KS (2008) Counting kinetochore protein numbers in budding yeast using genetically encoded fluorescent proteins. Methods Cell Biol 85:127–151. doi:10.1016/S0091-679X(08)85007-8
28. Coffman VC, Wu P, Parthun MR, Wu J-Q (2011) CENP-A exceeds microtubule attachment sites in centromere clusters of both budding and fission yeast. J Cell Biol 195:563–572. doi:10.1083/jcb.201106078

Chapter 3

Fluorescence Recovery After Photo-Bleaching (FRAP) and Fluorescence Loss in Photo-Bleaching (FLIP) Experiments to Study Protein Dynamics During Budding Yeast Cell Division

Alessio Bolognesi, Andrzej Sliwa-Gonzalez, Rupali Prasad, and Yves Barral

Abstract

The easiness of tagging any protein of interest with a fluorescent marker together with the advance of fluorescence microscopy techniques enable researchers to study in great detail the dynamic behavior of proteins both in time and space in living cells. Two commonly used techniques are FRAP (*F*luorescent *R*ecovery *A*fter *P*hoto-bleaching) and FLIP (*F*luorescence *L*oss *I*n *P*hoto-bleaching). Upon single bleaching (FRAP) or constant bleaching (FLIP) of the fluorescent signal in a specific area of the cell, the intensity of the fluorophore is monitored over time in the bleached area and in surrounding regions; information is then derived about the diffusion speed of the tagged molecule, the amount of mobile versus immobile molecules as well as the kinetics with which they exchange between different parts of the cell. Thereby, FRAP and FLIP are very informative about the kinetics with which the different organelles of the cell separate into mother- and daughter-specific compartments during cell division. Here, we describe protocols for both FRAP and FLIP and explain how they can be used to study protein dynamics during cell division in the budding yeast *Saccharomyces cerevisiae*. These techniques are easily adaptable to other model organisms.

Key words FLIP, FRAP, Dynamic, Fluorescent intensity, Fluorescent decay, Fluorescent loss, Recovery, Speed, Diffusion, Time

1 Introduction

During vegetative growth, the budding yeast *S. cerevisiae* divides asymmetrically into a bigger mother cell and a smaller daughter cell, the bud. This process starts with the bud appearing on the surface of the mother in S-phase, continues through mitosis when the duplicated chromosomes segregate into the two cells and ends with cytokinesis, which gives rise to two separated cells. All these steps rely on the function of many proteins whose behavior is often

Alberto Sanchez-Diaz and Pilar Perez (eds.), *Yeast Cytokinesis: Methods and Protocols*, Methods in Molecular Biology, vol. 1369, DOI 10.1007/978-1-4939-3145-3_3, © Springer Science+Business Media New York 2016

highly dynamic both in space and time. A remarkable example are septins, small GTPase proteins showing a highly dynamic behavior during cell division: upon bud emergence, at the beginning of S-phase, septins assemble in a fluid, ring-like structure at the base of the bud. At this stage, the septins diffuse easily within the ring and exchange with the cytoplasmic pool. Throughout the following phase of bud growth, the septin ring localizes to the mother–bud neck (the constriction between the mother and the future daughter cell) and forms a frozen, hourglass-like collar, within which septins are essentially immobile. Finally, during cytokinesis, when the septin ring splits and the mother and daughter cells separate from each other, septins become transiently dynamic again, before freezing in two rings, one on each side of the cleavage site [1–3].

During bud growth, the membrane proteins of the *e*ndoplasmic *r*eticulum (ER), which diffuse relatively rapidly in the plane of the membrane, soon stop exchanging between the mother and the daughter ER [4–6]. Similarly, the outer nuclear membrane proteins (such as the reporter NSG1) and the nuclear pores (as visualized by tagging core nuclear pore components such as Nup49 and Nup170) are also restricted in their ability to freely diffuse from the mother to the daughter part of the dividing nucleus [7, 8].

Over the years, fluorescence microscopy techniques (i.e. FLIP and FRAP) have been developed to look at these spatio-temporal dynamic processes and this has allowed scientists to obtain insights into a wealth of cellular processes. Both FLIP and FRAP rely on tagging a specific protein of interest with a genetically encoded fluorescent tag (e.g. GFP, *G*reen *F*luorescent *P*rotein) to visualize the protein of interest in living cells. In a typical FRAP experiment, a laser is pointed to a specific region of the cell to bleach the fluorescent signal of the protein in that area; the bleaching is performed only once and subsequently the recovery of the signal in the bleached area as well as its intensity profile in the surrounding areas are monitored over time. The fluorophore intensity profile in the bleached area is then usually fit to a mathematical model (e.g. an exponential recovery, *see* Subheading 3 for details): this enables to obtain both kinetic information (e.g. diffusion speed of the molecule of interest) as well as non-kinetic information (e.g. the amount of mobile versus immobile molecules). On the other hand, in a typical FLIP experiment, bleaching is repeated constantly during the entire time window of the experiment and the decay of the signal from the surrounding areas is monitored over time. The decay profiles from different regions of the cell are subsequently fit to a model (e.g. one- or two-phase exponential decay, *see* Subheading 3 for details): on one side, this enables obtaining kinetic information such as the diffusion speed of the molecule of interest; but, most importantly, the intensity profiles obtained in FLIP experiments can be used also to extract valuable information about the structural organization of the compartment where the tagged molecule is found: for example, a speed of decay which, upon constant bleaching, stays the same all over the entire compartment would indicate that the compartment is a continuous structure; on the other hand, a difference in speed of decay between different areas of the compartment would suggest the presence of different domains that are not freely communicating with each other.

Here, we present detailed protocols for both FRAP and FLIP experiments which enable the analysis of such dynamic phenomena during cell division in *Saccharomyces cerevisiae*. For practical reasons, we divided the "Materials" and the "Methods" sections in different subheadings. The subheadings 2.1, 2.2 and 2.3 refer to the "Materials" section; the subheadings 3.1, 3.2 and 3.3 refer to the section "Methods". More specifically:

- Subheading 2.1 describes the features of the yeast strains we used in this work as well as how they are constructed.
- Subheading 2.2 lists the composition of all the required media and reagents.
- Subheading 2.3 lists the needed microscopy equipment.
- Subheading 3.1 explains the sample preparation procedures.
- Subheading 3.2 explains the imaging settings. This section is divided into four subsections, respectively: "laser settings" to adjust the laser power; "time series settings" where the time resolution of the experiment is set; "bleaching settings" where the intensity and frequency of the bleaching is set, this subsection is where the main difference between FLIP and FRAP experiments is found; the last subsection, "start experiment," explains how to start the imaging. Note that most of the settings described in Subheading 3.2 are the same between FLIP and FRAP experiments and, unless otherwise stated, they apply to both techniques.
- Subheading 3.3 explains in details the data analysis procedures. We decided to divide this section into two subsections: one relative to FRAP experiments and the other relative to FLIP experiments. Both FRAP and FLIP subsections explain first how to perform the intensity measurements and then how to fit a mathematical model to the experimental data and how to retrieve biological information out of the model. It is important to stress out that in the data analysis section we will refer to specific examples shown in the different figures. We believe that by understanding these examples and adapting what we explain to his/her specific needs, the experimenter will be easily able to perform FRAP and FLIP in his/her condition of interest.

2 Materials

Prepare all materials in deionized water. Unless otherwise specified, only the growth media are sterilized. There is no need of sterile water. Unless otherwise stated, store all reagents at room temperature. The following reagents-media are the same for FLIP and FRAP experiments (*see* **Note 1**).

2.1 Yeast Strains

1. A list of strains used in this work is shown in Table 1. All strains are isogenic to either W303 or S288C (Table 1) and were constructed according to standard genetic techniques [9].

2.2 Media

All the following indicated amounts are for 1 l of total volume.

1. YPD solid-rich media (plates): Yeast Extract 10 g, bactopeptone 20 g, 2 % glucose, 20 mg adenine, Agar Agar SERVA high gel-strength 20 g.
2. Selective liquid –TRP synthetic media: 2.5 g Ammonium sulfate, 850 mg Yeast Nitrogen Base without both amino acids and ammonium sulfate, 2 % glucose, 10 mg adenine, 74 mg drop out –TRP (tryptophan) mix containing all amino acids but tryptophan.
3. Selective solid –TRP synthetic media: 2.5 g Ammonium sulfate, 850 mg Yeast Nitrogen Base without both amino acids and ammonium sulfate, 2 % glucose, 10 mg adenine, Agar Agar SERVA high gel-strength 20 g, 74 mg drop out –TRP (tryptophan) mix containing all amino acids but tryptophan.

2.3 Microscopy Equipment

1. 25 mm diameter round glass coverslips (Thermo Scientific) (Fig. 1).
2. Metallic circular Attofluor Cell chamber for microscopy (to mount the coverslip and the agar pad, *see* Subheading 3.1 and *see* **Note 3**, Fig. 1 for details).

Table 1
***S. Cerevisiae* strains used in this work**

Strain number	Use	Genotype	Background	Mating type
6077	Wild-type strain expressing GFP-Cdc12	GFP-Cdc12:URA3; leu2-3,112; trp1-1; can1-100; ura3-1; ade2-1; his3-11,15	W303	a
7828	Wild-type strain expressing Nup49-GFP	Nup49-GFP:His3; his3Δ1; leu2Δ0; ura3Δ0; met15Δ0	S288C	a
10411	Wild-type strain expressing Sec61-GFP	Sec61-GFP::hpnNT1; his3Δ0; leu2Δ0; met15Δ0; ura3Δ0	S288C	a

1

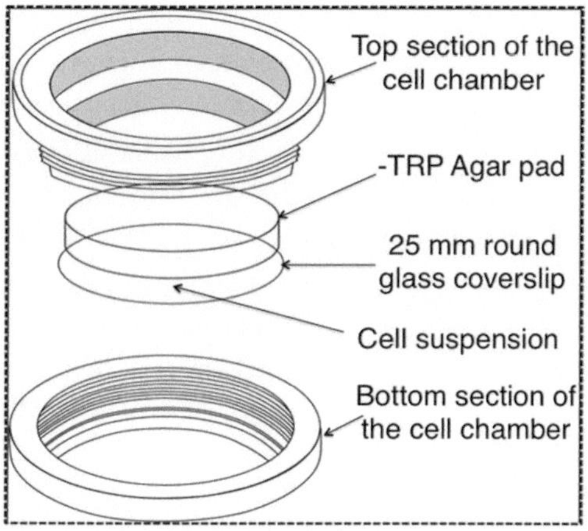

Fig. 1 Attofluor cell chamber used for FLIP and FRAP microscopy experiments. The cell suspension is placed as indicated between the round glass coverslip and the synthetic medium agar pad. The cell chamber is closed by screwing together its top and bottom sections. Another round glass coverslip (not depicted) is dropped on top of the cell chamber. See text for details about the sample preparation

3. Cylindrical metallic support to cut small circular agar pads (*see* Subheading 3.1 for details) (*see* **Note 3**).
4. Confocal microscope (e.g. Carl Zeiss NLO 780) with a Plan-Apochromat 63X/1.4 NA (*n*umerical *a*perture) oil immersion objective, an argon laser (488 nm line or other line of interest, depending on the used fluorophore), a photomultiplier detector system (PMT), an appropriate filter (e.g. 505 nm long-pass filter for GFP) to select for the desired fluorophore and a temperature-controlled imaging chamber.

3 Methods

Unless otherwise indicated, perform all the described steps at room temperature.

Note that, unless otherwise stated, the following steps apply to both FLIP and FRAP experiments.

3.1 Sample Preparation

1. Before the imaging, prepare a culture by streaking some yeast cells on YPD solid media (*see* **Note 2**). If willing to analyze mitotic cells, let the cells grow until exponential phase.
2. Unscrew the Attofluor cell chamber and mount into it a circular 25 mm glass coverslip (Fig. 1 and *see* **Note 3**). Do not touch the coverslip on its surface.
3. Screw the metallic cell chamber back.
4. Drop in the middle of the mounted coverslip not more than 10 μl of –TRP liquid media (*see* **Note 4**).

5. With a pipette, take a few cells previously plated on YPD solid media (**step 1** above) and resuspend them in the 10 μl of –TRP previously dropped on the coverslip.
6. With any available tool, cut a circular agar pad of –TRP solid media and put it on top of the cell suspension. Then put another 25 mm round glass coverslip on top of the agar pad.
7. The sample is now ready to be imaged: if desired, bring the microscope chamber up to 30 °C, then put abundant immersion oil on top of the 63X/1.4 NA microscope objective and fix the sample above the objective.

3.2 Imaging

3.2.1 Laser Settings

1. Launch the imaging software and set the argon laser excitation line wavelength to the desired value (for the FRAP–FLIP experiments shown in this work the wavelength was set to 488 nm, *see* **Note 5**).
2. Set the laser output (in the FRAP and FLIP experiments shown here we used 40–45 %; 2–3 % of this output was then used for imaging, *see* **Note 6**).

3.2.2 Time Series Settings

1. Set the desired time interval. In FRAP experiments, we acquired one frame every 10 s; this interval was 3–6 s for FLIP experiments (*see* **Notes 7** and **9**).
2. Set the number of cycles. This defines the entire duration of the movie. In FRAP–FLIP experiments shown in this work we set 40–60 cycles (*see* **Note 8**).

3.2.3 Bleaching Settings

1. Set the laser intensity to be used for bleaching. For the FRAP–FLIP experiments shown here, we used 80–100 % of the laser output set in the laser settings (*see* Subheading 3.2.1 above). *See* **Note 9** for further details.
2. Set the number of iterations: this number defines how many times the laser will go through the selected region in order to bleach the signal. In this work, this number was set on 80–110 (*see* **Note 9**).
3. Set when the bleaching must start and how often it has to be performed. For FRAP experiments, we performed only one single bleaching at the beginning of the movie. For FLIP experiments, a bleaching-acquisition cycle was performed for each time frame (*see* **Note 10**).

3.2.4 Start Experiment

1. Once all the imaging settings are defined, chose a good field where a cell of interest (*see* **Note 11**) and at least two control cells are present (*see* **Note 16**).
2. If necessary, zoom in to reduce the field of view to a window including only the cells of interest.
3. Using a drawing tool, draw on the cell of interest a small region where the bleaching will be performed (Fig. 2a, b; *see* **Note 9**).

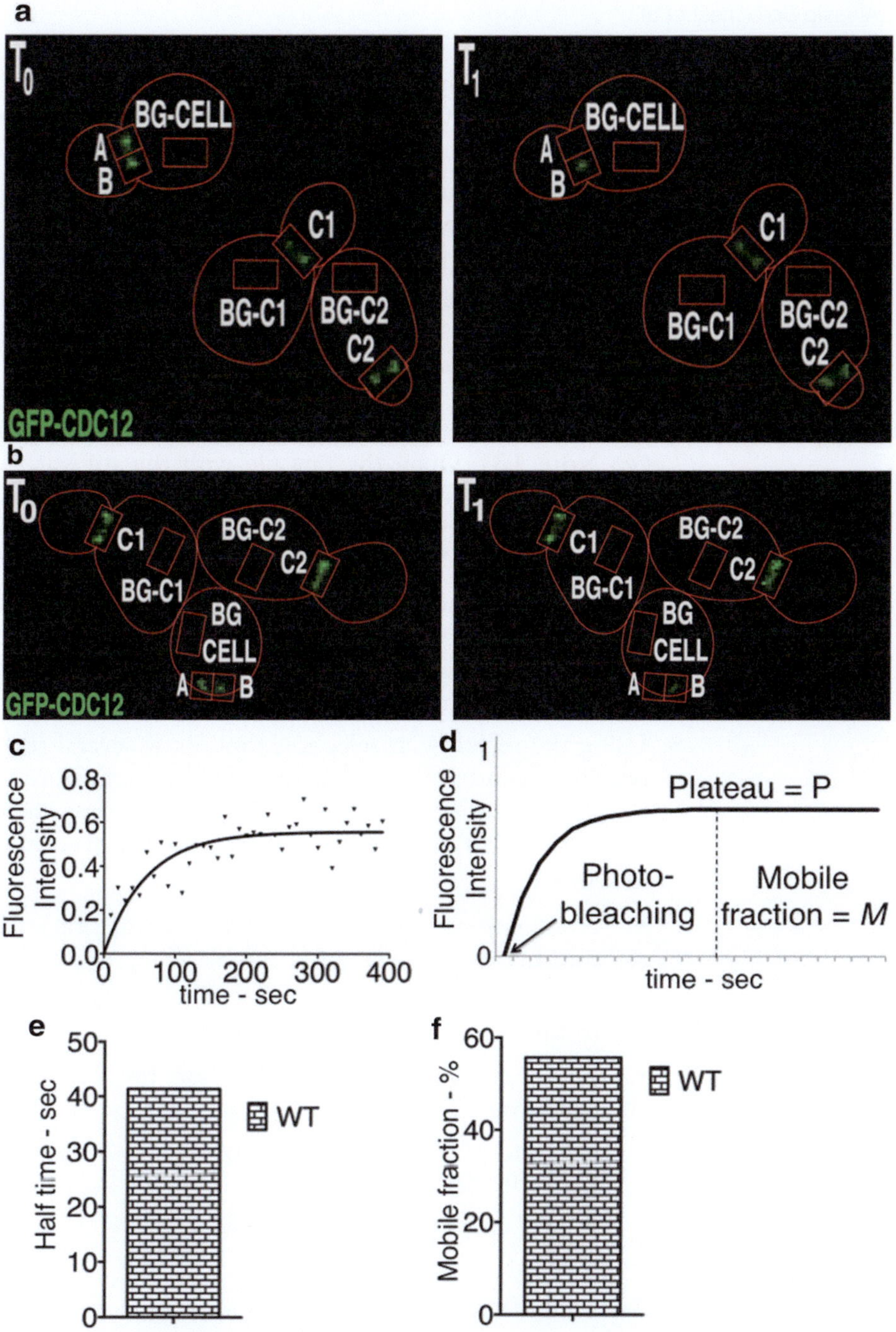

Fig. 2 Wild-type mitotic (**a**) and G1/S (**b**) cells expressing the septin ring component CDC12 tagged with GFP. T_0 and T_1, respectively, refer to the first and the second time frame of the movie. The single photo-bleaching was performed in the region "A" and the fluorescence intensity was measured in the red-labeled rectangular-like regions (see text for details). (**c**) Graph showing the fluorescence intensity recovery in the region "A" from a wild-type cell where the photo-bleaching was performed on half of the septin ring as in Fig. 2b. The fluorescence intensity is plotted on the *Y* axis over time (*X* axis). The data points were fit to a one-phase association function (exemplification in 2D). (**d**) Graph showing the one-phase association equation and its relative graph (see text for details). The fluorescence intensity recovery is plotted against time and the plateau (P), the fraction of mobile molecules (M = P) and the time of photo-bleaching (second time frame, T_1) are indicated. (**e**, **f**) Half time (measured as Ln2/*K*) and mobile fraction (in percentage) measured from the one-phase association model fit to the data in C (see text for details)

4. For each cell and for each time point, acquire both the fluorescent (e.g. GFP) and the transmission images. The transmission channel can be helpful in following the cell cycle stage.
5. Start the experiment and repeat the same procedure to acquire the desired number of cells (*see* **Note 12**).
6. Collect at least 30–40 data sets for each strain in three different experiments preferably on different days. A fraction of data sets may be unusable because of the focal plane shifted or the cell moved (*see* **Notes 12** and **13**).

3.3 Data Analysis

3.3.1 FRAP

1. Discard cells where the focal plane shifts and/or the cells move (*see* **Note 13**). Open the raw file with the software Image-J (FIJI can also be used) and split the two channels (transmission and GFP) apart (for Image-J follow the path: menu "Image" > "color" > "split channels") (*see* **Note 14**).
2. In the Image-J menu "Analyze" > "set measurements," click on either "integrated density" or "mean gray value"; in the FRAP experiments shown here we always measured the mean gray value and we kept the selection area constant (*see* **Note 15**).
3. Using the rectangular selection tool, draw several rectangular regions around the areas where you want to measure the fluorescence intensity (examples in Fig. 2a, b). For each movie (with the cell of interest and at least two control cells) measure as follows:
   - *A* = Bleaching region (half of the septin ring). Here we measure the recovery after the bleaching event.
   - *B* = Non-bleached region: here we monitor what happens to the non-bleached half of the ring.
   - *BG-CELL* = Background signal for the cell of interest (*see* **Note 17**).
   - *C1* and *C2* = Signal intensity for two control cells (*see* **Note 16**).
   - *BG-C1* and *BG-C2* = Background signal of the two control cells.
4. For each of the time frame of the movie and for each single cell of interest, compute as follows:
   - *A–BG-CELL* (to subtract the background signal from the recovery measured in the bleached region).
   - *B–BG-CELL* (to subtract the background signal from the recovery measured in the non-bleached region).
   - The difference *A–BG-CELL* at $t = 0$ (first time frame) will be $(A\text{–}BG\text{-}CELL)t_0$. The same applies to the other time frames and to the difference *B–BG-CELL*.

- *C1–BG-C1* and *C2–BG-C2* to subtract the background signal from the intensity profile measured for the two control cells.
- The difference *C1–BG-C1* at $t=0$ (first time frame) will be (*C1–BG-C1*)t_0. The same applies to the second control cell and to the other time frames.
- Decay of the control cell 1: (*C1–BG-C1*)$t_0 \times 100$/(*C1–BG-C1*)t_0. For the second time frame it will be: (*C1–BG-C1*)$t_1 \times 100$/(*C1–BG-C1*)t_0. And so on for the other time frames. The same applies to the second control cell.
- Decay of the control cells: This is computed by simply averaging for each single time frame of the movie the decays of the two single control cells. For example, this will give us (Average Decay Control)t_0 for the first time frame. Repeat the same for the other time frames.
- Recovery of the bleached region: for the first time frame this will be: ((*A–BG-CELL*)$t_0 \times 100$/(*A–BG-CELL*)t_0)/(Average Decay Control)t_0. For the second time frame: ((*A–BG-CELL*)$t_1 \times 100$/(*A–BG-CELL*)t_0)/(Average Decay Control)t_1. Repeat this for the other time frames to get the complete profile of fluorescence recovery.
- Recovery of the non-bleached region: for the first time frame this will be: ((*B–BG-CELL*)$t_0 \times 100$/(*B–BG-CELL*)t_0)/(Average Decay Control)t_0. For the second time frame you will compute: ((*B–BG-CELL*)$t_1 \times 100$/(*B–BG-CELL*)t_0)/(Average Decay Control)t_1. Repeat this for the other time frames to get the complete profile of fluorescence recovery.

5. Plot singularly all the recovery profiles measured for the bleached and the non-bleached region for each single cell. In Fig. 2c, we show an example of the recovery measured in the bleached area after bleaching half of the septin ring as in Fig. 2b.
6. Once you obtain the recovery profiles, a good way to extract biological information from them is to fit these experimental data with a model. Deciding which model to use is a key step for the analysis (*see* **Note 18**). For the FRAP experiments shown here, we used a one phase association model described by the equation: $\Upsilon = \Upsilon_0 + (P - \Upsilon_0) \times (1 - \exp(-KX))$ where Υ_0 is the Υ value at $t=0$ (first time frame), P is the plateau value and K is the rate constant, expressed in reciprocal of the X axis time unit (*see* Fig. 2d for an exemplification of a one-phase association).
7. Once the model is chosen, define which constrain to apply to the key parameters (e.g. Υ_0, P and K, *see* **Note 19**).

8. Fit the data for each single cell and go through the results to verify the goodness of the fit (*see* **Note 20**).
9. For each cell, take the half time and average this among all the cells to get a single half time. Compute the error on this average (e.g. as standard deviation) (Fig. 2e, *see* **Note 21**).
10. For each cell, take the plateau and average this among all the cells to get a single plateau value. Compute the error on this average (e.g. as standard deviation) (Fig. 2f, *see* **Note 22**).
11. Compare the different conditions among each other. When comparing the mean, a Student's *t*-test can be used to determine the statistical difference among different conditions (*see* **Notes 12, 23** and **24**).

3.3.2 FLIP

1. In order to extract the relevant fluorescence information from the acquired image, images should be analyzed by eye to discard any images where parts of the organelle/cells of interest are out of focus (*see* **Note 13**).
2. Open the raw file with the software Image-J and split the two channels (transmission and GFP) apart (for Image-J follow the path: menu "Image" > "color" > "split channels") (*see* **Note 14**).
3. Using the ImageJ polygonal selection tool, select and save six regions of interest. These regions should include the organelles/cells of interest during the entire course of the acquisition process (e.g. the nucleus growing in the bud during nuclear division, Fig. 3a, b) and should include the following:
   - Compartment 1 (mother compartment; M).
   - Compartment 2 (daughter–bud compartment; B).
   - Control cell 1 (c1).
   - Control cell 2 (c2).
   - Control cell 3 (c3).
   - Background (BG).
     In this case the organelle of interest is the dividing yeast nucleus/ER, which is further divided into the mother and daughter "halves."
4. Under "Analyze" > "Set Measurements", click on either "Mean grey value" or "Integrated Density" (*see* **Note 15**).
5. For each movie (with the cell of interest and at least two control cells) and for each time frame, measure the intensity in the selected areas to obtain raw intensities for the mother and daughter nuclei/ER as well as for the control cells.
6. Normalize the data by the background fluorescence. To do this, simply subtract the background fluorescence from the raw mother and daughter nuclei/ER intensities (*M* and *D*), as well

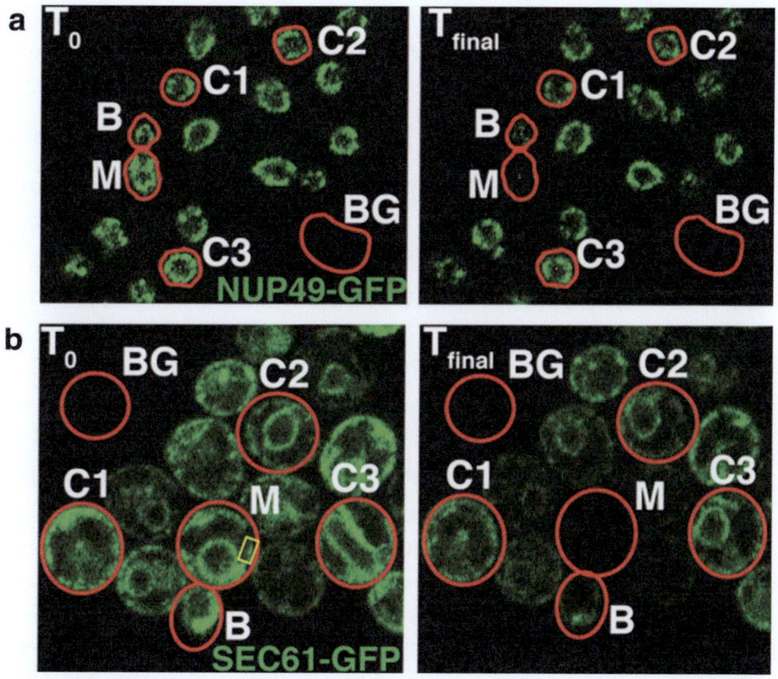

Fig. 3 (**a**) Wild-type anaphase cells expressing the nuclear pore component Nup49 tagged with GFP. T_0 and T_{final} indicate, respectively, the first and the last time frame of the movie. The fluorescence intensity was measured in the red-labeled polygonal regions (see text for details). The red-labeled polygonal regions include the entire area over which the nuclei move over the course of the experiment. Note that "B" in the *left panel* (T_0) is larger than the region of the actual daughter nucleus, but the daughter nucleus grows to the size of Bn at T_{final}. (**b**) Wild-type metaphase cells expressing the ER integral membrane protein Sec61 tagged with GFP. T_0 and T_{final} indicate the first and the last frame of the movie, respectively. The *yellow*-labeled region indicates the area where the photobleaching is performed. The fluorescence intensity was measured in the indicated red-labeled regions (see text for details)

as the three control cells (*C1*, *C2*, and *C3*) obtained in 5. Note that this must be done individually for every time point; for example at time t for the M compartment you have:

- M^t = raw $M^t - BGt$.

7. Photo-bleaching simply due to the acquisition process must also be corrected for. To control for this, the mother or daughter nuclear compartments (*M* and *D*) are divided by the average of the corrected control cells (after BG subtraction):

- Corrected $M^t = M^t / \langle C^t \rangle$ where $\langle C^t \rangle = (C1^t + C2^t + C3^t)/3$.

8. At this point, we have for a single cell and for all the compartments of interest (e.g. M and D) corrected fluorescence intensities profiles. Being interested in the fluorescence loss relative

to the initial fluorescence measurement (M^{t0} and D^{t0}), normalize each time frame of these profiles by its corresponding initial fluorescence measurement:

- Final $M^{t(i)}$ =corrected $M^{t(i)}$/corrected M^{t0}.

9. Repeat this for all the time points and for all the compartments of interests. This will yield the final fluorescence decay profiles.
10. In order to extract biological information from the decay profiles, we must first fit the data according to an appropriate model (*see* **Note 18**).
11. Once the model is defined, define which constrain to apply to the key parameters (e.g. Υ_0, P, and K). For FLIP experiments, we constrained Υ_0 to 100 and the plateau > 0 (*see* **Note 19**).
12. Take the FLIP decays for the M and the D compartments (each containing all the analyzed cells) and fit these data as two ensemble measurements (mother and bud) with one single fit (*see* **Note 20**). In the experiments shown here, we used for both the ER and the nucleus a one-phase decay (for the D) and a two-phase decay (for the M) to fit the decay profiles. An example of such decay is shown in Fig. 4a, b.

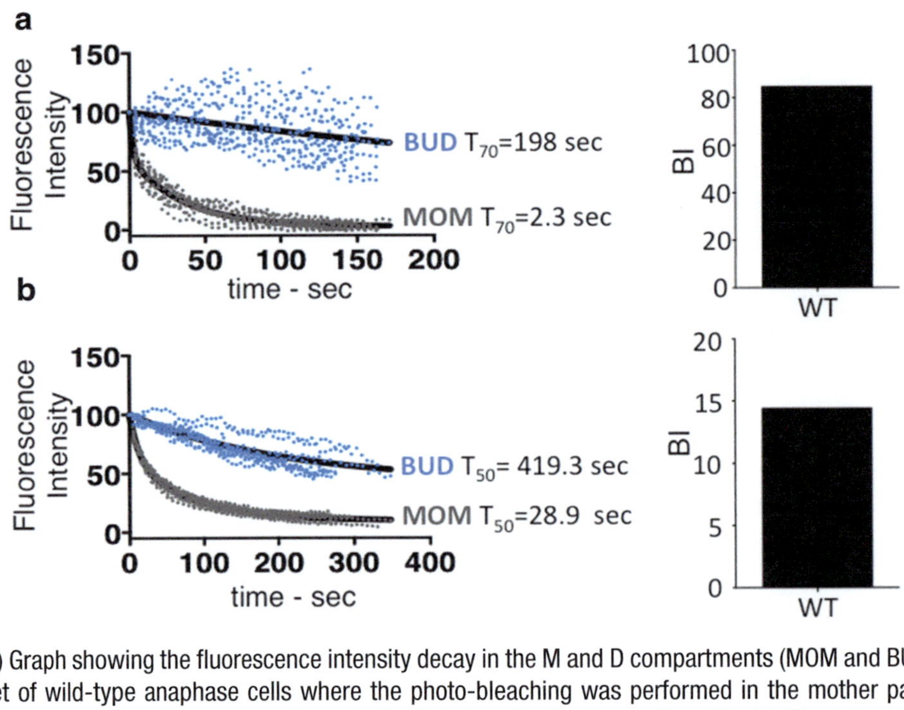

Fig. 4 (**a**) Graph showing the fluorescence intensity decay in the M and D compartments (MOM and BUD), from a dataset of wild-type anaphase cells where the photo-bleaching was performed in the mother part of the nucleus as in Fig. 3a. The fluorescence intensity is plotted on the *Y* axis over time (*X* axis). The data points were fit to a two-phase decay for the M and to a one-phase decay for the D compartments to obtain the BI indicated on the *right* (see text for details). (**b**) As in (**a**), but for the ER membrane protein SEC61-GFP in metaphase cells

13. One information we can retrieve from the fitting is to ask if the M and D regions we monitored are part of a continuous compartment or if they rather define two compartments not freely communicating with each other. To assess this we can measure what we refer to as *B*arrier *I*ndex (BI). The higher the BI the more restricted is the diffusion of the fluorophore (molecule of interest) between the M and D compartments. A BI value comprised between 1 and 2 is indicative of free exchange. Compute the BI as follows:

    - Compute the T_{70} (T_{50} in case of ER), which represents the time it takes to lose 30 % (50 % for the ER) of the original fluorescence intensity for both M and D compartments.
    - For both these compartments (valid for both ER and nucleus FLIP experiments), decide an arbitrary Υ value (e.g. 70, *see* **Note 25**) and draw a line from this Υ value along the X axis of the fitted curve. Stop when you reach the curve itself and then draw a line down, along the Υ axis, until you reach the X axis. The X value you find represents the T_{70}, which is the time at which 30 % of the fluorescence is lost.
    - Once the T_{70} has been calculated for both D and M compartments, the BI is the quotient $T_{70}{}^{\text{bud}}/T_{70}{}^{\text{mother}}$. For ER, BI would be $= T_{50}^{\text{bud}} / T_{50}^{\text{mother}}$ (Fig. 4a, b). The BI can be alternatively measured by a numerical approach, which follows the same principle described above (*see* **Note 26**).
    - Compare the degree of compartmentalization between mother and bud organelles for different conditions (*see* **Note 13–27**).

4 Notes

Here, we provide tips and suggestions concerning sample preparation, imaging settings, and data analysis. Note that unless clearly stated, the following notes are valid for both FLIP and FRAP experiments.

1. All solutions can be stored at 4 °C for several months, if not contaminated.
2. If willing to analyze mitotic yeast cells, it is important to prepare accordingly the culture before imaging (*see* Subheading 3.1). For this purpose, streak some cells on YPD solid media and let them grow either at 30 °C for 5 h or overnight at 25 °C. We observed that growth on solid medium ensures optimal oxygenation of the cells and oxidation of the fluorophores, which is an essential step in their maturation.

3. Other supports rather than the described Attofluor metallic cell chamber can be used to image the cells. The crucial aspect is to have the cells growing in contact with an agar pad in order to provide ideal conditions for their growth at least for 1–2 h, until the agar pad is dry. Before this happens, change the sample and use a fresh agar pad. Note that, in order to reduce the background signal and have a good signal/noise ratio, it is essential to image the cells in synthetic media (e.g. –TRP); this applies to both the media in which the cells are suspended on the coverslips as well as to the agar pad. Finally, note that any available tool can be used to cut the agar pad of –TRP solid media, but obviously, this has to be of the right diameter size compared to the cell chamber in use.
4. We observed that when putting too many cells on the coverslip, they have the tendency to move during the acquisition, preventing data analysis. On the other hand, having too few cells can result in spending too long before finding cells of interest. Also, using a too big volume of medium can result in cells moving during the acquisition of the movie. We found 10 μl to be a good volume of synthetic media (e.g. –TRP) in which to resuspend the cells before imaging.
5. For this work, we used the ZEN 2009 imaging software. Any other imaging software that enables to follow the described steps can be used. The set wavelength for the argon laser differs depending on the excitation and emission spectra of the used fluorophore. GFP (here we refer to the GFP derived from the jellyfish *Aequorea Victoria*) excitation and emission spectra, respectively, peak at 488 and 509 nm. Therefore, we used the laser line set to 488 nm. Adjust this value according to the used fluorophore.
6. The output of the laser in % highly depends on how well the laser itself performs: as long as the bleaching in the selected area is efficient and specific (such that only the region of interest is affected), then the selected output should be good. Please take into consideration that using a high output (usually higher than 45–50 %) can reduce quite quickly the lifespan of the laser. Therefore, it is advisable to reach these values only if the bleaching is not efficient. In addition, the experimenter should try to use as little laser power as possible also to prevent toxicity effects due to excessive exposition. Also, modify pinhole and detector gain for maximal fluorescence signal and minimal pixel saturation. Excessive saturation has to be avoided: once the signal is saturated, even if present, it is not possible to determine differences in fluorescence intensity.
7. The time interval between each frames needs to be adjusted depending on the desired time resolution. But, be aware that

the closer to each others the frames are acquired, the higher is the non-specific bleaching due to exposition; of course, this will cause premature weakening of the signal in the entire field.

8. The number of cycles (i.e. the entire duration of the movie) also depends on the process under investigation and needs to be adjusted accordingly.
9. The laser output used to bleach, the number of iterations, and the size of the bleaching region represent key parameters for efficient bleaching. These values need to be defined case by case, depending on the process of interest. As a general guideline, good values for these parameters should guarantee both rapid and specific bleaching, enabling the researcher to extensively bleach the signal after the first few frames and at the same time affecting only the area of interest. Note that by increasing too much the size of the bleaching region and/or the number of iterations and/or the used output, the bleaching usually becomes slower and more efficient but also too strong with the consequence that the surrounding areas are also affected over time. In addition, increasing the number of iterations and the size of the bleaching area artificially decreases the time resolution: this happens because the system takes longer to bleach a high number of iterations and/or a bigger area: if longer means longer than the set time interval, the system will automatically ignore the set time interval and it will use the time required for bleaching as the real-time interval. Remember, at the same time, to minimize the exposition to the laser in order to prevent photo-toxicity.
10. In FRAP experiments, the photo-bleaching is performed only once in the area of interest. In FLIP experiments (see below), the photo-bleaching is repeated constantly throughout the entire duration of the movie.
11. We found that one of the most critical step toward reproducible data is the cell cycle stage at which the cells are imaged. Therefore, the stage under investigation should be kept identical between different samples.
12. When comparing different strains among each other, it is important to keep the bleaching settings (number of iterations, laser output used to bleach, number of bleaching cycles, size and position of the bleaching area, etc.) and the imaging setting (e.g. laser output) constant between the different strains.
13. Note that if the cell migrates or the focal plane shifts during the course of the experiment, this can bias the data. For example, any decrease in fluorescence intensity due to a change in the focal plane is clearly not a biological phenomenon. Avoid considering cells where this happens.

14. As long as the steps described in the section data analysis are followed, any other image analysis software different from ImageJ and FIJI can be used.
15. The integrated intensity represents the sum of the values of all pixels in the selected area while the mean gray value equals the integrated density divided by the area of the selection. Therefore, the integrated density measures the intensity independently of the area of the selection; on the other hand, the mean gray value changes according to both the signal intensity and the area of the selection. This means that measuring the mean gray value of two different areas having the same signal intensity will give two different values as if the intensity in the selected areas were different. One way around this is to measure the integrated density. Another possibility is to simply measure the mean gray value but keeping the area of the selection always constant. It is imperative to be aware of this distinction and decide accordingly.
16. In order to exclude from the intensity measurements the contribution of any non-specific decay simply due to the constant exposition of the sample to the laser, at least two to three control cells should be included in each field with the cell of interest. Ideally, the signal of these control cells is measured at a stage where it is stable on its own such that any observed decay reflects specifically and exclusively the decay due to the laser exposition. For example, in the FRAP experiments shown here, we used mitotic cells as control cells because in this stage the signal of septins is known to be stable.
17. In order to exclude from the intensity measurements the contribution of any non-specific signal, a background region should be included in the measurements of both the cell of interest and the control cells and its signal should be deducted from the signal of interest (see text for details).
18. It is imperative that the model with which to fit the experimental data is carefully chosen, based on what biological phenomenon the experimenter is looking at and what is the model that best describes that process. Looking at the residual plot of the data (they should spread randomly around the central line) can be helpful in deciding the model to use (also *see* **Note 20**). In the present work, we used the software Prism 6 (GraphPad Software Inc., La Jolla, CA, USA) to perform nonlinear regression fitting and statistical analysis but any other appropriate software can be used as well.
19. Any set constrain can have a big impact on the results of the fitting. As a general rule, the key parameters of the model (e.g. Υ_0, P, and K for the one-phase association model used for FRAP experiments) should be constrained to specific values only if there is a scientific reason to do so. For example, if the

experimenter performs FRAP and decides to plot the recovery of the bleached region starting from the time frame after bleaching and if the first plotted value for all the cells is 0 then Y_0 could be constrained to 0; on the other hand, if there are scientific reasons to think that the recovery should approach a specific Y value, then this plateau value can be used as a constrain for the plateau parameter. In the case of FLIP experiments, since we want to monitor the proportion of fluorescence at each time point relative to the fluorescence at the initial time point (t_0), the initial value could be constrained to 100 %; furthermore, if we are fitting the data to an exponential decay model with a plateau (i.e. the compartment does not decay to 0), it is also important to specify whether we wish to constrain this parameter. This is also important since in some cases, the line of best fit will decay to values below 0, which in physical terms makes no sense. Also, in some conditions, the fluorescence decay of a FLIP experiment does not decay below 70 % in the bud, and hence, it is impossible to calculate a T_{70}. In such scenarios, we have found that imaging shorter time periods (corresponding to a smaller anaphase time window) and constraining the plateau to 0 in the bud can be used to extract bud T_{70}'s. By fitting the data in this way, we are assuming that early anaphase lasts an indefinite amount of time and that if we kept bleaching the cell in this stage, the fluorescence in the bud would eventually decay to 0. In our ensemble analysis, we did not constrain this parameter since the fluorescence in the bud decays to values below 70 % (or 0.7). However, fitting shorter time-windows with a plateau constrained to 0 can be used for single cell analysis, such that a T_{70} can be calculated for all buds. In this case, it is of course important to keep the same constrains among different conditions.

20. As a general rule, the closer the curve is to the data points, the more accurate and biologically significant are the best fit values we obtain from the fit. The R^2 value tells exactly how close the curve is to the data point. The closer the R^2 is to 1, the closer to the data points the curve will be. But it is important not to overestimate the importance of the R^2 value and not to judge the fit only based on this parameter. In fact a value of 1 does not mean anything on its own if for example the best-fit values make no sense (e.g. a negative rate constant) or the confidence intervals are too wide. In most cases, the entire point of nonlinear regression is to determine the best-fit values of the parameters in the model. The confidence intervals are another important parameter to take into account as they tell exactly how tightly you have determined these best-fit values. If a confidence interval is very wide, your data do not define that parameter very well.
21. The half time is the first important best-fit values obtained out of the fitting in a FRAP experiment. It is computed as Ln2/K

and it tells the time at which 50 % of the total signal is recovered. This means that the half time provides with kinetic information about how fast the molecules are diffusing (for example in the bleached area where we measure).

22. The plateau is the second important best-fit values obtained out of the fitting in a FRAP experiment. It represents the mobile fraction of the molecules pool. In other words, it does not provide with any kinetic information but it simply tells, independently of their speed, the amount of mobile molecules (mobile fraction) versus the immobile ones (immobile fraction). The immobile fraction can be of course computed as $1-P$.
23. It is important to stress that when fitting the cells singularly, as described in the text for FRAP experiments, it might happen that some of them will not fit the model. These cells are usually outliers and it often makes scientifically sense to exclude them from the analysis. But this decision has to be based on scientific reasons. If there are reasons to believe that all the cells should be taken into account then an alternative approach is to measure mobile fraction and half time also from one single fit performed on all the cells of a given condition. In this case all the cells are fitted to the model and one single value for the parameters of interest is obtained. This was the method of choice for FLIP experiments, although in this case an individual cell analysis can also be performed; this would mean that M and D compartments are fitted individually for each individual cell, and a BI is computed for that particular cell. Afterward, one can look at the distribution of BIs and compute the relevant statistics. We found that while the choice of how to do this analysis is up to the user, it is recommended that both types of analyses (one single fit and single cell analysis) are carried out, since ensemble measurements can mask phenotypic variability (e.g. two phenotypic groups can be masked by their average).
24. Note that for FRAP experiments, the recovery we measured in the bleached region after bleaching half of the septin ring in G1-S cells (as shown in Fig. 2b) is in principle the result of both lateral diffusion of molecules from the non-bleached part of the septin ring and then molecules diffusing from the cytoplasm. The recovery from the cytoplasm can be assessed by bleaching completely the ring of G1/S cells and then measuring the recovery in this bleached area; then, if willing to analyze specifically the contribution of lateral diffusion only, a more precise measurement can be performed by subtracting for each cell the recovery values obtained in the bleached area after bleaching the entire ring from the recovery obtained in the bleached region after bleaching of half of the ring.
25. The BI, or the Barrier Index, is a way of quantifying the relative compartmentalization of a molecule (in our case, Nup49-GFP

or SEC61-GFP) between two compartments. This is computed by first calculating the T_{70}, or the time it takes to lose 30 % of the initial fluorescence, for bud and mother, after which we take the quotient of their T_{70}'s (bud T_{70}/mother T_{70}). While we decided to use the time it takes to lose 30 % of the fluorescence for the nucleus and 50 % for the ER, this is an arbitrary value and it is up to the user to decide, depending on the process in analysis and what part of the decay the experimenter wants to look at.

26. To calculate the T_{70} for a one-phase decay curve, we have to solve the respective one-phase equation for both M and D compartments:
    - $\Upsilon = (\Upsilon 0 - \text{Plateau}) \times \exp(-K \times t) + \text{Plateau})$ for t where $\Upsilon = 70\,\% \times \Upsilon_0$.
    - Solving this equation yields: $T_{70} = \text{Ln}\left[(0.7 \times \Upsilon 0 - \text{Plateau}) / (\Upsilon 0 - \text{Plateau})\right] / -k$.
    - Note that if the T_{50} is calculated, then we simply solve the same equation, but for $\Upsilon = 0.5 \times \Upsilon_0$. For the M compartment where a two-phase decay equation was used, it is still possible to extract a T_{70} by numerical methods.
27. A decrease in the BI can be influenced by a faster or slower decay in the M compartment. Therefore, when comparing the BI for different conditions, it is important to check the decays of the M compartments before concluding anything about the compartmentalization index.

Acknowledgements

The authors thank the former members of the Barral laboratory: Jeroen Dobbelaere, Cosima Luedeke, Stéphanie Buvelot Frei, Zhanna Shcheprova, Sandro Baldi, Barbara Boettcher, Annina Denoth-Lippuner and Lori Clay, who have played an important role in establishing these techniques in the analysis of protein dynamics and cellular compartmentalization during yeast division. This work was supported by funding from the European Research Council and from ETH.

References

1. Dobbelaere J, Gentry MS, Hallberg RL, Barral Y (2003) Phosphorylation-dependent regulation of septin dynamics during the cell cycle. Dev Cell 4:345–357. doi:10.1016/S1534-5807(03)00061-3
2. Merlini L, Fraschini R, Boettcher B, Barral Y, Lucchini G, Piatti S (2012) Budding yeast dma vdle position checkpoint by promoting the recruitment of the Elm1 kinase to the bud neck. PLoS Genet 8:e1002670. doi:10.1371/journal.pgen.1002670
3. Lippincott J, Shannon KB, Shou W, Deshaies RJ, Li R (2001) The Tem1 small GTPase controls actomyosin and septin dynamics during cytokinesis. J Cell Sci 114:1379–1386

4. Clay L, Caudron F, Denoth-Lippuner A, Boettcher B, Buvelot Frei S, Snapp EL, Barral Y (2014) A sphingolipid-dependent diffusion barrier confines ER stress to the yeast mother cell. Elife. doi:10.7554/eLife.01883
5. Luedeke C, Frei SB, Sbalzarini I, Schwarz H, Spang A, Barral Y (2005) Septin-dependent compartmentalization of the endoplasmic reticulum during yeast polarized growth. J Cell Biol 169:897–908. doi:10.1083/jcb.200412143
6. Chao JT, Wong AKO, Tavassoli S, Young BP, Chruscicki A, Fang NN, Howe LJ, Mayor T, Foster LJ, Loewen CJ (2014) Polarization of the endoplasmic reticulum by ER-septin tethering. Cell 158:620–632. doi:10.1016/j.cell.2014.06.033
7. Shcheprova Z, Baldi S, Frei SB, Gonnet G, Barral Y (2008) A mechanism for asymmetric segregation of age during yeast budding. Nature 454:728–734. doi:10.1038/nature07212
8. Denoth-Lippuner A, Krzyzanowski MK, Stober C, Barral Y (2014) Role of SAGA in the asymmetric segregation of DNA circles during yeast ageing. Elife. doi:10.7554/eLife.03790
9. Guthrie C, Fink G (1991) Guide to yeast genetics and molecular biology: methods in enzymology, vol 194. Academic press, San Diego, NLM ID: 9201639

Chapter 4

High-Speed Super-Resolution Imaging of Live Fission Yeast Cells

Caroline Laplante, Fang Huang, Joerg Bewersdorf, and Thomas D. Pollard

Abstract

We describe a step-by-step method for high-speed fluorescence photoactivation localization microscopy (FPALM) of live fission yeast cells. The resolution with this method is tenfold better than spinning disk confocal microscopy.

Key words Super-resolution, Fluorescence microscopy, FPALM, Live cells, Fission yeast

1 Introduction

Imaging cellular structures with a tenfold improvement over classical light microscopy techniques has been achieved using single-molecule switching nanoscopy (SMSN) techniques called PALM, FPALM, or STORM [1–3]. These methods localize single fluorescent molecules in a field with nanometer precision as they are stochastically switched on and off. Precise localization of thousands of separately blinking emitters over thousands of consecutive camera frames achieves 25- to 40-nm resolution.

Until recently, data acquisition was too slow for live cell imaging. The readout speed of EMCCD cameras limited the rate of localizing molecules. Furthermore, the photoactivatable or photoswitchable fluorescent proteins available to tag proteins of interest had some limitations, such as oligomerization and/or fluorescence properties. Two recent developments opened the possibility for fast SMSN of live cells: ultrafast acquisition with the more sensitive scientific complementary metal–oxide–semiconductor (sCMOS) cameras [4, 5] combined with brighter monomeric photoconvertible fluorescent proteins such as mEos3.2 [6].

We show how to combine recent advances in high-speed FPALM [4] with endogenous gene targeting of the fission yeast

Alberto Sanchez-Diaz and Pilar Perez (eds.), *Yeast Cytokinesis: Methods and Protocols*, Methods in Molecular Biology, vol. 1369, DOI 10.1007/978-1-4939-3145-3_4, © Springer Science+Business Media New York 2016

Schizosaccharomyces pombe to achieve super-resolution images of endogenously mEos3.2-tagged proteins in live cells on a second time scale.

2 Materials

Prepare all solutions using ultrapure (deionized) water and store them at room temperature (unless indicated otherwise by the manufacturer). Filter-sterilize (0.22 μm pore size filter) all solutions to be used for FPALM imaging.

2.1 Gene Targeting Components

We followed the general protocols described by Bähler et al. [7].

1. High fidelity polymerase such as Phusion (New England BioLabs) and accompanying buffers and solutions. Store and handle the enzyme according to the manufacturer's instructions.
2. Long primers resuspended to 50 μM in ultrapure water [7] (Table 1).
3. Fission yeast transformation vectors containing the mEos3.2 DNA sequence to use for generating insertion DNA fragments.
4. dNTPs resuspended in ultrapure water at 10 mM.
5. PCR product purification kit such as spin column DNA purification kit by Qiagen.
6. 10× TE: 100 mM Tris–HCl, pH 7.5, 10 mM EDTA. Adjust pH to 7.5 with HCl and filter-sterilize [7].
7. 10× LiAc: 1 M lithium acetate (102 g/mol). Adjust the pH to 7.5 using acetic acid and filter-sterilize [7].
8. PEG/LiAc/TE: 40 % (w/v) PEG 3350 (Sigma P4338), 1× LiAc, 1× TE. Filter-sterilize [7].
9. 1× TE: Dilute 10× TE with ultrapure water.
10. 1× LiAc: Dilute 10× LiAc with ultrapure water.

Table 1
Sequences of long insertion primers

Vector	Forward primer sequence	Reverse primer sequence
pFA6a-mEos3.2-KanMX6	ATGAGTGCGATTAAGCCAGAC	GAATTCGAGCTCGTTTAAAC
PGOI-pFA6a-mEOS3 .2	GAATTCGAGCTCGTTTAAAC	TCGTCTGGCATTGTCAGGCA

The sequences listed in this table anneal to the vector only. The sequences that anneal to the genomic sequences are generated by the Bähler website (http://www.bahlerlab.info/resources/). Use GGAGGTGGAGGT as forward 4-glycine linker sequence for forward primers (C-terminal tagging) and use ACCTCCACCTCC as reverse 4-glycine linker sequence for reverse primers (N-terminal tagging), GGA and GGT being the two most abundant glycine codons in the fission yeast genome [15]

2.2 Yeast Growth Media

For both growth media, we added the five supplements (adenine, L-histidine, L-leucine, uracil, and L-lysine) for the auxotrophic markers expressed by our strains. Forsburg gives recipes for each [8] (http://www-bcf.usc.edu/~forsburg/media.html).

1. Rich yeast medium (YE): Forsburg gives the recipes for YE liquid medium and agar plates [8].
2. Edinburgh minimal medium (EMM) used for imaging: (*see* **Note 1**) Forsburg gives the recipe for EMM [8]. Follow the protocol for the ingredients (*see* **Note 2**).
3. 50 mL Erlenmeyer flasks with deep baffles and loose caps to ensure proper oxygenation of the cell culture.
4. Wood applicators (Puritan). Autoclave in stainless steel canisters (*see* **Note 3**).

2.3 Imaging Materials

1. n-propyl gallate to prevent photobleaching: Make a 100× stock (10 mM) solution by adding 0.22 g *n*-propyl gallate (Sigma P3031) to 100 mL EMM5S. Dilute to 10× (1 mM) with EMM5S, filter-sterilize, make 1 mL aliquots and freeze at −20 °C. Use at 0.1 mM final concentration in EMM5S imaging medium.
2. Coverslips thickness number 1.5.
3. Glass microscope slides 25 × 75 mm × 1 mm thick.
4. Gelatin pads: Add 0.25 g porcine gelatin (Sigma G2500, gel strength 300, Type A) to 1 mL EMM5S + 0.1 mM *n*-propyl gallate to obtain a 25 % (w/v) gelatin mix (*see* **Notes 4** and **5**). Heat to 65 °C for 10–15 min. Pipette 30 μL of gelatin onto a clean microscope glass slide and immediately place another glass slide on top to flatten the gelatin into a thin disk about 10 mm in diameter [9] (*see* **Note 6**, Fig. 1). Store gelatin pads in a sealed, humid container lined with wet paper towels at 4 °C for up to 2 days to prevent drying of the pad. Wrap the slides loosely in aluminum foil to avoid contact with the paper towels.
5. VALAP: Combine equal parts by weight of Vaseline, Lanolin, and Paraffin wax (flakes work best) in a disposable heat-resistant container (*see* **Note 7**). Heat the mixture to ~70 °C in a

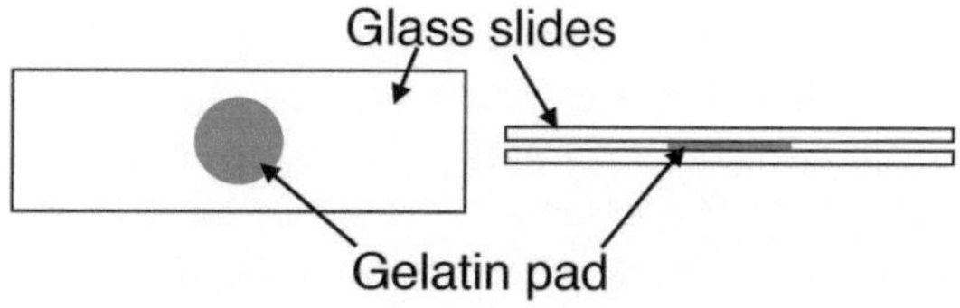

Fig. 1 Diagram of gelatin pad. A 30 μL aliquot of hot gelatin sandwiched between glass microscope slides flattens into a pad about 10 mm in diameter [9]

fume hood until the ingredients mix uniformly. Aliquot VALAP into 1 mL microcentrifuge tubes, cool to room temperature, cap and store at room temperature.

3 Methods

Among the currently available photoactivatable or photoswitchable fluorescent proteins for tagging fission yeast proteins we have experience with mEos2 and mEos3.2, a version of mEos2 mutated to eliminate oligomerization [6] (*see* **Note 8**).

3.1 Synthesize Long PCR Product

1. Design long primers specific to the gene to be tagged. Long primers for insertions are made of two parts; the first section anneals to the vector used as template (Table 1 for sequences to use, Fig. 2 for organization of vector template) and the second section anneals to the genomic sequence flanking the gene of interest (*see* **Note 9**). Order oligonucleotides and have them purified by polyacrylamide gel electrophoresis (PAGE).
2. Design and order checking primers. Checking primers are used to confirm the insertion of the mEos3.2 tag in the genome. Use Table 2 and the Bähler website (http://www.bahlerlab.info/resources/) to design your primers.

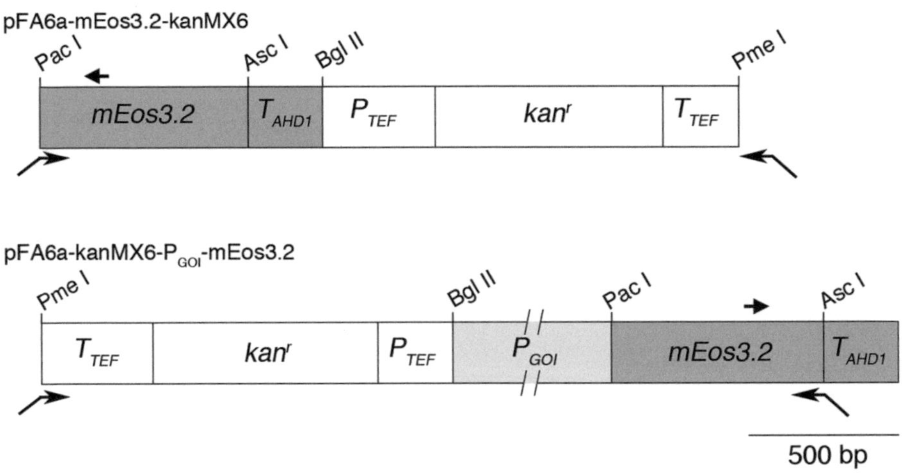

Fig. 2 Schematic of cassettes used as PCR templates to generate the fragments to be used to tag a protein coding sequence in the genome on the C-terminus (*top*, pFA6a-mEos3.2-kanMX6) or N-terminus (*bottom*, pFA6a-kanMX6-P_{GOI}-mEos3.2). *Thin arrows* within the *boxes* show directions of transcription. *Large arrows* outside the *boxes* indicate PCR primers for amplifying the insertion cassette (*bottom*) and checking primers to verify insertion (*top*). The primers are not to scale. *See* Table 1 for primer sequences annealing to the vector and Table 2 for checking primer sequences. For the N-terminal tagging cassette, insert the promoter for the gene of interest into the Bgl II-Pac I site. The breaks in the box for *PGOI* (Promoter of Gene of Interest) indicate that promoters vary in length

Table 2
Sequences for the checking primers of genes tagged at their endogenous gemomic loci

Checking primer sequences	Forward primer	Reverse primer
C-terminal tagged genes	Design using Bähler website	GCATAACTGGACCATTGGCGG
N-terminal tagged genes	GGTGTCAAGTTACCAGGCCGG	Design using Bähler website

These primers anneal to the mEos3.2 gene sequence. The primers that are not provided anneal to the sequence of the tagged gene and can be generated using the Bähler website or generated using the gene sequence

3. Use 0.25–0.5 ng of vector as amplification template DNA per 50 μL PCR reaction (*see* **Note 10**).
4. Synthesize the PCR products following the polymerase instructions and using buffers provided by the manufacturers (Fig. 2) (*see* **Notes 11** and **12**). Keep 1–2 μL of the unpurified PCR product to run by agarose gel electrophoresis.
5. Purify the PCR products using a purification kit with spin column (Qiagen) to remove contaminants from our PCR products. Elute the purified DNA in ultrapure water (*see* **Note 13**). Keep 1–2 μL of purified PCR product for DNA gel electrophoresis.
6. Run the samples of unpurified and purified PCR product by DNA gel electrophoresis to ensure the recovery of an amplified PCR product of the expected size.

3.2 Gene Targeting in Fission Yeast

1. Follow Bähler's protocol for fission yeast transformation [7]. We use 50 μL of PCR amplified DNA for each transformation.
2. Plate the transformed cells onto YE5S agar plates, incubate at 25 °C for 24 h.
3. Replica plate the transformed cells onto selective media YE5S plates, incubate at 25 °C for 3–4 days or until colonies are visible.
4. Select colonies and re-streak onto selective YE5S agar plate (*see* **Note 14**).

3.3 Selection of the Transformed Strains

1. Verify insert by PCR. Using a sterile wood stick, collect yeast cells (a tip full of yeast cells from a healthy culture growing on a YE5S agar plate) and resuspend into 50 μL of ultrapure water, vortex well to mix. Boil for 15 min and spin down at max speed in a tabletop centrifuge for 2 min. Take 10 μL of the supernatant genomic DNA for a 50 μL diagnostic PCR reaction using the mEos3.2 "checking" primers (Table 1 and Bähler website). Run the PCR products by agarose gel electrophoresis. Strains

with an insert should have amplified a PCR product of the desired size.

2. Verify expression of the fusion protein by fluorescence microscopy. An abundant protein tagged at its endogenous locus with mEos3.2 can be detected by fluorescence microscopy to confirm its expression. Image the green, non-photoconverted, fluorescent species of mEos3.2 using excitation at 488 nm and the emission filter used to detect GFP. This works well for abundant proteins, especially if they concentrate in a specific organelle or structure (such as the cytokinetic contractile ring), but the green fluorescence of mEos3.2 bleaches rapidly, making detection of low abundance and/or diffusely distributed proteins difficult.

3.4 Cell Preparation for Microscopy

Use standard sterile techniques to prevent contamination. Grow cells in liquid media in the dark to prevent photobleaching of the fluorescent protein. Solutions used for imaging are filter-sterilized (size of pore 0.22 μm) to remove particles that cause fluorescent background.

1. Grow cells in loosely capped Pyrex Erlenmeyer flasks (volume 50 mL) in a shaking water bath set to 25 °C in the dark for 36–48 h in 10 mL of YE5S medium prior to imaging. Dilute the cultures twice daily to maintain the cells at OD_{595nm} 0.05–0.5 (1×10^6 to 1×10^5 cells/mL) (*see* **Note 15**).
2. Cell cycle synchronization (Optional): Cross the strain expressing the tagged protein with a *cdc25-22* strain, a temperature-sensitive mutation that arrests the cell cycle at the G2-M transition at the non-permissive temperature of 36 °C. Select the resulting progeny to express your tagged protein and the *cdc25-22* mutation. Cells with the *cdc25-22* mutation grow very long when incubated at 36 °C and this phenotype can be used for selection. The day of imaging, dilute your *cdc25-22* mutant cells to OD_{595} 0.2 (4×10^6 cells/mL) and incubate in a shaking water bath at 36 °C for 4 h to arrest cells at the G2-M transition. Prior to imaging, return the arrested cell cultures to room temperature (22–25 °C) and proceed expeditiously with the following steps to prepare the cells for imaging as they enter mitosis synchronously.
3. Collect 1 mL of cells (OD_{595} 0.2–0.4 (4×10^6 to 8×10^6 cells/mL)) by centrifugation at 2348 × *g* (5000 RPM in a tabletop microcentrifuge) for 30 s with a tabletop centrifuge and discard the supernatant (*see* **Note 16**).
4. Wash the cells by resuspending the pellet in 1 mL EMM5S. Collect the cells as above. Discard the supernatant.
5. Wash the cells by resuspending the pellet in 1 mL EMM5S with 0.1 mM n-propyl-gallate. Collect the cells as above. Discard the supernatant.

6. Resuspend the pellet in 10–50 μL EMM5S with 0.1 mM n-propyl-gallate depending on the desired density of the cell suspension.
7. Pipette 5 μL of cell suspension onto the gelatin pad of a prepared microscope slide (*see* **Note 17**). Cover the cells with a coverslip and seal with VALAP.
8. Image the cells immediately.

3.5 FPALM Setup and Cell Imaging

Huang et al. describe the FPALM microscope illustrated in Figs. 3 and 4 [4]. The cost of the two lasers, acousto-optical tunable filter (AOTF), sCMOS camera, optical components, an optical bench and a computer is currently about 100,000 USD. With help from someone knowledgeable about optics and programming, a cell biology laboratory should be able to modify its own fluorescence microscope for FPALM imaging.

1. Focus on cells of interest using bright field illumination, preferably DIC or phase-contrast optics. Once the cells are in focus, turn off the bright field illumination.
2. Set the power of the 561 nm imaging laser at the sample plane to 1-2 kW/cm^2 (*see* **Note 18**).

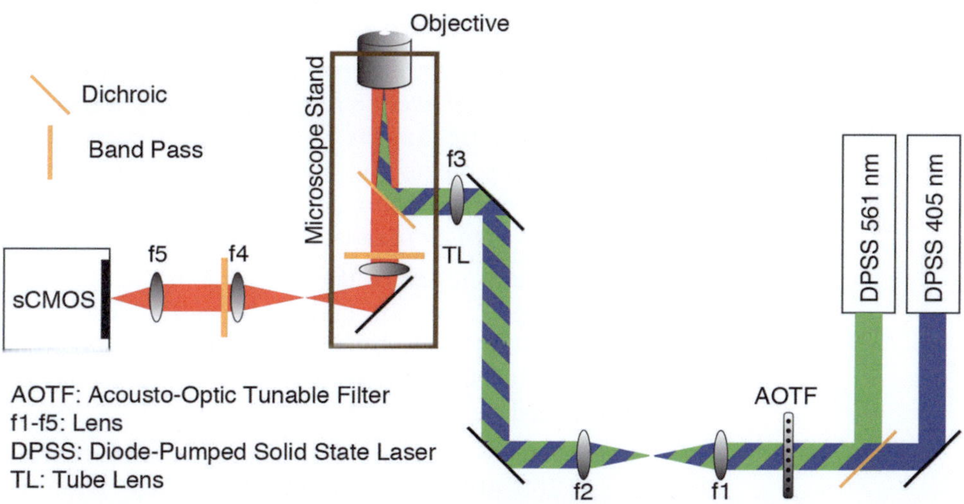

Fig. 3 Simplified schematic of custom-built FPALM setup with a sCMOS camera modified from [4]. Beams from a 405 nm laser (CrystaLaser model DL-405-50; 50 mW output power) and a 561 nm laser (MBP Communications model F-04306-102, 500 mW output power) are combined and sent through an acousto-optical tunable filter (AOTFnC-400.650-TN, AA Opto-Electronic). Lenses f1 and f2 expand the beam. Lens f3 focuses the laser beams into the back focal plane of the objective (alpha PlanApochromat 100×/1.46 Oil DIC, Zeiss) on an inverted microscope stand (Axio Observer D1, Zeiss) for wide field illumination. The emitted fluorescence from a single fluorophore such as mEos3.2 is collected by the objective and separated from the excitation light by a dichroic mirror (Di01-R405/488/561/635, Semrock; for a setup limited to imaging of the mEos3.2 detection channel only, a single-edge dichroic should be used) and two bandpass filters (FF01-446/523/600/677 and BLP01-635R-25, Semrock) before being focused on the sensor chip of the sCMOS camera (Orca Flash 4.0, Hamamatsu). Relay optics (lenses f4 and f5) magnify the image so that the camera pixel size corresponds to ~103 nm in the sample

Fig. 4 Photograph of FPALM microscope with labeled components and the light paths from the two lasers traced in blue and green. The laser clean-up filter provides high transmission of the desired wavelengths while blocking unwanted wavelengths. Visible components from the schematic in Fig. 3 are labeled

3. Use the AOTF to set the 405 nm photoactivation laser intensity to 0 W/cm^2 and then increase the power at the imaging plane by 3-5 W/cm^2 every 5 s, increasing the photoconversion rate to compensate for the progressive, irreversible photobleaching. Ramping up the photoconversion power is more reproducible with an automated software program (for example, LabView) than by manual adjustments.
4. Acquire images until all of the mEos3.2 in the field has been photoconverted, imaged, and irreversibly bleached. Then move on to a new field. Active mEos3.2 "blinks" on and off a few times by temporarily entering a dark state before being photobleached (*see* **Note 19**).
5. Set frame rate. We acquire data with a Hamamatsu sCMOS camera at 200–400 Hz. Other frame rates may be more suitable depending on the properties of the sample. If using our data analysis method and a Hamamatsu ORCA-Flash4.0 sCMOS camera, disable the automatic pixel correction in the acquisition software (check for similar settings on cameras from other manufacturers) (*see* **Note 20**).

3.6 FPALM Data Analysis

Huang et al. [4] describe the steps to localize the central position of each blinking molecule in each image (Fig. 5). Each pixel of a sCMOS camera has its own noise, so the key to using a sCMOS camera for FPALM is to characterize the camera noise. Use either

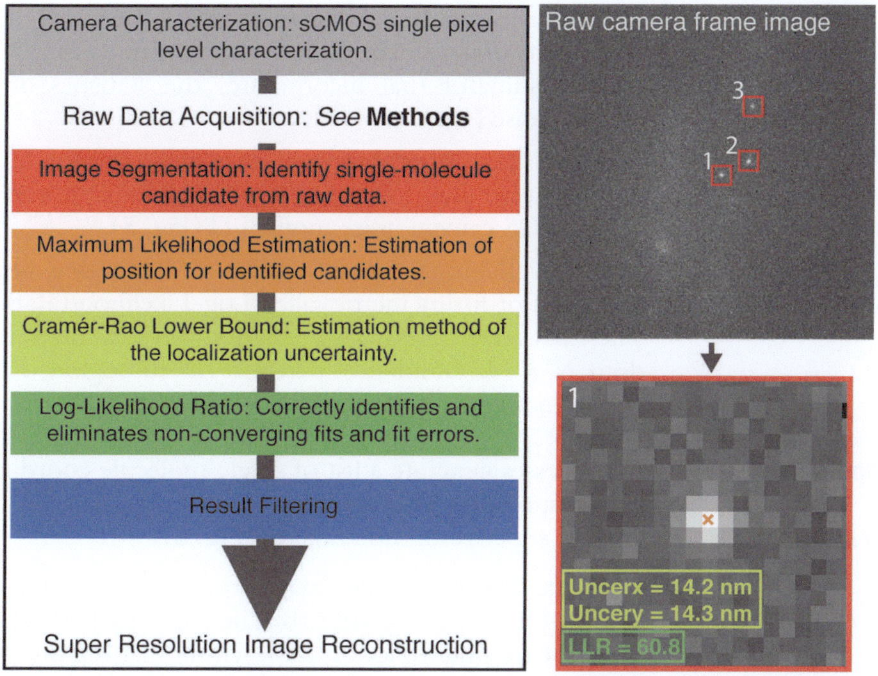

Fig. 5 Analysis and reconstruction of super-resolution images. *Left*, flow chart of image processing steps. *Right*, analysis of a single sCMOS camera frame with three single molecule emitters. The image analysis steps, color-coded in the flow chart, are highlighted on the picture using the same color code. The key to using a sCMOS camera is to generate the pixel-dependent noise map that includes the mean (offset), variance, and the amplification gain of each pixel [4]. This map is then applied at each step required to localize single-molecule emitters and generate a cumulative coordinate map of emitter locations for the final image reconstruction

GPU or CPU processors for the analyses depending on the complexity of the computation. We implemented all algorithms in MatLab, using CUDA-C-MatLab code for the GPU algorithms [4].

1. Characterization of sCMOS camera noise: Take a series of dark images with zero expected incident photon (camera aperture capped and lights turned off in the room) to characterize the offset and variance of each pixel on the camera. Following this step, take a series of images over a range of light levels ranging from 20 to 200 photons per pixel to characterize the gain of each pixel. Huang et al. describe a more detailed protocol [4]. Incorporate these statistics in the single-molecule detection, position estimation, uncertainty estimation, and rejection processes described below [4].
2. Image Segmentation (IS): Filter the raw, unprocessed sCMOS images, using a series of uniform filters to reduce the peak of the background fluorescence noise. Use a maximum filter to identify local maxima in the filtered image and isolate regions of potential single molecule emitters in square subregions of the raw images [10] (Fig. 5).

3. Maximum Likelihood Estimation (MLE): Fit each subregion with a two-dimensional (2D) Gaussian using a maximum likelihood estimator that incorporates the statistics of sCMOS camera noise [4].
4. Estimate the uncertainties of the single molecule position using Cramér-Rao lower bound (CRLB) taking into account the sCMOS camera noise model [4] (Fig. 5).
5. Reject overlapping emitters and non-converging fits using a goodness-of-fitting metric called Log-Likelihood Ratio (LLR). The LLR metric follows a χ^2 distribution with degrees of freedom N–K (N: number of pixels in the subregion; K: number of parameters in the fitting) [4, 11] (Fig. 5).

3.7 Reconstruction of Super-Resolution Images

These processes generate a list of single molecule coordinates and their corresponding uncertainties for use in quantitation or to reconstruct super-resolution images (Fig. 6).

1. Generate 2D histograms images on a fine pixel map (typical pixel size 5 nm) using MatLab. The count for each pixel represents the number of localization within that region.
2. To aid with visualization, convolve the 2D histogram images with a 2D Gaussian [σ=7.5 nm] with each localization and display using a color map such as "Gray" or "Hot" in MatLab (Fig. 6b).
3. We use the "Jet" color map in MatLab to color code each localization position according to its time of appearance during

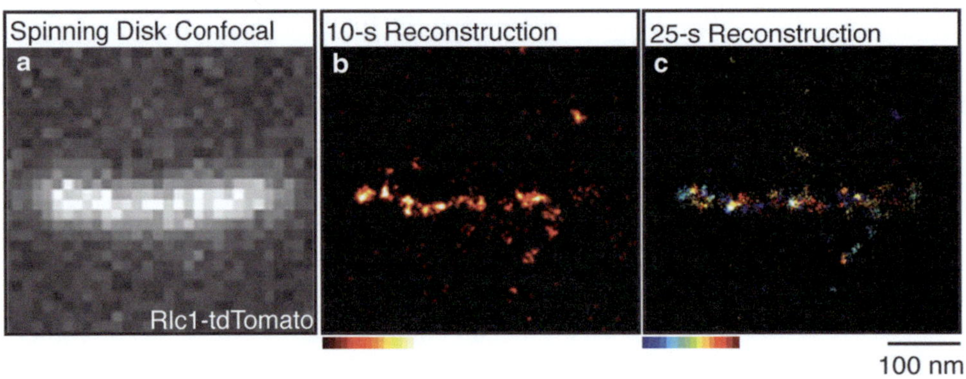

Fig. 6 Comparison of confocal and FPALM super-resolution fluorescence micrographs of the fission yeast cytokinetic ring expressing the myosin-II regulatory light chain Rlc1p fused to fluorescent proteins. (**a**) Spinning disk confocal micrograph of single focal plane of a cell expressing Rlc1p-tdTomato. The cytokinetic ring is pixelated and blurry with no obvious structural details. (**b**, **c**) FPALM reconstruction images of a cell expressing Rlc1p-mEos3.2. (**b**) An image reconstructed from 2000 camera frames (acquired at 200 fps for 10 s) and color coded for intensity with MatLab Hot map showing structural features of the cytokinetic ring. (**c**) An image reconstructed from 5000 camera frames (acquired at 200 fps for 25 s) and color coded for time with the MatLab Jet map. This image shows movements of features inside the ring. Scale bar: 100 nm

acquisition (Fig. 6c). The same information can be used to reconstruct images as time-lapse movies by combining frames acquired over 1–2 s and playing them in succession.

3.8 Quantitation and Simultaneous Observation of Two Tagged Proteins

Single-molecule switching nanoscopy can be used to count fluorescent emitters [12, 13] and should be valuable in the future for counting absolute numbers of proteins tagged with photoconvertible fluorescent proteins in live cells. In our experience, the numbers of detected single molecules of proteins tagged with mEos3.2 in nodes (precursors of the cytokinetic contractile ring) is proportional to their numbers measured by quantitative fluorescence microscopy [14].

The best photoactivatable proteins have similar emission wavelengths, so they are not useful for two-color live cell imaging. However, we have localized pairs of proteins tagged with mEos3.2 in cytokinesis nodes by comparing the distributions of single molecule detections in cells expressing each protein separately and the two proteins together.

4 Notes

1. Do not use rich growth medium YE5S for imaging. It exhibits autofluorescence in the range of emission wavelengths used for imaging and will cause an undesired increase in the fluorescence background of your datasets.
2. For EMM5S, we autoclave the dextrose separately from the rest of the ingredients. We make a 20 % dextrose solution by dissolving 20 g of dextrose into 100 mL, final volume, of ultrapure water. In a separate flask, we mix the remainder of the ingredients in 900 mL ultrapure water. We autoclave the two solutions separately, mix them together and then add the vitamin stock solutions.
3. We autoclave wood sticks in stainless steel metal canisters made for sterilizing glass pipets.
4. Some sources of gelatin are autofluorescent. Sigma gelatin G2500 has minimal to no autofluorescence at the emission wavelengths used for FPALM.
5. Immediately after adding the gelatin to the liquid medium, cap the tube and vigorously flick the tube to mix the gelatin and liquid together. The gelatin creates a "plug" on top of the liquid if not mixed immediately.
6. When making the pad, the gelatin will spread out too thinly between the glass slides if it is too hot and not enough if it is too cool. This takes a bit of practice.
7. These items can all be purchased at a pharmacy.

8. Some fission yeast strains with genes tagged with mEos2 were sick or inviable, but were healthy when tagged with monomeric mEos3.2.
9. Use the Bähler website (http://www.bahlerlab.info/resources/) to obtain the sequences of the primers that anneal to the genomic sequence.
10. We made mEos3.2 vectors by replacing the coding sequence of GFP [6] with the mEos3.2 sequence in the vectors described by Bähler et al. [7]. Our vectors express *Kan* marker for selection in yeast and *Amp* marker for selection in bacteria.
11. We use 0.5 kb/min to calculate the duration of the polymerization step.
12. We use 4 PCR reactions of 50 μL each for one transformation. We combine the products of 4 PCR reactions together, purify them following the manufacturer's instructions and resuspend the purified DNA in a final volume of 100 μL of ultrapure water.
13. We elute the DNA twice from the spin column by reapplying the first flow through to the column a second time and centrifuging. Some water volume is lost in the spin column, so the recovery volume is less than 100 μL. Each transformation reaction takes 50 μL of DNA and diagnostic gel electrophoresis requires 1–2 μL of DNA.
14. This second round of selection eliminates false-positive colonies.
15. Allow the cells to adapt to the liquid media by growing them for 36–48 h in a shaking water bath set to 25 °C in YE5S liquid medium prior to imaging. Failing to do so may result in phenotypic artifacts.
16. Centrifuging fission yeast cells at higher speeds or for longer times causes the nucleus to sediment to one pole of the cell.
17. Insert a clean razor blade between the two glass slides and in one sharp motion separate the slides, exposing the gelatin pad on one surface. Use the exposed gelatin pad immediately to prevent drying of the pad.
18. We set the power to 1.2 kW/cm^2 at the sample plane to achieve an average of 200 photons per emitter per frame.
19. The intensity of the 561 nm imaging laser photoconverts mEos3.2 at a high enough rate (without 405 nm illumination) to be useful for focusing on the region of interest of the cells (surface or cross-section through the cell).
20. Huang et al. describe the method to correct for the pixel-dependent noise [4].

Acknowledgements

This work was supported by the National Institute of General Medical Sciences of the National Institutes of Health under award number R01GM026132 to TDP, a grant 095927/A/11/Z from the Wellcome Trust to JB, a James Hudson Brown—Alexander Brown Coxe Postdoctoral Fellowship to FH and a HFSP Long-Term Fellowship to CL. The content is solely the responsibility of the authors and does not necessarily represent the official views of the National Institutes of Health. JB discloses significant financial interest in Bruker Corp. JB and FH disclose significant financial interest in Hamamatsu Photonics K.K.

References

1. Betzig E, Patterson GH, Sougrat R, Lindwasser OW, Olenych S, Bonifacino JS, Davidson MW, Lippincott-Schwartz J, Hess HF (2006) Imaging intracellular fluorescent proteins at nanometer resolution. Science 313(5793):1642–1645. doi:10.1126/science.1127344
2. Hess ST, Girirajan TP, Mason MD (2006) Ultra-high resolution imaging by fluorescence photoactivation localization microscopy. Biophys J 91(11):4258–4272. doi:10.1529/biophysj.106.091116
3. Rust MJ, Bates M, Zhuang X (2006) Sub-diffraction-limit imaging by stochastic optical reconstruction microscopy (STORM). Nat Methods 3(10):793–795. doi:10.1038/nmeth929
4. Huang F, Hartwich TM, Rivera-Molina FE, Lin Y, Duim WC, Long JJ, Uchil PD, Myers JR, Baird MA, Mothes W, Davidson MW, Toomre D, Bewersdorf J (2013) Video-rate nanoscopy using sCMOS camera-specific single-molecule localization algorithms. Nat Methods 10(7):653–658. doi:10.1038/nmeth.2488
5. Ma X, Kovacs M, Conti MA, Wang A, Zhang Y, Sellers JR, Adelstein RS (2012) Nonmuscle myosin II exerts tension but does not translocate actin in vertebrate cytokinesis. Proc Natl Acad Sci U S A 109(12):4509–4514. doi:10.1073/pnas.1116268109
6. Zhang M, Chang H, Zhang Y, Yu J, Wu L, Ji W, Chen J, Liu B, Lu J, Liu Y, Zhang J, Xu P, Xu T (2012) Rational design of true monomeric and bright photoactivatable fluorescent proteins. Nat Methods 9(7):727–729. doi:10.1038/nmeth.2021
7. Bahler J, Wu JQ, Longtine MS, Shah NG, McKenzie A 3rd, Steever AB, Wach A, Philippsen P, Pringle JR (1998) Heterologous modules for efficient and versatile PCR-based gene targeting in Schizosaccharomyces pombe. Yeast 14(10):943–951. doi:10.1002/(SICI)1097-0061(199807)14:10<943::AID-YEA292>3.0.CO;2-Y
8. Forsburg SL, Rhind N (2006) Basic methods for fission yeast. Yeast 23(3):173–183. doi:10.1002/yea.1347
9. Maddox PS, Bloom KS, Salmon ED (2000) The polarity and dynamics of microtubule assembly in the budding yeast Saccharomyces cerevisiae. Nat Cell Biol 2(1):36–41. doi:10.1038/71357
10. Huang F, Schwartz SL, Byars JM, Lidke KA (2011) Simultaneous multiple-emitter fitting for single molecule super-resolution imaging. Biomed Opt Express 2(5):1377–1393. doi:10.1364/BOE.2.001377
11. Wilks SS (1938) The large-sample distribution of the likelihood ratio for testing composite hypotheses. Ann Math Stat 9(1):60–62
12. Lee SH, Shin JY, Lee A, Bustamante C (2012) Counting single photoactivatable fluorescent molecules by photoactivated localization microscopy (PALM). Proc Natl Acad Sci U S A 109(43):17436–17441. doi:10.1073/pnas.1215175109
13. Rollins GC, Shin JY, Bustamante C, Presse S (2015) Stochastic approach to the molecular counting problem in superresolution microscopy. Proc Natl Acad Sci U S A 112(2):E110–E118. doi:10.1073/pnas.1408071112
14. Wu JQ, Pollard TD (2005) Counting cytokinesis proteins globally and locally in fission yeast. Science 310(5746):310–314. doi:10.1126/science.1113230
15. Forsburg SL (1994) Codon usage table for Schizosaccharomyces pombe. Yeast 10(8):1045–1047. doi:10.1002/yea.320100806

Chapter 5

Monitoring Chitin Deposition During Septum Assembly in Budding Yeast

Irene Arcones and Cesar Roncero

Abstract

The synthesis of the septum is a critical step during cytokinesis in the fungal cell. Moreover, in *Saccharomyces cerevisiae* septum assembly depends mostly on the proper synthesis and deposition of chitin and, accordingly, on the timely regulation of chitin synthases. In this chapter, we will see how to follow chitin synthesis by two complementary approaches: monitoring chitin deposition in vivo at the septum by calcofluor staining and fluorescence microscopy, and measuring the chitin synthase activities responsible for this synthesis.

Key words Chitin, Chitin synthase, Chitin ring, Primary septum

1 Introduction

In fungal cells the process of cytokinesis is necessarily coordinated with the synthesis and assembly of new cell wall (CW) components into a septum, which physically separates mother and daughter cells. *S. cerevisiae* has proved to be a fairly straightforward model to study such coordination because septum synthesis mostly depends on a single CW component, chitin [1]. This chitin, later linked to other CW components [2], assembles into a primary septum (PS) that physically separates mother and daughter cells at the time of cytokinesis. The PS, together with the rest of chitin and other cell wall components, forms a conspicuous structure, the bud scar, which remains in the mother cells after cell separation. This relies on the enzymatic action of chitinases, whose exclusive expression from the daughter cell side is directed by the cell cycle-controlled RAM network (review in [3]).

In *S. cerevisiae* chitin synthesis depends on three different chitin synthase (CS) activities, whose coordinated action results in proper septum formation. CSIII synthesizes the majority of cellular chitin [4, 5], this forming a ring that performs scaffold functions in the assembly of the septum. Although not essential, the chitin

Alberto Sanchez-Diaz and Pilar Perez (eds.), *Yeast Cytokinesis: Methods and Protocols*, Methods in Molecular Biology, vol. 1369, DOI 10.1007/978-1-4939-3145-3_5, © Springer Science+Business Media New York 2016

ring has a critical homeostatic function against any alteration in septum assembly [6, 7]. CSII performs an essential function in the synthesis of the primary septum (PS) and its activity has been linked directly to the proper contraction of the actomyosin ring [8, 9]. This coordination between PS synthesis and AMR contraction is probably conserved across fungi. The additional CSI activity seems to be involved in repair functions at the time of cytokinesis by synthesizing minutes amount of chitin, but its exact function remains poorly understood [10].

Chitin is an extremely insoluble polymer that needs to be synthesized directly into the periplasmic space from the plasma membrane (PM) [11]. Accordingly, CS activity is tightly regulated, mostly at the post-translational level [12]. The three Chs1, Chs2, and Chs3 chitin synthase proteins are transported to the yeast PM and activated in a timely fashion, and hence several different methods need to be combined in order to study chitin synthesis regulation. Chs2 and Chs3 localize to the neck during cytokinesis, but the protocols for the visualization of proteins during cytokinesis are extensively covered in the different chapters of this book and will not be discussed here. However, the precise localization of these proteins only gives a partial view of the story since it does not necessarily reflect the levels of their activities. In order to address this issue, here we shall describe two different experimental approaches to monitor chitin synthesis, either in vivo or in vitro, by means of determining chitin levels or chitin synthase activity respectively.

It should be recalled that in *S. cerevisiae* three different CS activities with different functions coexist. CSIII is responsible for the synthesis of more that 90 % of the cellular chitin in vivo, while in vitro CSI activity accounts for more than 90 % of the enzymatic activity measured, despite its almost null contribution to chitin synthesis. The protocols developed over the years will allow CS redundancy to be circumvented, providing consistent experimental results in vivo and in vitro.

2 Materials

2.1 All Reagents Are Dissolved in Water

1. Calcofluor White MR2 (10 mg/ml) (Fluorescent brightener 28, Sigma) (*see* **Note 1**).
2. Formaldehyde 16 % (Polysciences, Inc.).
3. Glass beads (0.45–0.6 mm, Sigma).
4. 50 mM Tris–HCl pH 7.5.
5. 50 mM Tris–HCl pH 8.0.
6. 50 mM Tris–HCl pH 6.5.
7. 0.8 M *N*-Acetylglucosamine (Sigma).
8. Co^{2+} 50 mM ($CoCl_2$, Sigma).

9. Ni^{2+} 50 mM ($(CH_3COO)_2Ni \cdot 4H_2O$, Merck).
10. Mg^{2+} 40 mM ($MgCl_2$, Sigma).
11. Trypsin (10 mg/ml) (Sigma).
12. Trypsin Inhibitor (10 mg/ml) (Sigma).
13. Trichloracetic Acid (TCA) 10 % (Merck).
14. Borosilicate tubes (10×70 mm).
15. 25 mm Glass microfiber filters GF/C (Whatman).
16. [^{14}C] 10 mM UDP-*N*-Acetylglucosamine (Substrate). For preparation of *1 ml of substrate*: 200 μl 50 mM UDP-N-Acetylglucosamine (Sigma), 72 μl [^{14}C] UDP-N-Acetylglucosamine (Amersham 235 mCi/mmol), and 728 μl water. The radioactive substrate should contain around 20,000 cpm/5 μl (400 cpm/nmol), however precise calibration should be performed in each substrate batch by measuring cpm levels directly in 5 μl of substrate.
17. Yeast growth media:

    YEPD medium (1 % yeast extract, 2 % peptone, 2 % glucose/dextrose).

    SD medium (0.67 % Yeast Nitrogen Base w/o amino acids, 2 % glucose, required amino acids).
18. Superspeed centrifuge (Sorvall RC5 or equivalent).
19. Filtration tower equipment.
20. Scintillation counter.
21. Fluorescence microscope with a UV filter (Chroma 4900-ET-DAPI or equivalent).

3 Methods

3.1 Qualitative Determination of Chitin Synthesis by Calcofluor Staining

Calcofluor binds rather specifically to β(1–4) glucans, and hence over the years it has been used for the specific staining of chitin along the yeast and fungal cell walls [13, 14]. In *S. cerevisiae*, calcofluor can be used directly on fixed cells, staining pre-existing chitin [13]. However, if added to growing cells the dye will bind much more efficiently to nascent chitin chains, providing much stronger staining at the sites where chitin is being actively synthesized [14]. The use of both protocols provides complementary answers to obtain a more detailed picture of the in vivo chitin synthesis occurring in a specific strain (*see* Fig. 1).

The two protocols described will mainly stain the chitin ring and any additional accumulation of chitin formed by CSIII activity [4, 15], thereby masking the staining of chitin accumulated at the primary septa which is synthesized by CSII. In order to

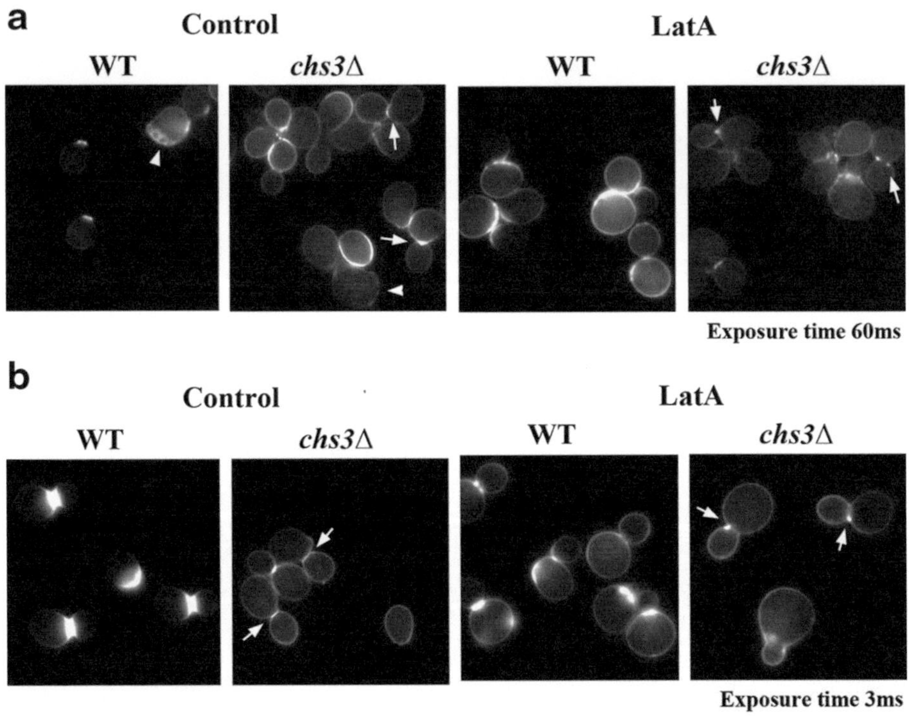

Fig. 1 Calcofluor staining of the indicated strains performed following the experimental protocols described in the text: (**a**) Calcofluor staining on fixed cells, (**b**) Calcofluor life cell staining. All images were processed in parallel to preserve relative fluorescence levels. However, note the 20-fold longer exposure time used on fixed cells. The WT images highlight the chitin derived from CSIII activity and accordingly the chitin rings are visible in some images (*arrowheads*). However, appreciate the absence of chitin rings in the bud scars of the *chs3Δ* mutant cells (*arrowheads*). Moreover, the images obtained from the *chs3Δ* strain highlighted the chitin associated with CSII activity and therefore chitin deposition at the PS (*arrows*). Latrunculin A was added directly to cell cultures at a concentration of 30 mM and cells were collected after 120 min. For life cell staining, Calcofluor was added 30 min after the addition of LatA. Note the alteration in chitin deposition after the blockade of endocytosis, including the collapse of some PSs in the *chs3Δ* strain (*arrows*)

observe PS staining directly it is necessary to use strains lacking CSIII activity [16] or inhibiting this activity by nikkomycin [6]. The use of *chs3Δ* cells allows the precise study of PS formation during cytokinesis, including potential alterations caused by the deregulation of CSII (Fig. 1; [17]; Sanchez-Diaz and Roncero unpublished observations).

3.1.1 Calcofluor Staining on Fixed Cells

1. Grow cells in your favorite medium to early logarithmic phase ($1–2 \times 10^7$ cells/ml).
2. Take 2 ml of culture and fix cells in by adding 500 μl of 16 % formaldehyde directly. Incubate at room temperature (RT) for 30 min in a roller.

3. Collect cells by centrifugation at 3500 × *g* for 2 min and wash them with 2 ml of water.
4. Resuspend them in 0.5 ml of 50 μg/ml of calcofluor in water; then incubate for 5 min in a roller.
5. Spin cells down and wash cells with 2 ml of water. Resuspend them in 100 μl of water. Cells can be maintained at 4 °C in the darkness.
6. Observe cells under the fluorescence microscope using the appropriate UV filter.

3.1.2 Calcofluor Staining on Growing Cells (Life Cell Staining)

1. Add calcofluor at a final concentration of 50 μg/ml to a YEPD culture of yeast cells growing in the early logarithmic phase (*see* **Note 2**).
2. Incubate for an additional 90 min and concentrate cells by centrifugation if necessary. Observe cells directly under the microscope with the appropriate UV filter.

3.1.3 Imagining Processes

Calcofluor staining can be usted as a semi-quantitative way of measuring chitin synthesis levels by carefully adjusting the microscopic/imaging processes. To do so:

1. Short exposure times should be used, since calcofluor provides a very strong degree of staining (*see* **Note 3**).
2. Use identical exposure times for all images, regardless of the staining intensity.
3. Process all images obtained in parallel in order to preserve the relative intensity values.
4. Avoid automatic machine set-ups for exposure times or brightness/contrast levels.

The images shown in Fig. 1 compare both experimental protocols for staining WT and *chs3*Δ cells. Note the different level of staining in fixed cells (A) or after life cell staining (B), even after the 20-fold longer exposure times in the fixed-cell images. The images highlight the alterations in chitin deposition after the blockade of endocytosis, and how it affects both the assembly of the chitin ring and the PS.

3.2 Measurement of Chitin Synthase Activity

Chitin synthase activity in yeast was described many years ago [18] and soon after it was recognized as an enzymatic activity associated with cellular membranes [11]. Accordingly, CS determination should include detailed protocols for the isolation of cellular membranes. However, *S. cerevisiae* contains three different CS activities encoded by three different genes [12] and the precise measurements of CS awaited the construction of individual and double mutants in the CS genes in order to design specific protocols for the assay of the different CS activities [19]. The protocols described here are based in the seminal work carried out at Enrico Cabib's

Lab and rely in the use of different cations and pHs and in the zymogenic nature of CSI and CSII, but not of CSIII, activity.

3.2.1 Preparation of Membrane Extracts

Membranes were prepared essentially as described by Orlean [20] with slight modifications (*see* **Note 4**).

Samples and reagents should be maintained continuously on ice.

1. Grow a 200–300 ml culture at 28 °C to a cell density of 1.5–2.0×10^7 cells/ml. Chill the culture down on ice.
2. Pellet cells by centrifuging them for 7 min at $6000 \times g$ at 4 °C. Discard the supernatant.
3. Wash cells with 20 ml of cold 50 mM Tris–HCl pH 7.5. Centrifuge for 7 min at $6000 \times g$. Discard the supernatant.
4. Resuspend cells in 200–500 μl of 50 mM Tris–HCl pH 7.5 and transfer them to 1.5 ml microcentrifuge tubes.
5. Add glass beads (0.45–0.6 mm) until the liquid is almost covered.
6. Break up the cells. Cells can be broken by different protocols. In a Fast-Prep (MPBio), using 15-s pulses at a speed of 5.5. Repeat three times, placing the tubes in ice after each interval. By Vortex, using glass tubes. Typically 8 pulses of 30 s at maximum speed. The tubes should be cooled down for 30 s between vortex pulses. Alternatively, a Braun homogenizer or a French Press can be used to break yeast cells as long as the temperature of the samples is maintained around 4 °C.

   Regardless of the method used, the degree of cell breakage should be checked by light microscopy.
7. Make a hole in the bottom of each microcentrifuge tube with a hot needle and place the tubes on top of new tubes. Spin for 15 s at $6000 \times g$ in a benchtop centrifuge at 4 °C in order to harvest cell extracts.
8. Dilute cell extracts up to 8 ml of 50 mM Tris–HCl pH 7.5. Centrifuge for 7 min at $6000 \times g$.
9. Transfer the supernatant to new centrifuge bottles. Centrifuge for 35 min at $40{,}000 \times g$ maintaining temperature between 0 and 6 °C. Discard all the supernatant.
10. Resuspend the membranes in 200–800 μl of cold 50 mM Tris–HCl pH 7.5, 33 % glycerol with the aid of a glass stick. Absolute homogenization is required in order to obtain reproducible results. Measure protein concentration by the Bradford method. The final extract should contain between 7 and 12 mg/ml of protein.

    Freeze extracts at −80 °C or proceed with chitin synthase reactions.

3.2.2 Chitin Synthase Assay

The general principle for CS measurement is based on the incorporation of a fully soluble radioactive precursor, [^{14}C] UDP-N-Acetylglucosamine, into a rather insoluble product: [^{14}C] chitin. Selective precipitation of chitin will allow the amount of radioactivity incorporated during the reaction time to be detected (*see* **Note 5**).

Owing to the in vitro zymogenic nature of some CSs, measurements are always performed under two different conditions, with and without trypsin, in order to measure total and basal CS activity respectively. The proteolytic activation of CS by trypsin was reported many years ago [21].

Reactions are always performed in 50 ml mixtures placed in 10 × 70 mm borosilicate tubes for effective temperature transfer. Mixtures are maintained continuously on ice just after the incubation period, performed at 30 °C in a water bath. All reactions, including blanks, should be performed in duplicate. The reaction conditions are based directly on the work of Cabib's Lab [19], but adapted over the years. We present three different chitin synthase assays to determine CSIII, CSII, and CSI.

CSIII Determination

This is based on the fact that CSIII activity is maximum at pH 8.0, using Co^{2+} as the activating cation. In addition, CSIII activity is not inhibited by Ni^{2+}. Thus, with these conditions only CSIII activity should be measured, since the potential contaminant CSII will be inhibited by the addition of Ni^{2+}.

1. *Reaction A*, without trypsin (Basal activity).

   Each tube should contain:

   5 μl of 50 mM Co^{2+} ($CoCl_2$).

   2 μl of 0.8 M *N*-Acetylglucosamine.

   5 μl of 10 mM [^{14}C] UDP-*N*-Acetylglucosamine.

   5 μl of 50 mM Ni^{2+} ($(CH_3COO)_2Ni \cdot 4H_2O$).

   10 μl of water.

   20 μl of membrane extracts (*see* Subheading 3.2.1).

   Incubate 90 min at 30 °C. Stop the reaction adding 2 ml of TCA.

2. *Reaction B*, with trypsin (Total activity):
   Reaction is performed in two consecutive steps. You should need to prepare three different tubes (reaction *B.1*, *B.2*, and *B.3*) per sample containing:

   3 μl of 50 mM Tris–HCl pH = 8.0.

   5 μl of 50 mM Co^{2+} ($CoCl_2$).

   5 μl of 10 mM [^{14}C] UDP-*N*-Acetylglucosamine.

   6 μl of H_2O.

   5 μl of 50 mM Ni^{2+} ($(CH_3COO)_2Ni \cdot 4H_2O$).

20 µl of membrane extracts.

Add trypsin as follow:

Reaction *B.1*) 2 µl of 1 mg/ml Trypsin.

Reaction *B.2*) 2 µl of 2 mg/ml Trypsin.

Reaction *B.3*) 2 µl of 3 mg/ml Trypsin.

Incubate tubes for 15 min at 30 °C.

Add to each tube:

Reaction *B.1*) 2 µl of 1.5 mg/ml Trypsin Inhibitor.

Reaction *B.2*) 2 µl of 3 mg/ml Trypsin Inhibitor.

Reaction *B.3*) 2 µl of 4.5 mg/ml Trypsin Inhibitor.

Finally, 2 µl of 0.8 M *N*-Acetylglucosamine to all of them.

Incubate 90 min at 30 °C. Stop the reaction adding 2 ml of TCA.

3 µl of 50 mM Tris–HCl pH 8.0.

CSII Determination

There are no specific reaction conditions for measuring CSII alone, and hence the joint CSIII/CSII activity is measured first, as indicated below, and CSII activity is determined by subtracting the CSIII activity measured as describe above from this value.

CSIII/CSII activity is measured in exactly the same way as CSIII but omitting Ni^{2+} in the assays and adding the corresponding volume of water. When using any strain devoid of CSIII activity, such as *chs3*Δ mutants, these reaction conditions will provide CSII activity directly.

1. *Reaction A*, without trypsin (Basal activity)
   Each tube should contain:

   3 µl of 50 mM Tris–HCl pH 8.0.

   5 µl of 50 mM Co^{2+} ($CoCl_2$).

   2 µl of 0.8 M *N*-Acetylglucosamine.

   5 µl of 10 mM [^{14}C] UDP-*N*-Acetylglucosamine.

   15 µl of water.

   20 µl of membrane extracts.

   Incubate for 90 min at 30 °C. Stop the reaction by adding 2 ml of TCA.

2. *Reaction B*, with trypsin (Total activity):

   3 µl of 50 mM Tris–HCl pH 8.0.

   5 µl of 50 mM Co^{2+} ($CoCl_2$).

   5 µl of 10 mM [^{14}C] UDP-*N*-Acetylglucosamine.

   11 µl of water.

   20 µl membrane extracts.

   Add trypsin as follow:

Reaction *B.1*) 2 μl of 1 mg/ml Trypsin.

Reaction *B.2*) 2 μl of 2 mg/ml Trypsin.

Reaction *B.3*) 2 μl of 3 mg/ml Trypsin.

Incubate tubes for 15 min at 30 °C.

Add to each tube:

Reaction *B.1*) 2 μl of 1.5 mg/ml Trypsin Inhibitor.

Reaction *B.2*) 2 μl of 3 mg/ml Trypsin Inhibitor.

Reaction *B.3*) 2 μl of 4.5 mg/ml Trypsin Inhibitor.

Finally, 2 μl of 0.8 M *N*-Acetylglucosamine to all of them.

Incubate for 90 min at 30 °C. Stop the reaction adding 2 ml of TCA.

CSI Determination

The protocol is identical to that described above but using different cations and buffers. Note that the differences are underlined.

1. *Reaction A*, without trypsin (Basal activity).
   Each tube should contain:

   3 μl of 50 mM Tris–HCl pH 6.5.

   5 μ of 40 mM Mg^{2+} ($MgCl_2$).

   2 μl of 0.8 M *N*-Acetylglucosamine.

   5 μl of 10 mM [^{14}C] UDP-*N*-Acetylglucosamine.

   15 μl of water.

   20 μl of membrane extracts.

   Incubate for 30 min at 30 °C. Stop the reaction by adding 2 ml of TCA.
2. *Reaction B*, with trypsin (Total activity):

   3 μl of 50 mM Tris–HCl pH 6.5.

   5 μl of 40 mM Mg^{2+} ($MgCl_2$).

   5 μl of 10 mM [^{14}C] UDP-*N*-Acetylglucosamine.

   11 μl of water.

   20 μl of membrane extracts.

   Add trypsin as follow:

   Reaction B.1) 2 μl of 1 mg/ml Trypsin.

   Reaction B.2) 2 μl of 2 mg/ml Trypsin.

   Reaction B.3) 2 μl of 3 mg/ml Trypsin.

   Incubate tubes for 15 min at 30 °C.

   Add to each tube:

   Reaction B.1) 2 μl of 1.5 mg/ml Trypsin Inhibitor.

   Reaction B.2) 2 μl 3 mg/ml Trypsin Inhibitor.

Reaction B.3) 2 μl 4.5 mg/ml Trypsin Inhibitor.

Finally, 2 μl of 0.8 M N-Acetylglucosamine to all of them.

Incubate for 30 min at 30 °C. Stop the reaction by adding 2 ml of TCA.

3.2.3 Filtration Procedures

Incorporated radioactivity is measured after filtration of TCA precipitates into glass fiber filters. To do so the content of each tube containing the CS reaction is poured on a 25 mm GF/C filter placed in a filter tower equipment (*see* Fig. 2). After application of vacuum, the filter is washed twice with 2 ml of 10 % TCA and once with 100 % ethanol. Filters are later retired from the tower and dried at RT before adding scintillation liquid for radioactivity measurements in a scintillation counter.

For the final calculation of CS activity, the cpm counts determined as indicated are transformed into nanomoles of *N*-Acetylglucosamine incorporated into the insoluble material, after which CS activity is usually expressed in mU (incorporated nanomoles/hour/milligram protein). For a standard CSIII reaction using 20 μl of membrane and 90 min of incubation you can calculate CS activity as indicated in Fig. 3:

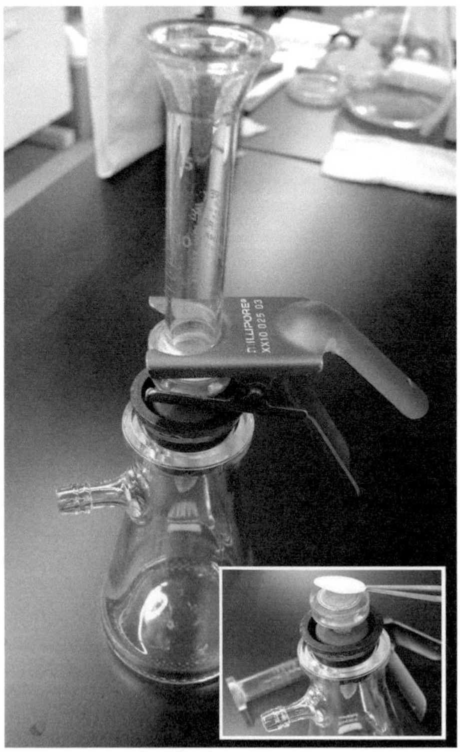

Fig. 2 Filtration equipment. Individual samples can be processed using a simple filtration equipment as depicted in the Figure. 25 mm Fiber glass filters are routinely used

Time — Protein — N-Acetylglucosamine incorporated

$$\text{CS activity (mU/h or nmol/hour/mg)} = \left[\frac{1}{1.5}\right] \times \left[\frac{50}{\text{mg/ml}}\right] \times \left[\frac{\text{CPM}^{\text{REACTION}} \times 50}{\text{CPM}^{\text{SUSTRATE}}}\right]$$

1.5	Reaction time expressed in hours
mg/ml	Protein concentration in cell extracts
$\text{CPM}^{\text{REACTION}}$	cpm counted in the filter corresponding to each reaction
$\text{CPM}^{\text{SUSTRATE}}$	cpm of the substrate added in each reaction (5ml)

Fig. 3 Calculation of chitin synthase activity

1. *An alternative method for CS measurement.* A non-radioactive protocol has been described for CS measurement [22]. This protocol has been used in HTS assays [23], but its use in the characterization of CS has not been extensively tested. In our experience it does not work as well as the radioactive method described here, lacking sufficient sensitivity for use in the measurement of the CSII activity involved in PS formation (Foltman and Sanchez-Diaz, unpublished results).

   Multiple adaptations of the different protocols described here have been used to measure chitin synthase activity in different fungi such as *Candida albicans* or *Aspergillus* sp. However, to date the multiplicity of chitin synthase genes present in fungi has made it extremely difficult to link in vitro CS activities to chitin synthesis in vivo in any organism other than *S. cerevisiae.*

2. *The physiological relevance of CS measurements.*

   The in vitro zymogenic nature of some CS makes it difficult to interpret the levels of CS activities in physiological terms. It is unclear whether basal or total activities correspond to the real action of the CS in the cell. This interpretation has proved to be even more complicated since the physiological activator/s of the different CSs has not been defined.

   CSIII is only modestly zymogenic, because trypsin treatment only increases activity by a factor of 1.7. Moreover, increases in the levels of Chs3 at the PM have been correlated with higher basal activities and increases in chitin synthesis [15, 24]. Therefore, it can be assumed that CSIII basal activity correspond with the actual levels of chitin synthase in vivo.

   CSII is clearly zymogenic, and trypsin treatment increases CSII activity several-fold. Only very recently has it been possible to link the increase in CS basal activity to a higher in vivo functionality of CSII [17, 25]. Although the molecular mechanism that lead to CSII activation remains largely unknown, these results strongly support a direct relationship between

CSII basal activity and the in vivo synthesis of chitin at the PS, expanding our understanding of how CSII might be regulated at the time of cytokinesis.

4 Notes

1. *Calcofluor preparation*: Calcofluor is prepared as a stock solution at 10 mg/ml concentration in water. Only the salt form of calcofluor is soluble in water, and hence depending on the manufacturer small amounts of NaOH should be added to the solution for complete dissolution. Calcofluor should be sterilized by filtration though a 0.22 μm filter. Once clear, the solution can be maintained at 4 °C in the dark for months. Sigma provides calcofluor under the name of fluorescent brightener 28. This compound is fully soluble in water at 10 mg/ml and does not require the addition of NaOH.
2. *Life cell staining with Calcofluor*: *Growth of yeast cells in* YEPD medium is preferred for life staining because a much more uniform staining of the cells can be achieved. However, if selective media is required, the SD medium should be buffered with 50 mM Phthalate, pH 6.3, in order to prevent calcofluor precipitation at acidic pH. Longer or shorter incubation times with calcofluor can be used if desired. Life cell staining provides a much higher levels of fluorescence than fixed staining, highlighting the sites of active chitin synthesis (*see* Fig. 1).
3. *Imaging procedures*: Be careful during images capturing, image overexposure is relatively frequent in un-experienced observers. If some of the images are overexposed you should not be able to compare relative intensity values. Exposure times should be adjusted depending on the microscopy equipment used.
4. *Measurement of CS activity*: The preparation of membrane extracts is the most critical step in chitin synthase measurement since partial inactivation of the samples may occur during processing. It is therefore recommended to include control cells in each experiment as a reference. Cell extracts are stable for weeks If maintained frozen at −80 °C.

   Alternative methods for preparation of crude cell extracts have been used for the measurement of CS activities. These methods rely on the direct permeabilization of cells, either by detergent treatment [26] or osmotic shock [27]. While these methods allow efficient measurement of CS activities, they would alter CS properties either directly or by altering CS intracellular distribution [15, 28]. Accordingly, permeabilized cells should be used with caution in order to translate the in vitro results to cell physiology.

   Researchers should be aware that CS activities change significantly, depending on the different phases of cellular growth and on the growth medium used. Therefore, it must always be assured that cells growing in the early logarithmic phase are used. The use of synchronized cultures in not practical owing to the large amount of cells required in extract.

Changes in the growth medium would alter not only CS activity but also, indirectly, other parameters able to influence CS measurement. For example, growth in galactose medium alters yeast cell wall composition, rendering cells highly resistant to mechanical rupture. Therefore, extended pulse times are required in either the vortexing step or the fast-prep breakage protocols.

5. *Chitin synthase assay*: The protocol described make use of a radioactive substrate, therefore you should be aware of the security rules regarding the use of this type of compounds, including not only the protection of the user but also the proper disposal of contaminated material.

Acknowledgments

We acknowledge the help of N. Skinner for language revision. We are indebted to E. Cabib for his seminal and continued work on chitin synthesis and to A. Duran for transferring his enthusiasm for this field of research to Salamanca. CR was supported by a grant GR31 from the Excellence Research program from the Junta de Castilla y Leon, and by grants BFU2010-18693 and BFU2013-48582-C2-1-P from the CICYT/FEDER program.

References

1. Cabib E (2004) The septation apparatus, a chitin-requiring machine in budding yeast. Arch Biochem Biophys 426:201–207
2. Cabib E, Arroyo J (2013) How carbohydrates sculpt cells: chemical control of morphogenesis in the yeast cell wall. Nat Rev Microbiol 11: 648–655
3. Roncero C, Sanchez Y (2010) Cell separation and the maintenance of cell integrity during cytokinesis in yeast: the assembly of a septum. Yeast 27:521–530
4. Roncero C, Valdivieso MH, Ribas JC, Duran A (1988) Isolation and characterization of *Saccharomyces cerevisiae* mutants resistant to Calcofluor white. J Bacteriol 170:1950–1954
5. Shaw JA, Mol PC, Bowers B, Silverman SJ, Valdivieso MH, Duran A, Cabib E (1991) The function of chitin synthases 2 and 3 in the *Saccharomyces cerevisiae* cell cycle. J Cell Biol 114:111–123
6. Cabib E, Schmidt M (2003) Chitin synthase III activity, but not the chitin ring, is required for remedial septa formation in budding yeast. FEMS Microbiol Lett 29:299–305
7. Gomez A, Perez J, Reyes A, Duran A, Roncero C (2009) Slt2 and Rim101 contribute independently to the correct assembly of the chitin ring at the budding yeast neck in *Saccharomycces cerevisiae.* Eukaryot Cell 8:1449–1559
8. Schmidt M, Bowers B, Varma A, Roh DH, Cabib E (2002) In budding yeast, contraction of the actomyosin ring and formation of the primary septum at cytokinesis depend on each other. J Cell Sci 115:293–302
9. VerPlank L, Li R (2005) Cell cycle-regulated trafficking of Chs2 controls actomyosin ring stability during cytokinesis. Mol Biol Cell 16:2529–2543
10. Cabib E, Sburlati A, Bowers B, Silverman SJ (1989) Chitin synthase 1, an auxiliary enzyme

for chitin synthesis in *Saccharomyces cerevisiae*. J Cell Biol 108:1665–1672
11. Duran A, Bowers B, Cabib E (1975) Chitin synthase zymogen is attached to the plasma membrane. Proc Natl Acad Sci U S A 72: 3952–3955
12. Roncero C (2002) The genetic complexity of chitin synthesis in fungi. Curr Genet 41: 367–378
13. Pringle JR (1991) Staining of bud scars and other cell wall chitin with calcofluor. Methods Enzymol 194:732–735
14. Roncero C, Duran A (1985) Effect of Calcofluor white and Congo red on fungal wall morphogenesis: in vivo activation of chitin polymerization. J Bacteriol 163:1180–1185
15. Reyes A, Sanz M, Duran A, Roncero C (2007) Chitin synthase III requires Chs4p-dependent translocation of Chs3p into the plasma membrane. J Cell Sci 120:1998–2009
16. Sanz M, Castrejon F, Duran A, Roncero C (2004) Saccharomyces cerevisiae Bni4p directs the formation of the chitin ring and also participates in the correct assembly of the septum structure. Microbiology 150:3229–3241
17. Oh Y, Chang KJ, Orlean P, Wloka C, Deshaies R, Bi E (2012) Mitotic exit kinase Dbf2 directly phosphorylates chitin synthase Chs2 to regulate cytokinesis in budding yeast. Mol Biol Cell 23:2445–2456
18. Keller FA, Cabib E (1971) Chitin and yeast budding. Properties of chitin synthetase from Saccharomyces carlsbergensis. J Biol Chem 10:160–166
19. Choi WJ, Cabib E (1994) The use of divalent cations and pH for the determination of specific yeast chitin synthetases. Anal Biochem 219:368–372
20. Orlean P (1987) Two chitin synthases in *Saccharomyces cerevisiae*. J Biol Chem 262: 368–372
21. Cabib E, Farkas V (1971) The control of morphogenesis: an enzymatic mechanism for the initiation of septum formation in yeast. Proc Natl Acad Sci U S A 68:2052–2056
22. Lucero HA, Kuranda MJ, Bulik DA (2002) A nonradioactive, high throughput assay for chitin synthase activity. Anal Biochem 305: 97–105
23. Magellan H, Boccara M, Drujon T, Soulié MC, Guillou C, Dubois J, Becker HF (2013) Discovery of two new inhibitors of *Botrytis cinerea* chitin synthase by a chemical library screening. Bioorg Med Chem 21:4997–5003
24. Jimenez C, Sacristan C, Roncero MI, Roncero C (2010) Amino acid divergence between the CHS domain contributes to the different intracellular behaviour of Family II fungal chitin synthases in *Saccharomyces cerevisiae*. Fungal Genet Biol 47:1034–1043
25. Devrekanli A, Foltman M, Roncero C, Sanchez-Diaz A, Labib K (2012) Inn1 and Cyk3 regulate chitin synthase during cytokinesis in budding yeasts. J Cell Sci 125: 5453–5466
26. Fernandez MP, Correa JU, Cabib E (1982) Activation of chitin synthetase in permeabilized cells of a *Saccharomyces cerevisiae* mutant lacking proteinase B. J Bacteriol 152:1255–1264
27. Crotti LB, Drgon T, Cabib E (2001) Yeast cell permeabilization by osmotic shock allows determination of enzymatic activities in situ. Anal Biochem 292:8–16
28. Valdivia RH, Schekman R (2003) The yeasts Rho1p and Pkc1p regulate the transport of chitin synthase III (Chs3p) from internal stores to the plasma membrane. Proc Natl Acad Sci U S A 100:10287–10292

Chapter 6

Imaging Septum Formation by Fluorescence Microscopy

Juan Carlos Ribas and Juan Carlos G. Cortés

Abstract

Fungal cleavage furrow formation during cytokinesis relays in the coordinated contraction of an actomyosin-based ring and the centripetal synthesis of both new plasma membrane and a special wall structure named division septum. Through transmission electron microscopy, the septum exhibits a three-layered structure with a central primary septum, flanked at both sides by the secondary septum. In contrast to the chitinous primary septum present in most of fungi, the fission yeast *Schizosaccharomyces pombe* does not contain chitin, instead it divides through the formation of a linear β(1,3)glucan-rich primary septum, which has been shown to be specifically stained by the fluorochrome Calcofluor white. Recent findings in *S. pombe* have revealed the importance of septum synthesis for the steady contraction of the ring during cytokinesis. Therefore, to study the molecular mechanisms that connect the extracellular septum wall with the other components of the cytokinetic machinery located in the plasma membrane and cytoplasm, new experimental approaches are needed. Here we describe the methods developed to image the septum structure by fluorescence microscopy, with a special focus in the analysis of septum progression by the use of time-lapse microscopy.

Key words Cell wall, Septum, Polysaccharide, β(1,3)-glucan, Calcofluor white, Glucan synthase, Cytokinesis

Abbreviations

B-BG	Branched β(1,3)-glucan
L-BG	Linear β(1,3)-glucan
CW	Calcofluor white

1 Introduction

The *S. pombe* septum is a special structure of wall material formed during the cytokinesis. Its synthesis is intimately coupled to the actomyosin ring contraction and the addition of new plasma membrane. Once completed, the septum physically separates the original cell in two identical new cells. Next, the controlled dissolution of the glucans in the septum edging and the primary septum in the

Alberto Sanchez-Diaz and Pilar Perez (eds.), *Yeast Cytokinesis: Methods and Protocols*, Methods in Molecular Biology, vol. 1369, DOI 10.1007/978-1-4939-3145-3_6, © Springer Science+Business Media New York 2016

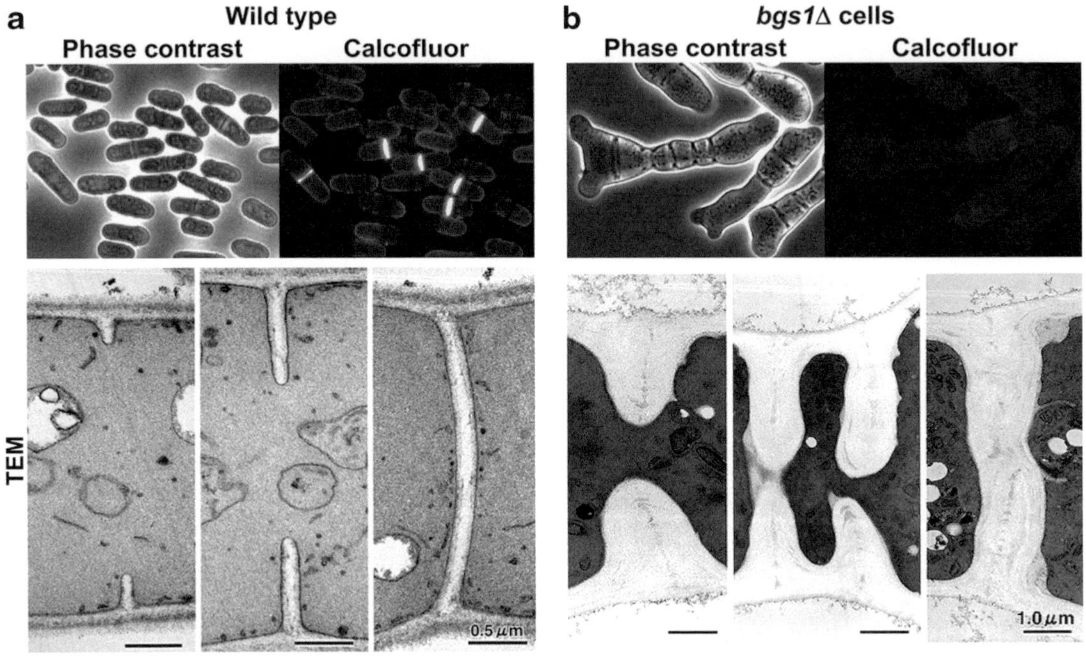

Fig. 1 In *S. pombe* cells, the fluorochrome Calcofluor white (CW) binds specifically and with high affinity to the L-BG of the primary septum. Cell morphology and septum structure of wild-type cells (**a**) and cells from germinated *bgs1*Δ spores (**b**) observed by phase-contrast, CW-stain fluorescence, and transmission electron microscopy (TEM). Differently from the wild-type septa, the *bgs1*Δ septa do not contain primary septum and L-BG, and are not stained with Calcofluor. Images adapted from [8]

complete septum structure allows cell separation and the end of cytokinesis [1–4]. When observed by transmission electron microscopy, the septum displays a three-layered structure (Fig. 1a), with a central disk of primary septum flanked at both sides by the secondary septum [5]. The septum is mainly composed of interlinked glucose polysaccharides [6]. Immunoelectron microscopy using colloidal gold-labeled antibodies specific against different β-glucans has helped to define the organization of these polysaccharides in the *S. pombe* septum [7]:

1. Linear β(1,3)-glucan (L-BG) is mostly found in the primary septum structure. This polysaccharide consists of β-(1,3)-linked glucose units forming linear chains in a conformation of single-helix with a small proportion of triple-helix structures. The L-BG has been shown to be responsible for the primary septum structure observed by electron microscopy [8]. The primary septum organization is similar to those found in plants, which is also formed by a L-BG named callose [9], or in *Saccharomyces cerevisiae*, which is formed by chitin. The single-helix conformation of L-BG makes the primary septum more susceptible than the secondary septum to degradation by

β(1,3)-glucanases [10]. The L-BG is recognized with higher affinity than the rest of *S. pombe* wall polysaccharides by the fluorochrome Calcofluor White (CW), and is responsible for the strong labeling of the septum wall with this fluorescent dye [8] (Fig. 1a, b).

2. Branched β(1,3)-glucan (B-BG) is located along all the septum structures, including the primary septum. It is made of β(1,3)-linked glucose units forming linear chains with β(1,6)-linked branches of β(1,3)-linked chains. Unlike L-BG, the B-BG chains form closed triple-helix structures that are more resistant than those of L-BG to degradation by β(1,3)-glucanases [11–13]. The B-BG is essential for cell integrity and for secondary septum formation. In addition, the B-BG plays a crucial role in cytokinesis, linking the cell wall to the plasma membrane and contractile actomyosin ring, which is needed to couple the septum synthesis with the plasma membrane and ring progression [4].
3. β(1,6)-glucan is only located in the secondary septum. It is a highly branched polysaccharide formed by a main chain of β(1,6)-linked glucose units and 75 % of β(1,3)-linked branches of β(1,6)-linked chains [14, 15].

Besides the β-glucans, the *S. pombe* septum also contains α-glucan. This polysaccharide has not been detected yet by immunoelectron microscopy, although it has been shown that the α-(1,3)-glucan is localized with the B-BG in both primary and secondary septum [3, 4]. It is absent in the budding yeast and in *C. albicans*, but is present in the cell wall of filamentous and dimorphic fungi [3, 4, 16, 17]. *S. pombe* α-glucan is a linear polysaccharide formed by chains of D-glucose bound by α-(1,3) linkages and with a small proportion of α-(1,4)-linked residues located at the reducing end [18]. Like B-BG, this polysaccharide is essential for cell integrity and for secondary septum formation. It also plays an essential role in the primary septum adhesion strength needed to support the physical force of the internal turgor pressure during cell abscission [3].

Recent studies have revealed the importance of septum synthesis for the proper actomyosin ring contraction and cleavage furrow formation in *S. pombe* [4, 19, 20]. Direct visualization of living cells is an important method for cytokinesis studies and fluorochromes can be used for cell wall and septum fluorescence microscopy analysis. The most commonly used is CW, also called Blankophor (Blankophor GmbH & Co.) or Fluorescent Brightener 28 (Sigma-Aldrich). Taking advantage of the higher affinity of CW for the L-BG of the primary septum, here we describe a method to follow the synthesis of the septum by time-lapse fluorescence microscopy (Fig. 2a). Besides, we describe additional methods developed to image the septum structure either in live or fixed cells.

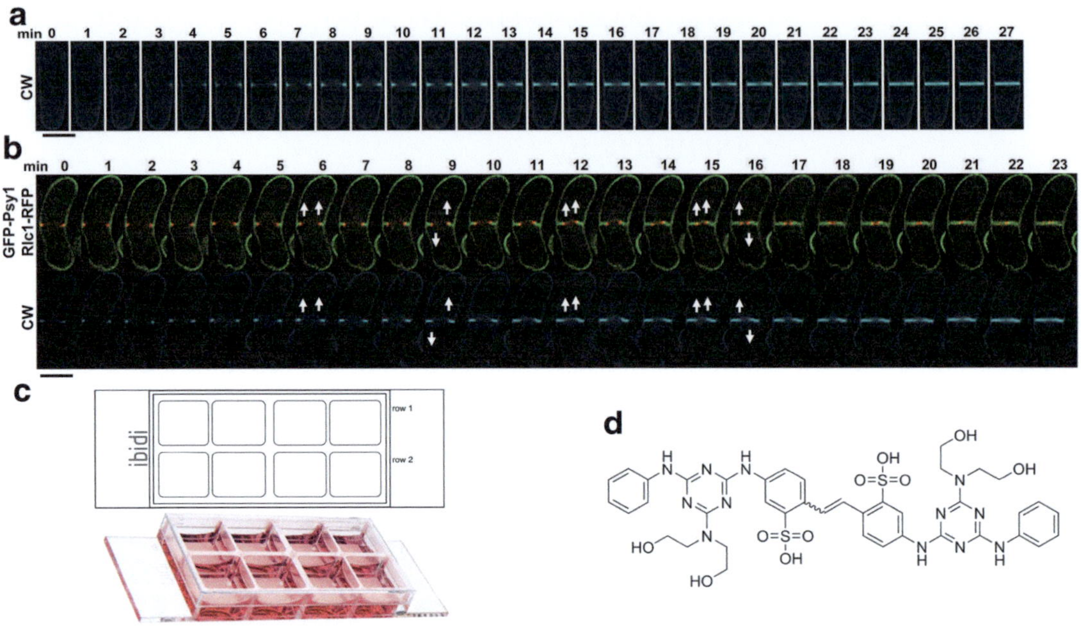

Fig. 2 Progression of septum formation followed by time-lapse microscopy of CW-staining fluorescence. (**a**) Time-lapse showing the progression of the primary septum formation in a wild-type cell by CW-staining fluorescence. Early log-phase wild-type cells growing in YES liquid medium at 28 °C were collected and processed for time-lapse CW-staining visualization. Images were captured every 1 min at 28 °C. (**b**) A defect in B-BG synthesis causes misdirected actomyosin ring contraction and septum progression (*arrow*). Fluorescence time-lapse of the progression of the primary septum (CW), the septum plasma membrane (GFP-Psy1), and the contractile actomyosin ring (Rlc1-RFP) of a cell affected in the synthesis of the B-BG. Early log-phase cells of the Bgs4 mutant allele *cwg1-1* growing in YES liquid medium at 37 °C were collected and processed as in (**a**). Images were captured every 1 min at 37 °C. Images adapted from [4]. (**c**) Scheme of the μ-Slide 8 well (Ibidi GmbH, http://ibidi.com) used for the time-lapses presented in (**a**) and (**b**). The dimensions of the μ-Slide 8 well are those of a standard slide and it can be mounted and the cells observed in any microscope with slide holder. Images adapted from http://ibidi.com. (**d**) Chemical structure of CW fluorescent brightener 28 (Sigma). Image adapted from http://www.sigmaaldrich.com

CW (Fig. 2d) is a water-soluble dye that interacts with various polysaccharides, but displays a higher affinity for chitin and cellulose. It intercalates into nascent chains of chitin, and at high concentrations this prevents chitin microfibrils assembly [21]. However, at very low concentrations it can be used as a live-cell stain. Concentrations of CW ranging between 10 and 100 μg/ml are sublethal and higher concentrations are lethal for *S. cerevisiae* cells [22], while *S. pombe* cells, which do not contain detectable chitin in their cell wall, are highly resistant to CW [21], growing in the presence of concentrations of up to 1.5 mg/ml of the dye [23–25]. Compared to these concentrations, our method to follow septum formation and progression by time-lapse fluorescence microscopy uses a highly reduced concentration of CW (1.25–5 μg/ml) which does not affect the physiology of *S. pombe*.

CW is excited with UV light and fluoresces in a spectrum near-UV light (optimum excitation wavelength of 347 nm and peak emission wavelength of 450 nm), close to the range of DAPI and Hoechst dyes used for nuclear staining. The low excitation/emission peaks allow the use of CW for multiple fluorescence labels (Fig. 2b) in combination with the different versions of the GFP and/or RFP fluorescence proteins, or with other vital dyes which have major excitation peaks to higher spectral wavelengths [26, 27].

Although here we mainly describe a time-lapse method based on the CW-staining fluorescence, this method can be adapted to other dyes used to stain the cell wall polymers after the previous testing of the doses that do not alter cytokinesis and permit to image the derived-septum fluorescence along the time. Aniline blue and its derivative Sirofluor (also named Water blue, Cotton blue, Poirriers blue, Methyl blue) (Sigma-Aldrich; Polysciences, Inc.; Merck Millipore; Kingston Chemistry; etc.) are also used to stain the primary septum, but the concentration to be used varies considerably depending on the supplier. This dye binds to the single-helix and open triple-helix conformations of the L-BG but not to the closed triple-helix conformation of the B-BG [28]. In budding yeast it has been assumed that this dye stains specifically the closed triple-helix B-BG of the secondary septum and not other polysaccharide of the septum [29]. Since in this yeast it has not been described the existence of a single-helix L-BG, the specific affinity of Aniline blue in the cell wall of budding yeast still must be determined. Congo red (Sigma-Aldrich) binds to the chitin fibrils in the fungal cell walls and it is thought to interfere with cell wall assembly by binding to chitin as CW does [21, 22]. Solophenyl Flavine 7GFE 500 (Ciba Specialty Chemicals) and Pontamine Fast Scarlet 4B (Bayer Corp.) have also been described as useful dyes of fungal cell walls, septa, and bud scars [26]. These two dyes fluoresce in the green and red spectrum respectively, so theoretically they could be used in combination with CW for a putative differential cell wall staining.

2 Materials

2.1 Reagents

- CW fluorescent brightener 28 (Sigma-Aldrich). A stock of CW to 10 mg/ml is made by adding 100 mg of the fluorescent brightener 28 in 10 ml of distilled water and filter-sterilized. The final stock can be stored in the dark at 4 °C for at least 2–3 months. Discard when the solution displays signs of precipitated material (*see* **Notes 1** and **2**).
- 10 % potassium hydroxide.
- Lectin from Glycine max, soybean (Sigma-Aldrich). Prepare 1 mg/ml by adding the corresponding volume of distilled water

to the original vial with the powder. Next the solution is divided into 100–200 μl aliquots and stored at −20 °C.

- Fresh standard fission yeast culture media: yeast extract plus supplements (YE4S), or Edinburgh minimal medium + supplements (EMM4S). Recipes can be found at: http://www-bcf.usc.edu/~forsburg/media.html.

2.2 Disposables and Equipment

- Uncoated plastic or glass 15 μ-Slide 8 well for Live Cell Analysis (Ibidi GmbH). These slides (Fig. 2c) allows standard immunofluorescence protocols and live imaging of cells growing in liquid medium over extended periods of time on an inverted microscope (*see* **Note 3**).
- Inverted fluorescence microscope.

3 Methods

3.1 Time-Lapse Fluorescence Microscopy of the Septum Formation in S. pombe

This method is basically as described in [3, 4], with some modifications and tips to improve its understanding and make it simpler. We prefer to image the cells in liquid growth medium instead of using agarose embedding medium (*see* **Note 4**). Using liquid medium allows to maintain the cells in the growth conditions more efficiently and without any stress. In addition, this method is more conservative because it does not apply pressure on the cells, contrarily to the pressure caused by the coverslips used in agarose embedding medium.

3.1.1 Preparing the Cells

Use early log-phase cells growing in the same liquid medium and temperature that are going to be used for imaging. Maintain the cells in early log-phase with the appropriate dilutions.

Incubate 4-ml of early log-phase cells in 25-ml flasks using an incubator shaker set to 200 rpm shaking speed, which allows a stable incubation temperature and the adequate aeration of the cells (*see* **Note 5**).

3.1.2 Coating the Slide

The uncoated slide must be coated with lectin to promote the cell adhesion to the bottom surface.

1. Add the lectin (1 mg/ml) 20–30 min before image acquisition onto the bottom surface of a well of the μ-Slide (5 μl for plastic or 2.5 μl for glass bottom slides).
2. Gently spread the lectin to cover uniformly the well bottom surface. The side of a clean pipette tip can be used for spreading. Let the lectin dry at room temperature and wash the well twice with 300 μl of growth medium each, before addition of cells (*see* **Notes 6** and **7**).

3.1.3 Seeding Cells

1. Prepare a fresh stock of growth medium with CW. Dilute the CW to the desired final concentration (1.25–5 μg/ml) in 10-ml of the same growth medium where the cells are growing.
2. Centrifuge 750 μl of early log-phase growing cells at 2348 × *g* during 1 min and resuspend them in 300–500 μl of liquid medium containing CW.
3. Apply 300 μl of the cell suspension into a well of the μ-Slide and leave it at room temperature for 1.5 min. Avoid shaking or moving the μ-Slide as this will result in a non-homogenous distribution of the cells (*see* **Note 8**).
4. After seeding the cells, carefully remove the medium and wash the well twice with 300 μl of liquid medium containing CW. The washes permit to remove the cells that are not attached to the bottom surface of the well.
5. Add 300 μl of liquid medium containing CW and cover the μ-slide with the supplied lid. Then the cells are ready to be observed and imaged on the inverted fluorescence microscope.

3.1.4 Time-Lapse Microscopy

Consider the time scale of the dynamic processes being imaged to specifically optimize the image acquisition intervals and the total time of cell imaging. In a wild-type *S. pombe* strain, septum formation takes approximately 20–30 min depending on the temperature and growth medium selected for the specific experiment. However, in diverse mutant strains, septum formation may take longer. Although the septum forms a disk structure, 3D acquisition over the time is not required. As an example, a typical capture uses: 100×/1.40 IX70 objective lens; the camera set up to 2 × 2 binning; 30–120 s interval between successive images; 0.250 s exposure time with 10 % of light transmission intensity; and 30–60 min of total cell imaging time (*see* **Notes 9** and **10**).

3.2 Fluorescence Microscopy of the Septum Structure in *S. pombe*

3.2.1 CW-Staining of Live Cells

To image directly the primary septum of live cells, early log-phase cells grown in liquid medium can be visualized directly by adding a solution of CW to the sample (1.25–50 μg/ml final concentration):

1. Use precleaned/ready-to-use microscope slides. Always clean the surface of the slide with a coverslip to remove the remaining dust particles. This will help to image all the cells in focus, in a single focal plane. The single layer of cells between slide and coverslip permits to keep the cells stopped, without the cell movement that would be caused by convection of the liquid medium in a broader space.
2. Centrifuge 500–1000 μl of growing early log-phase cells at 2348 × *g* during 1 min. Discard the supernatant leaving 10–25 μl of the liquid medium, depending on the amount of cells to be imaged. Resuspend the cells and add CW to make 1.25–50 μg/ml final concentration. Alternatively, remove the entire

liquid medium and resuspend the pellet of cells in 10–25 μl of water, PBS, or liquid medium containing CW (1.25–50 μg/ml final concentration).

The standard concentration of CW used for septum staining is 50 μg/ml final concentration; but lower concentration can be useful for differential staining in non-saturated conditions (*see* **Notes 11** and **12**).

3. Deposit 1.75 μl of the cell suspension in the slide and place a clean coverslip over the sample. To obtain a monolayer of cells distributed uniformly, be sure that the cell suspension is covering all the surface of the coverslip. Depending on the time used for imaging cells in a single slide, the coverslip can be sealed with melted VALAP (1:1:1 vol mixture of vaseline, lanolin, and paraffin) to prevent the sample from drying.
4. Image the cells under a microscope equipped with a UV lamp and the appropriate filter (*see* **Note 13**).

3.2.2 CW-Staining of Fixed Cells

When acquisition of multiple label fluorescences is required, the cells may be occasionally fixed to preserve the fluorescence of a specific dye or tagged protein or its localization to a specific cellular structure. Several methods for *S. pombe* cells fixation have been described [30]. For CW-staining, cold 70 % ethanol is the most generally used:

1. Centrifuge 500–1000 μl of growing early log-phase cells at 2348 × *g* during 1 min.
2. Add 1-ml of cold 70 % ethanol and resuspend by vortexing the pellet of cells. In this step the samples can be stored at 4 °C.
3. Before microscope acquisition rehydrate the cells by removing the ethanol by centrifugation as above and adding 1 ml of water. Centrifuge the cells and resuspend the pellet in 10–25 μl of a solution containing CW and proceed as described for CW-staining of live cells (*see* Subheading 3.2.1).

Ethanol fixation can be used to image the nucleus (DAPI or Hoechst staining) and septum (CW-staining) simultaneously. CW fluorescence interferes with the signal from the DAPI or Hoechst staining. Thus proper CW concentration should be adjusted before imaging. Here, we briefly describe a protocol that works well for live and fixed cells. Centrifuge early-logarithmic phase cells, live or fixed with cold 70 % ethanol, and resuspend them in 20 μg/ml Hoechst or DAPI. Keep the cells at room temperature in the dark for several minutes, wash them to remove the Hoechst or DAPI, and add 10–25 μl of a 10 μg/ml CW solution. All these fluorophores are visualized with the same UV filter (*see* **Note 14**).

An alternative method using mounting media, CW, and DAPI can be found in [30].

3.2.3 Immunofluorescence of the Linear β(1,3)-Glucan of the Primary Septum

Since the primary septum contains L-BG, its structure can be studied through immunofluorescence by using a specific monoclonal antibody against the L-BG [7, 31]. *S. pombe* immunofluorescence method is essentially described in [32] with some modifications [8]. Briefly, cells are fixed with 100 % ethanol for 30 min at −20 °C, washed and incubated with 0.15–5.0 μg/ml of Novozyme-234, also named Glucanex (Sigma-Aldrich) for 45 min at 30 °C in PEMS buffer (100 mM Pipes pH 6.9, 1 mM EGTA, 1 mM $MgSO_4$, 1.2 M sorbitol) to allow a controlled and progressive partial digestion of the cell wall. Then labeling of the primary septum is performed using primary anti-L-BG-specific monoclonal antibody 400-2 (1:100 dilution; Biosupplies), and secondary Alexa Fluor 594-labeled anti-mouse antibody (1:400 dilution) (*see* **Note 15**).

4 Notes

1. CW is described as hazardous and potentially carcinogenic. Wear protective gloves, eye shields, and a dusk mask to prevent inhalation during stock preparation.
2. Depending on the supplier the CW cannot be directly dissolved in water so its dissolution may be aided by the addition of 10 N KOH dropwise, with mixing until the solution becomes clear.
3. In addition to the 15 μ-Slide 8 well, this method can be used with glass bottom dishes (P35G, MatTek), although we favor the use of the μ-Slides because they can be inserted in most of microscope stages and can be used for eight individual experiments.
4. As an alternative method to the live fluorescence imaging with lectin-coated μ-Slides or bottom-glass dishes the cells can be mounted in agarose pads, where the agar embedding growth medium keeps the cell in a immobilized position and provides an acceptable environment for sustained cell growth. Briefly, these agarose pads contain the appropriate yeast growth medium plus 2 % w/vol of agarose. Place approximately 100 μl of hot melted agarose onto a clean glass slide and immediately place a second glass slide on top of the agar drop. After a few minutes to let the agar to cool and solidify, remove the top glass slide from the agar pad by gently but quickly sliding apart both glass slides. Next add quickly 1–2 μl of cell suspension onto the solid agar pad, place a coverslip on top of the cells and seal it with melted VALAP (a 1:1:1 vol mixture of vaseline, lanolin, and paraffin). Then the cells are ready to be imaged for a long-term period [33].
5. To improve the cell wall and septum labeling, CW can be added to the growth medium and the cells can be incubated in

this medium for 12–24 h or longer times before image acquisition. Use the same concentration of CW (1.25–5 μg/ml) before and during image acquisition.

6. Coating can be prepared a day in advance, but it is preferable to make it the same day of use. Keep the coated μ-Slide at room temperature until its use. Be sure to keep track of which well is coated.
7. Poly-L-lysine or Concanavalin A-coated μ-Slides do not completely immobilize fission yeast cells; however, they can be used for budding yeast cells microscopy.
8. The cells must be handled and seeded carefully. The absence of adequate care can lead to non-flattened cells, resulting in regions of the cells out of focus and hence, with multiple focal layers. In our case, 1.5 min of seeding works well in most of the cell backgrounds displaying wild-type morphology. However the seeding time may vary depending on the concentration of lectin used to coat the well, the growth medium or the cell shape. For example, liquid medium containing sorbitol is denser and therefore the cells need longer times (2.5–3 min) to seed and attach to the well surface. On the contrary, longer mutant cells need less time to seed (0.5–1 min).
9. Phototoxicity and photobleaching are two major problems in live cell analysis. Although it has been described that CW fades rapidly when exposed to excitation wavelengths [26], in our experimental conditions we never detected any bleaching of the CW fluorescence over the time.
10. CW is excited with UV or near-UV light, phototoxicity can be also an issue. Phototoxicity describes a general class of damaging effects on live cells based on either long term or short but extreme exposures to a light source. Phototoxicity can be a result of DNA damage (UV light) or protein damage (infrared light). Besides, excessive excitation of a fluorophore can lead to oxide radical formation and may also negatively impact cell growth [34]. To avoid unexpected effects during acquisition, 3D microscopy of the septum over the time is not recommended unless necessary for a specific experiment. On the contrary, reducing the percentage of light transmission to 10 % and using 0.250 s of exposure time permit to extend the movie acquisition time up to at least 90 time points during 90 min using the CW fluorescent dye to follow septum formation.
11. CW signal depends on the concentration of cells in the sample. To more cells concentration, less CW signal will be detected.
12. We have observed that much lower concentrations of CW, such as 1.25 μg/ml, are still able to stain all the septa of a wild-type strain. Concentrations below 1.25 μg/ml can also be used by increasing the exposure time. The analysis with low CW

concentration can be useful for strains affected in the formation of the primary septum, where 50 μg/ml of CW can saturate and eventually mask a defective staining of the septum. This is the case of Bgs1-depleted cells, whose defects in the primary septum structure are clearly observed using CW at 1.25 μg/ml, but are not detected at the standard concentration of 50 μg/ml. Under conditions of low CW concentration, many Bgs1-depleted septa did not stain or displayed a very weak and irregular CW-staining [8].

13. CW only stains the L-BG and therefore, the primary septum structure. No fluorochrome has been described for B-BG, α-glucan, or secondary septum labeling. Lectins that bind the outer layer of cell wall galactomannoproteins have been used to image the cell outline, but they do not label the septum, probably because the outer layer of galactomannoproteins is absent of this structure. Therefore, to study the complete septum morphology by fluorescence microscopy, the only option available nowadays is the analysis of the septum thickness by using fluorescent tagged proteins or probes that localize to the septum plasma membrane (glucan synthases, acylated GFP, syntaxin Psy1, etc.) and therefore, that are surrounding the septum structure.
14. When imaging fixed cells notice that while some cellular structures are well preserved after rehydration, others can result affected. The last is the case for the septum structure. When *S. pombe* cells are fixed with ethanol (or methanol, formaldehyde, etc.) and imaged, the septa appear as thicker and curved structures with a very weak CW-stain when compared with the septa of live cells. This is probably due to alterations in the wall structure and/or linkage between polysaccharides, causing fluffy and expanded wall structures.
15. For optimal primary septum L-BG labeling, we observed that 1 μg/ml of Novozyme preserves the cell shape and allows labeling of all the septa in wild-type cells. However, this concentration could not be adequate for cells of mutants affected in the cell wall structure, and therefore in this case a range of Novozyme 234 or Glucanex concentrations should be previously tested.

References

1. Bathe M, Chang F (2010) Cytokinesis and the contractile ring in fission yeast: towards a systems-level understanding. Trends Microbiol 18(1):38–45. doi:10.1016/j.tim.2009.10.002
2. Sipiczki M (2007) Splitting of the fission yeast septum. FEMS Yeast Res 7(6):761–770
3. Cortés JCG, Sato M, Muñoz J, Moreno MB, Clemente-Ramos JA, Ramos M, Okada H, Osumi M, Durán A, Ribas JC (2012) Fission yeast Ags1 confers the essential septum strength needed for safe gradual cell abscission. J Cell Biol 198(4):637–656
4. Muñoz J, Cortés JC, Sipiczki M, Ramos M, Clemente-Ramos JA, Moreno MB, Martins IM, Pérez P, Ribas JC (2013) Extracellular cell wall β(1,3)glucan is required to couple septation to actomyosin ring contraction. J Cell Biol 203(2):265–282

5. Johnson BF, Yoo BY, Calleja GB (1973) Cell division in yeasts: movement of organelles associated with cell plate growth of *Schizosaccharomyces pombe*. J Bacteriol 115(1):358–366
6. Pérez P, Ribas JC (2004) Cell wall analysis. Methods 33(3):245–251
7. Humbel BM, Konomi M, Takagi T, Kamasawa N, Ishijima SA, Osumi M (2001) In situ localization of b-glucans in the cell wall of *Schizosaccharomyces pombe*. Yeast 18(5):433–444
8. Cortés JCG, Konomi M, Martins IM, Munoz J, Moreno MB, Osumi M, Durán A, Ribas JC (2007) The (1,3)β-D-glucan synthase subunit Bgs1p is responsible for the fission yeast primary septum formation. Mol Microbiol 65(1):201–217
9. Verma DP (2001) Cytokinesis and building of the cell plate in plants. Annu Rev Plant Physiol Plant Mol Biol 52:751–784
10. Pelosi L, Imai T, Chanzy H, Heux L, Buhler E, Bulone V (2003) Structural and morphological diversity of (1-3)-β-D-glucans synthesized *in vitro* by enzymes from *Saprolegnia monoica*. Comparison with a corresponding *in vitro* product from blackberry (*Rubus fruticosus*). Biochemistry 42(20):6264–6274
11. Gawronski M, Park JT, Magee AS, Conrad H (1999) Microfibrillar structure of PGG-glucan in aqueous solution as triple-helix aggregates by small angle x-ray scattering. Biopolymers 50(6):569–578
12. Kopecka M, Kreger DR (1986) Assembly of microfibrils *in vivo* and *in vitro* from (1-3)-β-D-glucan synthesized by protoplasts of *Saccharomyces cerevisiae*. Arch Microbiol 143(4):387–395
13. Saito H, Yoshioka Y, Yoloi M, Yamada J (1990) Distinct gelation mechanism between linear and branched (1-3)-β-D-glucans as revealed by high resolution solid state ^{13}C NMR. Biopolymers 29(14):1689–1698
14. Magnelli PE, Cipollo JF, Robbins PW (2005) A glucanase-driven fractionation allows redefinition of *Schizosaccharomyces pombe* cell wall composition and structure: assignment of diglucan. Anal Biochem 336(2):202–212
15. Sugawara T, Takahashi S, Osumi M, Ohno N (2004) Refinement of the structures of cell-wall glucans of *Schizosaccharomyces pombe* by chemical modification and NMR spectroscopy. Carbohydr Res 339(13):2255–2265
16. Bush DA, Horisberger M, Horman I, Wursch P (1974) The wall structure of *Schizosaccharomyces pombe*. J Gen Microbiol 81(1):199–206
17. Edwards JA, Alore EA, Rappleye CA (2011) The yeast-phase virulence requirement for α-glucan synthase differs among *Histoplasma capsulatum* chemotypes. Eukaryot Cell 10(1):87–97. doi:10.1128/EC.00214-10
18. Grun CH, Hochstenbach F, Humbel BM, Verkleij AJ, Sietsma JH, Klis FM, Kamerling JP, Vliegenthart JF (2005) The structure of cell wall α-glucan from fission yeast. Glycobiology 15(3):245–257
19. Proctor SA, Minc N, Boudaoud A, Chang F (2012) Contributions of turgor pressure, the contractile ring, and septum assembly to forces in cytokinesis in fission yeast. Curr Biol 22(17):1601–1608
20. Zhou Z, Munteanu EL, He J, Ursell T, Bathe M, Huang KC, Chang F (2015) The contractile ring coordinates curvature-dependent septum assembly during fission yeast cytokinesis. Mol Biol Cell 26(1):78–90. doi:10.1091/mbc.E14-10-1441, mbc.E14-10-1441 [pii]
21. Roncero C, Durán A (1985) Effect of Calcofluor white and Congo red on fungal cell wall morphogenesis: *in vivo* activation of chitin polymerization. J Bacteriol 163(3):1180–1185
22. Ram AF, Klis FM (2006) Identification of fungal cell wall mutants using susceptibility assays based on Calcofluor white and Congo red. Nat Protoc 1(5):2253–2256. doi:10.1038/nprot.2006.397, nprot.2006.397 [pii]
23. Carnero E, Ribas JC, García B, Durán A, Sánchez Y (2000) *Schizosaccharomyces pombe* Ehs1p is involved in maintaining cell wall integrity and in calcium uptake. Mol Gen Genet 264(1-2):173–183
24. Tajadura V, García B, García I, García P, Sánchez Y (2004) *Schizosaccharomyces pombe* Rgf3p is a specific Rho1 GEF that regulates cell wall β-glucan biosynthesis through the GTPase Rho1p. J Cell Sci 117(Pt 25):6163–6174
25. Martín V, García B, Carnero E, Durán A, Sánchez Y (2003) Bgs3p, a putative 1,3-β-glucan synthase subunit, is required for cell wall assembly in *Schizosaccharomyces pombe*. Eukaryot Cell 2(1):159–169
26. Hoch HC, Galvani CD, Szarowski DH, Turner JN (2005) Two new fluorescent dyes applicable for visualization of fungal cell walls. Mycologia 97(3):580–588
27. Shaner NC, Steinbach PA, Tsien RY (2005) A guide to choosing fluorescent proteins. Nat Methods 2(12):905–909
28. Young SH, Jacobs RR (1998) Sodium hydroxide-induced conformational change in schizophyllan detected by the fluorescence dye, aniline blue. Carbohydr Res 310(1-2):91–99
29. Onishi M, Ko N, Nishihama R, Pringle JR (2013) Distinct roles of Rho1, Cdc42, and Cyk3 in septum formation and abscission during

yeast cytokinesis. J Cell Biol 202(2):311–329. doi:10.1083/jcb.201302001, jcb.201302001 [pii]

30. Forsburg SL, Rhind N (2006) Basic methods for fission yeast. Yeast 23(3):173–183. doi:10.1002/yea.1347
31. Meikle PJ, Bonig I, Hoogenraad NJ, Clarke AE, Stone BA (1991) The location of (1-3)-β-glucans in the walls of pollen tubes of *Nicotiana alata* using a (1-3)-β-glucan-specific monoclonal antibody. Planta 185(1):1–8
32. Hagan IM, Hyams JS (1988) The use of cell division cycle mutants to investigate the control of microtubule distribution in the fission yeast *Schizosaccharomyces pombe*. J Cell Sci 89:343–357
33. Tran PT, Paoletti A, Chang F (2004) Imaging green fluorescent protein fusions in living fission yeast cells. Methods 33(3):220–225
34. Rines DR, Thomann D, Dorn JF, Goodwin P, Sorger PK (2011) Live cell imaging of yeast. Cold Spring Harb Protoc 2011(9). doi:10.1101/pdb.top065482, 2011/9/pdb.top065482 [pii]

Chapter 7

Visualization of Cytokinesis Events in Budding Yeast by Transmission Electron Microscopy

Franz Meitinger and Gislene Pereira

Abstract

In yeast cells, cytokinesis is accompanied by morphological changes due to cell wall growth during furrow ingression and abscission. The characteristics of the growing cell wall can be used as an indicator for the function of the contractile actomyosin ring, the Rho-GTPases Rho1 and Cdc42 and/or other factors that drive cytokinesis. The ultrastructural information of the cell wall can be easily acquired by transmission electron microscopy, which makes this technique an invaluable tool to analyze cell division in yeast cells. Here, we describe the process of embedding and staining budding yeast cells for transmission electron microscopic analysis of cytokinetic events.

Key words Budding yeast, Cell division, Cytokinesis, Primary septum, Secondary septum, Electron microscopy

1 Introduction

Cytokinesis is a complex process that involves the re-organization of the cytoskeleton, targeted membrane trafficking, and the formation of an actomyosin ring (AMR) that drives membrane ingression. In yeast, an additional structure composed of chitin, named the primary septum, is formed during AMR contraction. Primary septum formation and AMR contraction are followed by the deposition of the secondary septum, which is composed of extracellular matrix that is continuous with the cell wall [1].

Transmission electron microscopy (TEM) is a powerful tool to analyze the ultrastructure of yeast cells (Fig. 1) and in particular of the primary/secondary septa during cytokinesis (Fig. 2). Electron microscopes operate on the same principle as optical/light microscopes. However, the physical properties of electrons, which have a wavelength below those of visible light, allow us to obtain 1000 times higher magnifications than with a light microscope. The image is formed through the interaction of an emitted electron beam with the specimen under a high vacuum. The specimen

Alberto Sanchez-Diaz and Pilar Perez (eds.), *Yeast Cytokinesis: Methods and Protocols*, Methods in Molecular Biology, vol. 1369, DOI 10.1007/978-1-4939-3145-3_7, © Springer Science+Business Media New York 2016

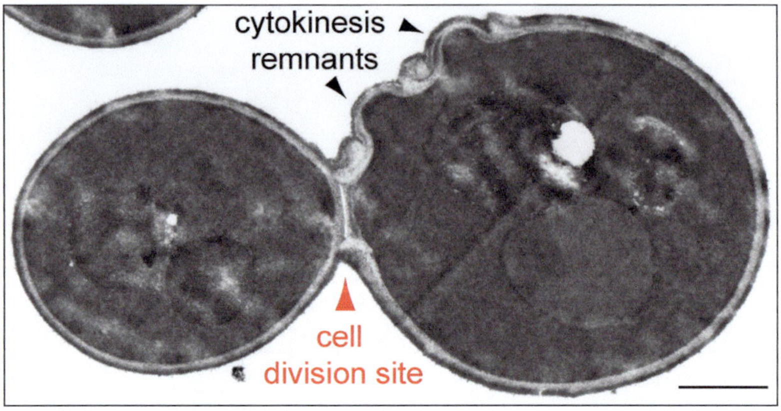

Fig. 1 Electron micrograph of a budding yeast cell. Yeast cells were embedded and stained as described in this protocol. The cross-section shows the bud neck region after completion of cytokinesis (*red arrowhead* points to the primary and secondary septa at the site of cell division) and cytokinesis remnants or bud scars (*black arrowhead*). Scale bar, 1 μm

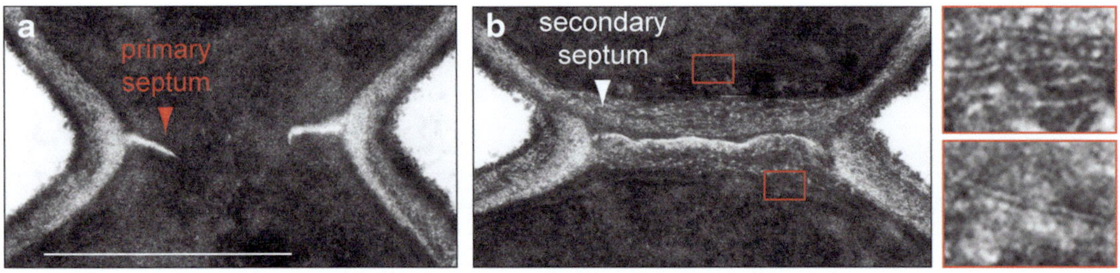

Fig. 2 Electron micrographs of the cell division site. (**a**, **b**) Electron micrographs show the cell division site during primary septum formation (**a**) and after secondary septum deposition (**b**). The enlarged regions (*red boxes*) show plasma membrane staining. Scale bar, 1 μm

therefore has to be free of water and resistant to vacuum. Biological samples like yeast cells have to be fixed and embedded into resin to meet the requirements for TEM. Special staining procedures allow the discrimination between cellular structures. Several different techniques have been developed to visualize different aspects of yeast cells [2]. Here, we describe how to fix and embed yeast cells for inspection of the cell wall and plasma membrane structures using TEM (Figs. 1 and 2). The method is based on a combination of potassium permanganate and uranyl acetate [3–5]. The protocol described here was successfully used to analyze the functions of genes in cell division [6–9] and cell polarization [10, 11].

2 Materials

Prepare all solutions with ultrapure distilled water. Hazardous materials should be handled and disposed according to local safety regulations. Unless otherwise specified, solutions should be stored at room temperature (23 °C).

2.1 Yeast Growth Media

1. YPDA medium: dissolve 10 g Bacto Yeast Extract (BD Biosciences), 20 g Bacto Peptone (BD Biosciences), 20 g glucose (BD Biosciences), and 0.1 g adenine (Sigma) in 1 l distilled water. Autoclave or filter-sterilize using a 0.22 μm filter unit (*see* **Note 1**) [12].

2.2 Fixation and Embedding

1. Glutaraldehyde 25 % (biological grade, Polyscience).
2. Potassium permanganate: 4 % solution in water (*see* **Note 2**).
3. Uranyl acetate: saturated solution in water (*see* **Note 3**). Store at 4 °C.
4. Ethanol absolute (≥99.8 %, Sigma) and dilutions (30, 50, 70, 85, and 95 %).
5. Propylene oxide (Sigma) (*see* **Note 4**).
6. Agar 100 resin kit (Agar Scientific) (*see* **Note 5**).
7. Agar100 mix1: 5 ml Agar100 epoxy resin, 3.5 ml MNA, and 1.5 ml DDSA (*see* **Note 6**).
8. Agar100 mix2: Agar100 mix1 + 0.15 ml DMP-30 (*see* **Note 6**).
9. BEEM® capsules 5.2 mm for EM-embedding (Serva Electrophoresis).
10. Tweezers for handling grids (e.g. Dumont Biology, Dumostar tweezer high precision grade, tip dimensions 0.05 × 0.01 mm).

2.3 Sectioning and Contrasting

1. Ultramicrotome.
2. Formvar film-coated copper grids (3.05 mm 100 mesh) (obtained from Electron Microscopy Sciences or any other company that supplies reagents for TEM).
3. Filter paper (e.g. Whatman® qualitative filter paper, grade 1 from Sigma or similar).
4. Uranyl acetate: 2 % in absolute methanol. Store at 4 °C protected from light. Allow to reach room temperature before use.
5. Methanol (analytical grade).
6. Lead citrate: mix 1.33 g Lead(II)nitrate ($Pb(NO_3)_2$) (Sigma) with 1.76 g sodium citrate dihydrate ($Na_3C_6H_5O_7 \cdot 2H_2O$) (Sigma) in 30 ml distilled water in a 50 ml volumetric flask. Stir for 30 min. Add 8 ml of 1 N NaOH and distilled water to

reach a final volume of 50 ml. Mix well. The pH of the solution was routinely found to be around 12. Faint turbidity, if present, can be easily removed by centrifugation. The staining solution, stored in glass or polyethylene bottles, is stable for a period of up to 6 months [13].

7. 0.05 N Sodium hydroxide (NaOH). Store in glass bottles at room temperature.
8. Nine 5 ml glass snap-cap bottles (approx. 35 mm H, 13 mm D).
9. Grid storage boxes (e.g. Gilder Grid Storage Box SB100, Gilder Grids Ltd., UK).

3 Methods

3.1 Embedding

1. Grow yeast cells in the appropriate growth medium until the cell suspension reaches an optical density of 0.5–1.0 at 600 nm (approximately $1\text{–}2 \times 10^7$ cells/ml). We normally use 10–20 ml of this cell suspension per embedding.
2. Fix cells in 2.5 % glutaraldehyde for 1 h at room temperature. For this, add 1 ml of 25 % glutaraldehyde to 9 ml cell suspension in a conical tube. Leave it on your bench and mix the cells every 10 min by inversion. After incubation at room temperature, the cell suspension can be placed on ice overnight or processed further.
3. Spin cells with a relative centrifugal force of $400 \times g$ or less to prevent clumping. Centrifugation steps are carried out at room temperature for 2 min. Discard the supernatant and gently resuspend the cell pellet in 1 ml of distilled water using a micropipette or by flicking the tube by hand (do not vortex). Use this condition for all subsequent washing steps.
4. Transfer the cell suspension into a 2 ml conical polypropylene tube (we normally use Eppendorf Safe-Lock tubes®). Wash cells once more with water. Centrifuge and discard supernatant.
5. Resuspend cells in 1 ml of 4 % potassium permanganate and incubate for 30 min at room temperature. Centrifuge and discard supernatant.
6. Wash cells twice with 1 ml of distilled water as described in **step 3**.
7. Resuspend cells in 500 μl of aqueous saturated uranyl acetate solution (filtered prior to use, *see* **Note 3**). Place tubes in the dark for 30 min at room temperature (tubes can be placed in a drawer or covered with aluminum foil). We normally do not mix the samples during this incubation.
8. Wash cells once with 1 ml of distilled water as described in **step 3**.

9. Dehydrate cells by sequentially incubating them with increasing concentrations of ethanol. For this, centrifuge cells (400 × *g*, 2 min), discard supernatant and gently resuspend the cells in 1 ml of 30 % ethanol by flicking the tubes by hand (do not vortex). Incubate for 5–10 min. Centrifuge, discard the supernatant, and repeat the incubation with fresh 30 % ethanol. Repeat this procedure with 50, 70, 85, 95 % and absolute ethanol solutions.
10. Wash cells twice with 1 ml of propylene oxide. After the last centrifugation step, aspirate off the supernatant leaving 100–200 μl of propylene oxide in the tube (*see* **Notes 4**, **7** and **8**).
11. Add 1 ml of Agar100 mix1 to the cells and incubate for at least 30 min at room temperature. Repeat this step two more times (three washes with Agar100 mix1 in total) (*see* **Notes 8–10**).
12. Wash once with 1 ml of Agar100 mix2. Transfer cells into BEEM®-capsules and incubate cells in Agar100 mix2 for 1–2 days at room temperature in the dark (*see* **Note 11**). Spin capsules using a swing out rotor at 4000 × *g* for approximately 20 min or until the cells are pelleted (*see* **Note 10** and Fig. 3a). Incubate cells for another 2 days at room temperature in the dark.
13. Label (*see* **Note 12** and Fig. 3a) and place the BEEM®-capsules in an incubator at 60–65 °C for 24 h to allow the solidifying polymerization process.

3.2 Sectioning and Contrasting Procedure

1. Ultrathin sections (50–90 nm) should be prepared with a diamond knife using an ultramicrotome (*see* **Notes 13** and **14**) [14].
2. The sections should be mounted on the Formvar-coated side of the grid (*see* **Note 15** and Fig. 3b).
3. For the post-embedding contrasting of the cells, prepare silicone tubes with slots to hold the grids (*see* Fig. 4 and **Note 16**).

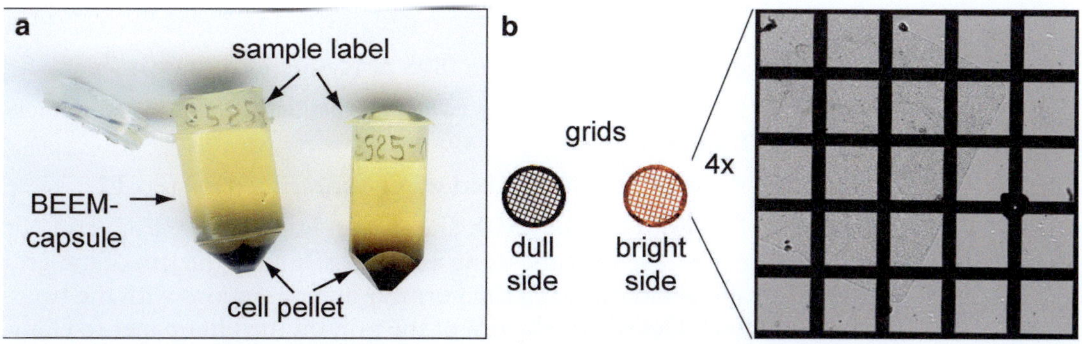

Fig. 3 Embedding of yeast cells and mounting on a grid. (**a**) Resin-embedded yeast cells in the BEEM®-capsule (*left*) or after removal of the BEEM®-capsule after resin polymerization (*right*). (**b**) Formvar film-coated copper grids (3.05 mm 100 mesh). The magnification on the *left* shows one mounted section on the grid (image taken using a DeltaVision RT microscope system with a 4× magnification objective lens)

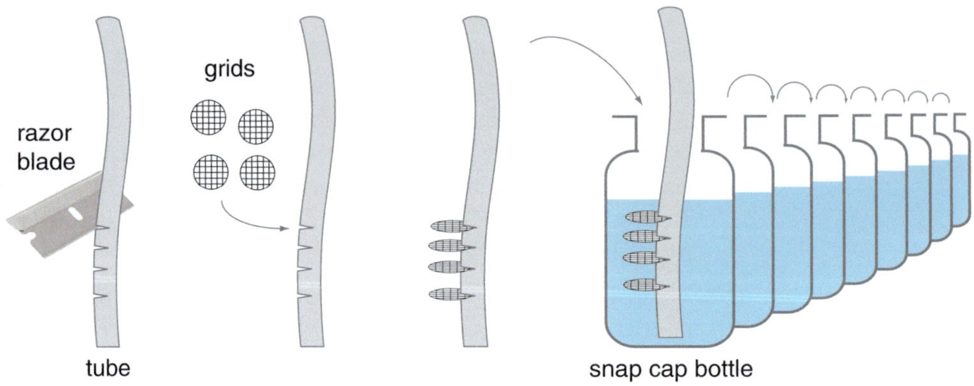

Fig. 4 Post-embedding contrasting procedure. Schematic representation of the different steps of the contrasting procedure

Attach grids (section-face down) at the silicone tube as depicted in Fig. 4. Also prepare nine glass snap-cap bottles as containers for the following incubation steps (see below). Add approximately 4–5 ml of the corresponding solution per container.

4. Place the silicone tube carrying the grids in container 1 filled with 2 % uranyl acetate solution in methanol. Incubate for 20 min (protected from light).
5. Immerse grids briefly in methanol in container 2.
6. Briefly immerse grids in a 1:1 mixture of methanol and distilled water in container 3.
7. Briefly immerse grids twice in distilled water in containers 4 and 5.
8. Using a wash bottle filled with distilled water, rinse the grids carefully with steady streams of water for approximately 30 s.
9. Incubate the grids for 5 min in lead citrate solution in container 6.
10. Briefly immerse grids in 0.05 N NaOH in container 7.
11. Briefly immerse grids twice in distilled water in containers 8 and 9.
12. Rinse grids with distilled water as described in **step 12**.
13. Dry grids carefully with filter paper. For this, take each grid out of the silicone tube using a tweezer (hold at the rim of the grid and avoid touching the Formvar film or sections with the tweezer). Gently tap the rim of the grid on the filter paper to eliminate the excess of water. Place the grid on a Petri dish coated with clean filter paper (section-face up) and allow the grids to dry. Store grids in a grid holder.

3.3 Analysis of Specimen

1. Analyze prepared sections at a transmission electron microscope (e.g. Zeiss EM900 or EM910; follow the manufacturer's instructions).
2. Find the bud neck region (cell division site) of dividing yeast cells. Use a magnification of 3,000–4,000.
3. Acquire images of the region of interest. Use a magnification of 20,000–40,000.
4. Quantifiable phenotypes include the symmetry or the directionality of primary septum formation (indicator for actomyosin ring function) [6, 8, 9], the thickness of the secondary septum (indicator for Rho1 activity) or the appearance of bud neck collars (indicator for impaired spatial regulation of Cdc42) [10, 11].

4 Notes

1. YPDA is a rich medium and it is used to grow cells when selection is not required. Use your choice of appropriate growth medium if your strains carry plasmids or genes under control of inducible promoters [12].
2. Prepare potassium permanganate solution fresh and filter it prior to use. Potassium permanganate ($KMnO_4$) is a strong oxidizer and forms a brown product (MnO_2) over time once in solution.
3. Note that uranyl acetate is toxic and radioactive and should be handled carefully. Uranyl acetate saturates at around 7 % in water. To prepare 100 ml of a saturated uranyl acetate solution, weigh 7 g of uranyl acetate in a glass beaker and add 100 ml distilled water. Stir the suspension overnight at room temperature and store at 4 °C in a glass bottle protected from light. Before use, allow the not-fully dissolved uranyl acetate to settle down, remove the supernatant, and filter it through a 0.22 μm syringe filter. Use uranyl acetate solution at room temperature. This procedure prevents crystallization of uranyl acetate during sample preparation.
4. Propylene oxide is a potential carcinogen and should be handled carefully inside the fume hood.
5. The Agar 100 resin kit contains Agar 100 epoxy resin, which is equal to Epon 812 but it has a more defined composition and is less viscous. The lower viscosity of Agar 100 in comparison to Epon 812 facilitates the handling and improves the infiltration of the resin into the specimen. In addition, the Agar 100 kit contains the hardeners dodecenylsuccinic anhydride (DDSA) and methyl nadic anhydride (MNA), and the polymerization accelerator Tris-(dimethyl amino methyl) phenol

(DMP-30). Safety data sheets and more detailed technical specifications of the Agar 100 kit components can be downloaded from the supplier (http://www.agarscientific.com/fr/agar-100-resin-kit.html). We strongly recommend a closer inspection of these files, as these substances are hazardous and should be handled with care. Also, follow the supplier's instructions to prepare the mixes.

6. Prepare Agar100 mixes 1 and 2 1 day before usage. For this, add the components into a 50 ml conical polypropylene tube. Place the tubes overnight on a roller mixer to obtain a homogenous mix. Avoid the formation of bubbles. The mixes can be stored at −20 °C. In this case, allow the mixes to reach room temperature before use.
7. Propylene oxide is an ethereal liquid and therefore difficult to pipette because of its high vapor pressure. To overcome this problem, pipette the liquid several times up and down using disposables or glass pipettes to adjust the pressure inside the pipette.
8. A critical step during embedding of the sample is the infiltration of the Agar100 mix into the cells. Infiltration can be improved by mixing the Agar100 mix1 with the leftovers of propylene oxide from the previous washing step. To further improve penetration of the resin into cells, the Agar100 mix1 can be incubated with the cells for up to 12 h in the second and third washing steps.
9. Please keep in mind that Agar100 mix1 will increase in viscosity over time. If cells are incubated for longer periods of time with this mix, they will become more difficult to be pelleted by centrifugation. It is therefore important to find a good balance. In our hands, a total incubation for up to 2 days with Agar100 mix1 worked well.
10. If cells are not pelleted, you can increase the centrifugation speed and time gradually until a good size pellet is obtained. Try to prevent high-speed centrifugations to avoid cell clumping.
11. Do not fill the BEEM®-capsules more than three-quarters full with the cell suspension in Agar100 mix2 at this step to leave some space for labeling. We normally prepare two BEEM®-capsules per sample strain.
12. Label the BEEM®-capsules before placing them in the 60–65 °C incubator. We prepare a small piece of paper, which is approximately as long as the perimeter of the BEEM®-capsules and 3–4 mm broad. Label the paper with a pencil and place it inside the BEEM®-capsule along the edge, so that the label faces the outside. Carefully fill up the BEEM®-capsule with Agar100-mix2 and place it in the incubator (Fig. 3a).

13. Making ultrathin sections with a microtome is technically difficult and requires a lot of practice and experience. Thus, we strongly recommend that you attend an appropriate training course or work in close collaboration with experienced researchers.
14. Thin sections (50–60 nm; these sections have a silver-gray color) tend to break and may have holes. Yeast cells especially tend to break out of the sections during cutting when embedding is not optimal. Therefore, it can be useful to also make thicker sections (70–90 nm; these sections have a bright silver-pale or gold color), which are less likely to break. These sections should however be thin enough to allow visualization of the desired structures. Mount thin and thick sections on the same grid.
15. The pre-coated grid usually has a bright dull side and a darker shiny side. The Formvar film is usually at the dull side (Fig. 3b).
16. Take a silicone tube with an approximate diameter of 5 mm and cut the slots for the grids with a razor blade or scalpel. Attach the grid with the rim to the tube, so that the specimen itself is not covered (Fig. 4).

References

1. Lesage G, Bussey H (2006) Cell wall assembly in Saccharomyces cerevisiae. Microbiol Mol Biol Rev 70(2):317–343
2. Osumi M (2012) Visualization of yeast cells by electron microscopy. J Electron Microsc (Tokyo) 61(6):343–365. doi:10.1093/jmicro/dfs082
3. Luft JH (1956) Permanganate; a new fixative for electron microscopy. J Biophys Biochem Cytol 2(6):799–802
4. Watson ML (1958) Staining of tissue sections for electron microscopy with heavy metals. J Biophys Biochem Cytol 4(4):475–478
5. Neiman AM (1998) Prospore membrane formation defines a developmentally regulated branch of the secretory pathway in yeast. J Cell Biol 140(1):29–37
6. Meitinger F, Boehm ME, Hofmann A, Hub B, Zentgraf H, Lehmann WD, Pereira G (2011) Phosphorylation-dependent regulation of the F-BAR protein Hof1 during cytokinesis. Genes Dev 25(8):875–888
7. Meitinger F, Palani S, Hub B, Pereira G (2013) Dual function of the NDR-kinase Dbf2 in the regulation of the F-BAR protein Hof1 during cytokinesis. Mol Biol Cell 24(9): 1290–1304
8. Meitinger F, Palani S, Pereira G (2012) The power of MEN in cytokinesis. Cell Cycle 11(2):219–228
9. Meitinger F, Petrova B, Lombardi IM, Bertazzi DT, Hub B, Zentgraf H, Pereira G (2010) Targeted localization of Inn1, Cyk3 and Chs2 by the mitotic-exit network regulates cytokinesis in budding yeast. J Cell Sci 123(Pt 11):1851–1861
10. Meitinger F, Richter H, Heisel S, Hub B, Seufert W, Pereira G (2013) A safeguard mechanism regulates Rho GTPases to coordinate cytokinesis with the establishment of cell polarity. PLoS Biol 11(2):e1001495
11. Meitinger F, Khmelinskii A, Morlot S, Kurtulmus B, Palani S, Andres-Pons A, Hub B, Knop M, Charvin G, Pereira G (2014) A memory system of negative polarity cues prevents replicative aging. Cell 159(5):1056–1069. doi:10.1016/j.cell.2014.10.014
12. Sherman F (1991) Getting started with yeast. Methods Enzymol 194:3–21
13. Reynolds ES (1963) The use of lead citrate at high pH as an electron-opaque stain in electron microscopy. J Cell Biol 17:208–212
14. Burghardt RC, Droleskey R (2006) Transmission electron microscopy. Current protocols in microbiology Chapter 2:Unit 2B 1. doi:10.1002/9780471729259.mc02b01s03.

Chapter 8

Visualization of Fission Yeast Cells by Transmission Electron Microscopy

Matthias Sipiczki

Abstract

This chapter deals with the preparation of fission yeast (*Schizosaccharomyces*) cells for ultrastructural examination. The structure of the cell must be preserved as close to the in vivo situation as possible. This can be achieved by either chemical or cryofixation; the latter will not be dealt with in this chapter. Aldehydes that cross-link proteins and permanganates that besides cross-linking also stain membranous and cell wall structures are used for chemical fixation. This step is followed by dehydration and embedding of the cells in epoxy or acrylic resin. Sectioning of the embedded material produces slices of the cells that have to be stained with heavy metals to increase contrast differences between different structures or can be used for immunodetection of antigens (polysaccharides or proteins) with specific primary antibodies and gold-conjugated secondary antibodies.

Key words Transmission electron microscopy, Immunoelectron microscopy, *Schizosaccharomyces*, Cell wall, Septum, Ultrastructure

1 Introduction

The major reason for using a transmission electron microscope as opposed to a light microscope is the vast increase in resolution of the images. Cellular and subcellular structures can be viewed in greater detail with an electron microscope than with a light microscope (general refs. 1, 2). Electron microscopy in combination with immunolabeling also allows the localization of macromolecules at high resolution in the structural context of the cell.

Transmission electron microscopy of fission yeasts requires the transformation of their cells from the living, hydrated state to a dry state that faithfully maintains the structural and biological relationships. To accomplish this, the cells must pass through a series of preparatory steps: fixation, embedding, sectioning, and contrasting.

The procedure starts with the stabilization of the cell structure by fixation. There are three main ways in which fission yeast cells can be processed to retain their structural organization for

Alberto Sanchez-Diaz and Pilar Perez (eds.), *Yeast Cytokinesis: Methods and Protocols*, Methods in Molecular Biology, vol. 1369, DOI 10.1007/978-1-4939-3145-3_8, © Springer Science+Business Media New York 2016

subsequent staining. These are fixation by cross-linking, fixation by precipitation, and fixation by freezing (cryofixation). Cryofixation is probably the best technique for cellular preservation [3] and has been efficiently used to study fission yeasts (e.g. [4–6]). The disadvantage of this method is that it requires specialized equipments usually unavailable in most laboratories. Therefore, I will not discuss it in this chapter. The most frequently used methods are based on treatment of the cells with specific chemicals called fixatives (general refs. 2, 7). Today, the most common primary fixatives for electron microscopy (EM) are aldehydes (e.g. glutaraldehyde, formaldehyde) that preserve the structure by cross-linking proteins and certain other types of macromolecules. Their application to fixation of fission yeast cells will be discussed in this chapter. Oxidizing compounds (e.g. osmium tetroxide, permanganates, and dichromates) form another group of fixatives. The way of the action of permanganate is not well known [8, 9]. It also may cross-link proteins but simultaneously causes denaturation of proteins much more extensively than aldehyde fixatives. The two types of fixatives are frequently used in combination (e.g. glutaraldehyde-fixed cells are post-fixed with potassium permanganate; e.g. [10]).

Potassium permanganate was introduced as an alternative fixative to OsO_4 by Luft [11] prior to the introduction of glutaraldehyde fixation and became popular because of the clarity of the membranes on the sections and the ease of its application. It has the advantage that it acts both as a fixative and as a stain because it not only preserves the membranes but also gives them high contrast (Fig. 1) by depositing MnO_2. Nowadays, it is sparsely used when internal structures are to be visualized but has proven very effective in visualization of the structure of the cell wall and the division septum (cell plate) in fungi and plants (e.g. [10, 12–15]). It appears that permanganate fixation reveals more details of the organization of the cell wall and septum of fission yeasts than any other method. It was the permanganate fixation that revealed that the division septum consists of three layers (a primary septum flanked by two secondary septa; Fig. 1c, d), the cell wall has a laminal structure and forms a specific structural element called "fuscannel" around the base of the centripetally growing septum (septum edging) (Fig. 1c, d, *arrows*) which persists after cytokinesis as a ring (Fig. 1a, b, *asterisk*) of the division scar (Fig. 1a–c, o, *asterisk*) visible on the surface of the growing daughter cell [12, 13]. Secretion vesicles are also clearly visible near the site of cytokinesis in the TEM images of permanganate-fixed cells [16] (Fig. 1o). Since building and degradation of wall structures and vesicle trafficking are important elements of the cytokinetic process, special attention will be devoted to permanganate fixation in this chapter.

The next step is the elimination of free water from the fixed cells [2, 7]. Dehydration is necessary because the resins used to

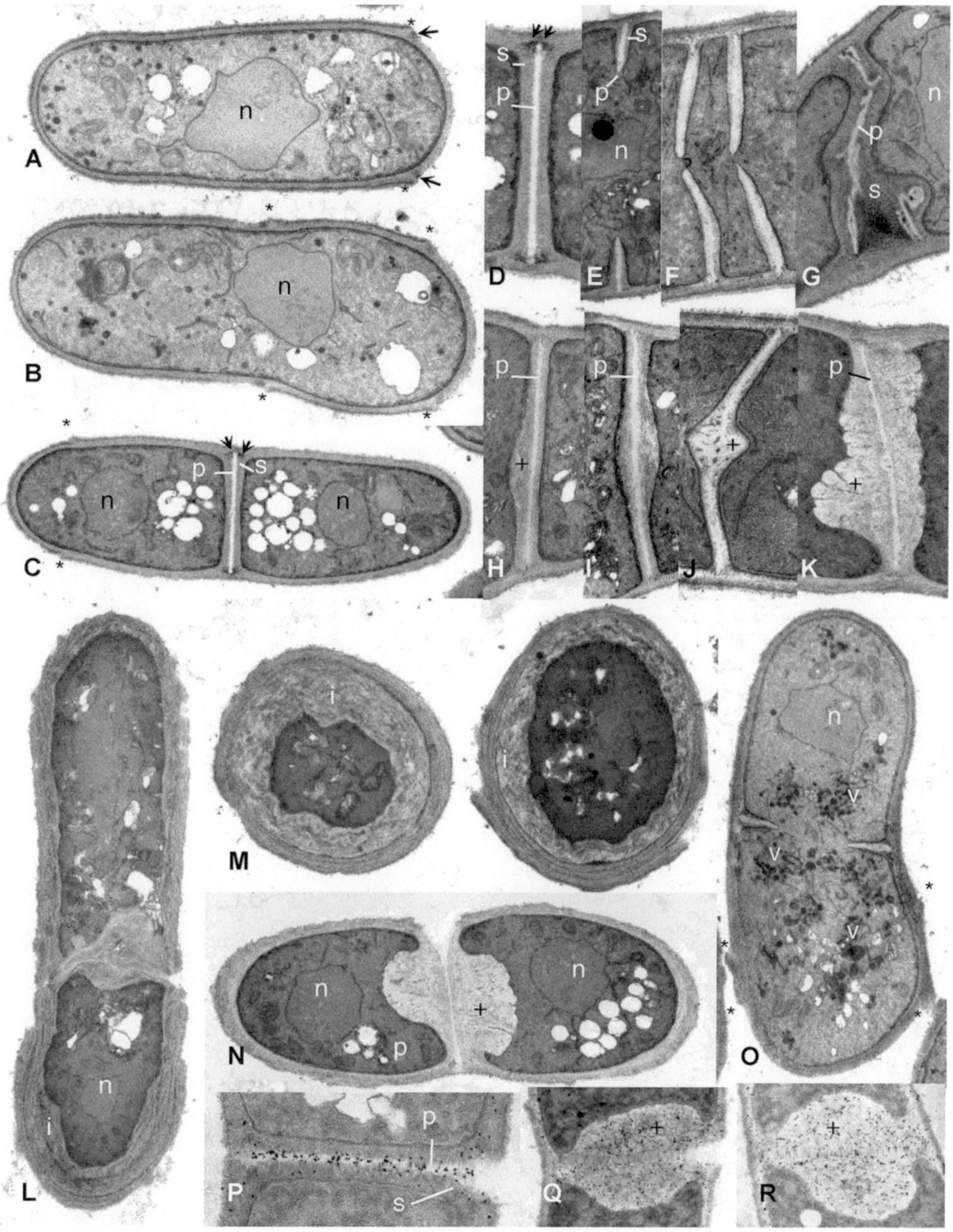

Fig. 1 Permanganate-stained *Schizosaccharomyces pombe* cells. (**a**) Young cell with one division scar. (**b**) Young cell with two division scars. (**c**) Dividing cell with complete septum. (**d**) The standard trilaminar structure of the septum consisting of a primary septum and two secondary septa. (**e**) A growing septum. (**f**) Twin septa in the mutant *sep2-SA2*. (**g**) Aberrant septum structure in a cdc3ts cell growing at the restrictive temperature 35 °C. (**h**, **i**) Septum swelling in rho⁻ cells. (**j**) Septum swelling in a *bgs4Δ* cell. (**k**) Overproduction of septal material in a mutant defective in the regulation of septum synthesis. (**l**) Multilayered cell wall in a *kin1Δ* cell. (**m**) Multilayered cell wall and secondary inwards wall synthesis in *kin1Δ exo70Δ* cells. (**n**) Septum cleavage in a mutant defective in the regulation of septum synthesis. (**o**) Accumulation of vesicles near the septum of a *sec8-1* cell. (**p**) Immunogold localization of 1,3-β glucan in the primary septum of a wild-type cell. (**q**, **r**) Demonstration by immunogold localization that the overproduced septal material of the mutant shown in (**k**) and (**n**) contains 1,3-β glucan. *, division scar; +, overproduced septal material; arrow, fuscannel; p, primary septum; s, secondary septum; i, inwards deposition of cell wall layers

infiltrate and embed the cells to make them suitable for sectioning are not miscible with water. The dehydrated cells are then infiltrated with a plastic monomer, most commonly an epoxy resin (Araldite, Epon) [17] or acrylic media (Lowicryl, LR White, LR Gold) [18]. Upon infiltration, the monomers are polymerized, resulting in chemically preserved yeast cells completely filled and surrounded by a hard plastic. The hardened resin blocks containing the yeast cells are thinly sectioned with an ultramicrotome, and the sections treated with solutions of heavy metal salts (lead citrate, uranyl acetate) in order to impart contrast [19, 20] or with specific antibody probes to detect and localize defined antigens (e.g. specific proteins or wall polysaccharides) (Fig. 1p–r). This chapter describes protocols for fixation, dehydration and embedding of fission yeast cells as well as contrasting and immunogold labeling of their ultrathin sections.

2 Materials

2.1 Consumables and Equipments

1. Glassware (Erlenmeyer flasks, beakers, cylinders, 10 mL tubes, etc.) including brown glass bottles (200 mL) with screw caps.
2. Microcentrifuge tubes and capped vials.
3. Rotary shaker (incubator).
4. Refrigerated centrifuge.
5. Laboratory oven.
6. Electron microscopy grade forceps.
7. Filter paper (Whatman #1).
8. Agarose or agar.
9. Polypropylene embedding capsules and holder.
10. Gelatin embedding capsules and holder.
11. Primary antigen with antigen specificity (e.g. mouse monoclonal linear β-1,3-glucan antibody).
12. Gold-conjugated secondary antigen ("gold probe"; e.g. ultra small colloidal gold-conjugated goat antimouse IG).
13. Silver enhancement kit (e.g. LI SILVER).
14. Ultramicrotome.
15. Glass or diamond EM knives.
16. Copper and nickel EM grids.

2.2 Fixatives

Fixatives are poisonous irritants; work in a fume hood and wear gloves.

1. *Potassium permanganate* (*PM*), 1 g of potassium permanganate crystals in 100 mL of distilled water (1 % PM). For postfixation of glutaraldehyde-fixed cells dissolve 1.2 g crystals in

the same volume of distilled water (1.2 % PM). Approximately 15–30 min afford ample time for complete dissolution of the crystals. Use an Erlenmeyer flask with screw lid to reduce exposure to air. Contact with air promotes precipitation. The stain solution should not be filtered in attempting to remove crystals, since precipitation of permanganate is augmented by aeration (*see* **Note 1**).

2. *Glutaraldehyde* (*GA*), GA fixatives of various concentrations (0.5, 1.0, or 2.5 %) can be prepared from electron microscopy grade solution (25 %) available in sealed glass ampoules by dilution with 0.2 M sodium phosphate buffer or PBS (for the composition of PBS *see* Subheading 2.4, **item 5**) (pH 7.2) and distilled water (final buffer concentration: 50 mM).
3. *Glutaraldehyde and paraformaldehyde* (*GA-PA*).

   A mixture of 0.5 % GA and 3 % paraformaldehyde in 0.1 M PBS (pH 7.2) can also be used as a fixative. However, paraformaldehyde adversely affects the preservation of ultrastructural details.
4. *Osmium Tetroxide* (*OT*)

   1 g of osmium tetroxide is dissolved in 50 mL of distilled water. Since the OT solution must be absolutely pure, it is better to prepare it from ampoules of EM-grade solution.

2.3 Embedding Resins

1. *Epoxy resin 812* (*see* **Note 2**) (*ER*), 30 mL Epoxy resin 812, 16 mL Dodecenylsuccinic anhydride, 8 mL NMA, 1.3 mL, BDMA (benzyldimethylamine). To achieve faster mixing (reduced viscosity) warm up the components separately in a 60 °C oven. After measuring the warmed components into a graduated disposable 50 mL centrifuge tube, tighten the cap and invert the tube end over end continuously for 5 min. Prepare the mixture several hours in advance.
2. *LR White* (*see* **Note 11**) (*LRW*), LR White is a ready-to-use embedding medium and can be purchased from suppliers of materials for electron microscopy. When using it, follow the manufacturer's instructions.

2.4 Other Reagents and Solutions

1. *Ethyl alcohol solutions* (*ethanol*). Dehydration solutions are prepared from absolute ethanol in 50, 70, 95, and 100 % concentrations.
2. *Propylene oxide* (*PO*). PO is used as a transitional solvent to replace the dehydration agent (ethanol) before the infiltration step. PO is carcinogenic and highly flammable; use a fume hood.
3. *Uranyl acetate* (*UA*), 4 g of uranyl acetate under the fume hood (wear protection) to 100 mL of near-boiling (CO_2-free) double-distilled water in an Erlenmeyer flask. Place the flask on a stirrer until the uranyl acetate dihydrate crystals are dissolved.

After dissolving, let the solution cool down to room temperature. Filter the solution through filter paper (Whatman #1) into a brown bottle for light protection and cap tightly. This stock solution can be stored for months at 4 °C. Uranyl acetate is extremely toxic due to the combined effect of chemical toxicity and mild radioactivity and there is a danger of cumulative effects from long-term exposure.

4. *Reynolds lead citrate* (*LC*), 30 mL freshly boiled (CO_2-free) and cooled double-distilled water in an Erlenmeyer flask with screw cap, 1.33 g of lead nitrate, 1.76 g of sodium citrate. Close the flask and shake vigorously for 1 min; continue intermittent shaking for 30 min. The solution should appear cloudy. Add 8 mL of 1 N NaOH solution (carbonate free) and invert slowly. The solution should appear clear. Add the solution to 50 mL of freshly boiled (CO_2-free) and cooled double-distilled water in a brown bottle (light protection) and cap tightly. This stock solution can be stored for about 1 month at 4 °C. The pH should be 12.0 ± 0.1. If the pH is too low add more NaOH to the clear solution (*see* **Note 3**).
5. *Blocking buffer for immunogold-silver labeling* (*PBS-BSA*), 1 % solution of bovine serum albumin in PBS. PBS buffer: 20 mM phosphate, 150 mM NaCl, pH 7.4.

3 Methods

3.1 Culture Preparation (See Note 4)

1. Inoculate 15–50 mL of yeast extract liquid medium (YEL) to an OD_{600} of 0.2–0.5 ($0.5–2 \times 10^6$ cells/mL), depending on the doubling time of the cells and incubate overnight in a rotary shaker at the proper temperature. For inoculation, use cells of overnight precultures. If the strain is not temperature-sensitive, 30 °C is an ideal temperature to obtain a log-phase overnight culture. If the cells are cultured in a selective minimal medium, transfer the overnight culture into YEL for 1 or 2 h before fixation.
2. Centrifuge 2 mL of log-phase culture at room temperature to generate a cell pellet (e.g. $800 \times g$ for 5–10 min). Extensive centrifugation must be avoided, but the pellet should be compact enough to remain a firm mass.
3. Carefully remove the supernatant by a pipette and gently resuspend the cells in distilled water (*see* **Note 5**).
4. Repeat washing two or three times.

3.2 Fixation

Fixation is the first and most important step in any EM study, since mistakes made at this point render the whole project useless. Cells can be fixed in one step with one fixative (primary fixation) or in two steps by treating the cells with two fixatives in succession (primary fixation + postfixation). Postfixation can take place before (Subheading 3.2.2) or after (Subheading 3.2.4) agarose/agar enrobement. Primary fixation can be performed with PM, GA, or GA-PA.

3.2.1 Primary Fixation of Cells with Potassium Permanganate (PM)

1. Resuspend pellet in 2 mL of 1 % freshly prepared PM fixative and keep them at 4 °C.
2. After 1 h, collect cells by centrifugation (e.g. 800 × *g* for 5–10 min).
3. Resuspend cells in 5 mL of distilled water.
4. After 10 min, collect cells by centrifugation.
5. Repeat **steps 3** and **4** three times.

3.2.2 Primary Fixation with Glutaraldehyde (GA)

1. Resuspend pellet in 2 mL of 2 % GA and keep them at 4 °C.
2. After 2 h, collect cells by centrifugation.
3. Resuspend cells in 2 mL of distilled water.
4. After 10 min, collect cells by centrifugation.
5. Repeat **steps 3** and **4** two times.

3.2.3 Primary Fixation with Mixture of Glutaraldehyde and Paraformaldehyde (GA-PA)

1. Resuspend cells in 2 mL of GA-PA and keep them at 4 °C.
2. After 2 h, collect cells by centrifugation.
3. Resuspend cells in 2 mL of PBS.
4. After 5 min, collect cells by centrifugation.
5. Repeat **steps 3** and **4** two times.
6. Resuspend cells in 2 mL of 1 % sodium metaperiodate and keep them at 4 °C.
7. After 15 min, collect cells by centrifugation.
8. Resuspend cells in 2 mL of 50 mM ammonium chloride.
9. After 30 min, collect cells by centrifugation and repeat **steps 3** and **4** two times.

3.2.4 Postfixation of Cells with Potassium Permanganate (See Note 6)

1. Resuspend cells fixed with GA or GA-PA in 2 mL of freshly prepared 1.2 % PM and keep them at 4 °C.
2. After 1 h, collect cells by centrifugation at 4 °C (e.g. 800 × *g* for 5–10 min).
3. Resuspend cells in 2 mL of 0.2 M sodium phosphate buffer (pH 7.2).
4. After 10 min, collect cells by centrifugation.
5. Repeat **steps 3** and **4** three times.

3.2.5 Agarose/Agar Enrobement

To prepare a firm pellet suitable for further processing, cells must be embedded in agarose or agar blocks.

1. Using a warmed pipette, quickly transfer approx 50 μL of molten agarose (or agar) onto the pellet of fixed cells and gently stir the cells with the tip of the plastic pipette to suspend the cells.
2. Place the tube in a refrigerator until the agarose solidifies (15 min).
3. Overlay the solidified agar with several mL of buffer and use a spatula to dislodge the agarose plug.
4. Transfer the plug into a Petri dish containing several mL of buffer and trim the plug into 1- to 2-mm cubes using a razor blade.

3.2.6 Postfixation of the Enrobed Cells (Agarose/Agar Cubes) with Osmium Tetroxide (See Note 7)

Cubes containing cells fixed with GA or GA-PA must be post-fixed with OT to impart contrast to the specimen.

1. Transfer the agarose/agar cubes into a small vial containing 1 mL of OT for 1 h at room temperature.
2. Remove OT with a pipette and rinse cubes by adding and removing 1 mL of distilled water two times.

3.3 Dehydration (See Note 8)

All types of cubes regardless of the mode of fixation and postfixation applied must be dehydrated before infiltration because most epoxy resins used for infiltration are not miscible with water. Thus, all free water from the cubes must be replaced with an organic solvent.

1. Transfer the cubes into a small capped vial containing 1 mL of 50 % ethanol, incubate at room temperature for 15 min.
2. Remove ethanol with a pipette.
3. Repeat **steps 1** and **2** with 70, 95, and 100 % ethanol.
4. Repeat dehydration with 100 % ethanol three times, each for 10 min.

3.4 Infiltration and Embedding (See Note 9)

3.4.1 Infiltration and Embedding with Resin (See Note 2)

1. Repeat dehydration **step 4** with propylene oxide (PO), if necessary (*see* **Note 10**).
2. Place 1 mL of 1:1 PO:ER embedding medium over the cubes and gently rotate the capped vial several times during a period of 20 min at room temperature. Vials should be closed so that moisture from the air does not enter them.
3. Replace the 1:1 mixture with 1 mL of 1:3 mixture of PO:ER and rotate the capped vial a few times during a period of 45 min.
4. Replace the 1:3 mixture with several mL of pure ER and rotate the vial a few times during a period of 90 min.

5. Fill several embedding capsules half-way with ER. Dislodge air bubbles from the mixture with a needle. Use a small spatula to transfer the specimen cubes from the vials into the capsules. Allow the cubes to sink to the bottom of the capsules which may take an hour or more.
6. Transfer the capsules into a 60 °C oven for 24 h until the epoxy resin is completely hardened.

3.4.2 Infiltration and Embedding in LR White (See Note 11)

1. Place 1 mL of LRW embedding medium over the ethanol-dehydrated cubes (*see* **Note 12**) and gently rotate the capped vial in a fume hood overnight at room temperature.
2. Replace the embedding medium with fresh medium and gently rotate the vials for 20 min at room temperature.
3. Repeat **step 2**.
4. Transfer the cubes from the vial into small gelatin capsules (*see* **Note 13**) completely filled with fresh LRW medium and then seal the capsules to keep oxygen from interfering with the polymerization process.
5. Place the loaded capsules in an oven adjusted to 60 °C for a period of 20–24 h (*see* **Note 14**).

3.5 Sectioning

Ultrathin sections from the hardened ER and LRW blocks are produced by sectioning with an ultramicrotome and are collected on copper or nickel grids. For producing ultrathin sections by ultramicrotomy, the reader is referred to recent methodological descriptions such as [21].

3.6 Contrasting (Poststaining of Sections on Grids)

To enhance differential contrast on ultrathin sections, the sections placed on grids are usually stained with heavy metal salts (*see* **Note 15**). The double-contrast method of ultrathin sections on grids with uranyl acetate and lead citrate is the standard contrasting technique.

1. Fill several small beakers with warm, freshly boiled deionized water.
2. Cut two pieces of parafilm and place them in two small Petri dishes.
3. In the first dish, create a CO_2-free area for staining: place several pellets of sodium hydroxide around the edge of the parafilm to quickly absorb CO_2 and place the lid on the Petri dish (*see* **Note 3**).
4. In the second dish, place 1 drop of UA for each grid onto the parafilm.
5. Place grids section side down on the UA drops and let them float for 10 min (*see* **Note 16**). Cover the dish with a larger aluminum-foil wrapped Petri dish lid to protect against light.

6. Carefully remove the grid from the UA drop and dip it in the water of the first beaker and swirl it around. Repeat the process in three other beakers.
7. In the first dish lift the lid only enough to insert the Pasteur pipette and place 1 drop of LC for each grid on the parafilm.
8. Place the UA-stained grids on the LC drops and cover the dish to reduce CO_2 contamination. Let the grids float for 5 min (*see* **Note 16**).
9. Carefully remove the grid from the UA drop and dip it in the water of the first beaker and swirl it around. Repeat the process in three other beakers.
10. Blot grids and forceps on filter paper (*see* **Note 17**).
11. Place the grids in a small Petri dish and let them dry at room temperature.

3.7 Immunogold-Silver Labeling of Sections (Immunolocalization)

Use cells in which the antigen has survived fixation, processing, and embedding in such a form as to be recognizable to the specific antibody. For detection of cell wall polysaccharides, both PM-fixed and aldehyde-fixed cells can be used because neither treatment destroys the ability of the polysaccharides to bind antibodies. If proteins are to be localized, do not use PM-fixed cells because permanganate denaturizes protein antigens much more extensively than aldehyde fixatives. Another problem is that the epoxy resin, which is needed to support the sectioned cells, prevents the antibody from gaining access to many antigens. Section etching with agents that cleave the ester bounds in the polymerized epoxides may produce a gel layer on the surface of the section, which may allow an antigen–antibody interaction. The new generation polyhydroxy acrylates (i.e. LR White) can be polymerized with some 5–30 % of water in the cells, which may help maintain the proteins in a state closer to their native conformation and allows better access for antibodies. Therefore, it is recommended to use LR-White-embedded sections for immunolocalization of proteins. Fixing with osmium tetroxide also may cause problems. When this compound reacts with cellular components, the resulting osmium metal coating will prevent the antibodies from binding to the antigens. To visualize the antigen–antibody interaction, "secondary" antibodies bound to colloidal gold particles are applied. These bind the gold particles to the "primary" antibodies specifically bound to the antigen to be detected. However, the small gold particles can hardly be recognized on the contrasted ultrastructure. For better visibility, they have to be enlarged, which can be achieved by the silver enhancement reaction. In this reaction, the gold particle catalyzes the reduction of silver ions to metallic silver on its surface which enlarges the particle. It must be completed before contrasting reagents such as osmium tetroxide, lead citrate, or uranyl acetate are applied, since the osmium and lead deposits can also nucleate silver deposition in the same manner as gold and produce non-specific staining.

1. Fill several small beakers with the PBS-BSA blocking buffer.
2. Cut several pieces of parafilm and place them on a large filter paper.
3. Dilute the primary antibody in PBS-BSA and centrifuge (*see* **Note 18**).
4. Put small drops of PBS-BSA and diluted antibody onto parafilm pieces.
5. To mask nonspecific binding site, place the grid section side down on a small drop of PBS-BSA on parafilm for 5 min (*see* **Note 19**).
6. Transfer the grids and float them section-side down on the antibody drops for 20–60 min.
7. Transfer the grids on drops of PBS-BSA for 2 min. Repeat washing five times (in five drops).
8. Dilute the colloidal gold probe to the working concentration in PBS-BSA and place drops on parafilm pieces.
9. Transfer the grids, section-side down, to the drops of gold probe (secondary antibody) and incubate for 10–60 min.
10. Transfer the grids on drops of PBS-BSA for 2 min. Repeat washing two times.
11. Transfer the grids on drops of PBS for 2 min. Repeat washing two times.
12. Prepare silver enhancement reagent (developer) by combining equal volumes of enhancer and initiator solutions and put drops on parafilm. Avoid skin contact since the silver enhancement reagents will stain skin.
13. Transfer the grids on developer drops for 1–4 min (*see* **Note 20**).
14. Transfer the grids on drops of deionized water for 5 min, twice.
15. If desired, contrast sections with UA and/or LC (*see* Subheading 3.6).

4 Notes

1. Potassium permanganate is presently not widely used as a fixative. Its disadvantages include capriciousness, contamination of sections by precipitation ("dirt"), granularity, and fragility of the section to the electron beam. Contamination of the section by precipitated material can be prevented/reduced by using freshly prepared solution and by excluding air from the solution. If dirt is already produced, a treatment of the specimen with a

0.025 % solution of citric acid can remove most of the precipitated material. Although Luft [11] originally prepared $KMnO_4$ in a buffer, unbuffered solutions yield equally satisfactory preservation and staining. Potassium permanganate has been shown to preserve some carbohydrates that are extracted by glutaraldehyde.

2. Epoxy Resin 812 is made by various manufacturers (Poly/Bed, EMbed, etc.). The Epon embedding media consist of epoxy resin (Epon 812 or substitute), the hardeners DDSA (dodecyl succinic anhydride) and NMA (nadic methyl anhydride), and an accelerant such as BDMA (benzyldimethylamine) [17]. It is possible to prepare resin mixtures by manipulating with the proportions of the components that will polymerize into plastics of various hardness. More NMA makes a harder resin, more DDSA makes it softer.
3. Lead salts are extremely toxic, exercise care in weighting the powder. Lead citrate reacts with CO_2 in the atmosphere to form lead carbonate precipitate, which appears as electron-dense particles of various sizes and shapes ("dirt") on the section. Another common cause of lead precipitates is incorrect pH. The correct pH for Reynolds led citrate is 12. A pH lower than 12 causes precipitation of lead carbonate.
4. The cells should be fixed during early log phase when the density of the cytoplasm appears minimal and the permeability of the cell wall appears maximal.
5. Because compounds can be present in media that interact with fixatives, cells have to be washed thoroughly. If mutants with fragile cell walls are to be fixed, use a sorbitol solution (0.8–1.2 M) for washing.
6. Primary fixation with GA or GA + PA, followed by a secondary fixation with PM, has become a routine procedure in fission yeast laboratories (e.g. [10, 22]). It has been found, through years of personal experience, that GA can be omitted and direct (primary) fixation with PM can result in images superior to those obtained with double fixation. The involvement of PA can reduce the clarity of certain intracellular structures. Formalin fixation yields unacceptable EM ultrastructure.
7. Osmium tetroxide has a relatively slow rate of penetration and for this reason is usually used as a secondary fixative. It also serves as a stain of membranes and lipid-containing bodies. It reacts poorly with proteins and carbohydrates. It should be avoided when immunolabeling is to be performed as it almost totally destroys antigenicity of reactive sites.
8. The aim of the dehydration process is to replace the water in cells with a fluid that acts as a solvent between the aqueous environment of the cell and the hydrophobic embedding

media. Water is removed stepwise by passing the cubes through a series of solutions of increasing concentration of an organic solvent (ethanol, acetone, or acetonitrile). Ethanol is preferred to acetone as a dehydrating agent because anhydrous acetone absorbs water from the atmosphere and is a more powerful extractor of lipids within the cell. Many dehydration schedules exist.

9. Infiltration is the process by which dehydrants (e.g. ethanol) and transition fluids (e.g. PO) are replaced by resin monomers, most of which are very viscous. For infiltration, acrylic media (Lowicryl, LR White, LR Gold) or epoxy resins (Araldite, Epon) are used.

10. Replacement of ethanol by another intermediary solvent that is highly miscible with the plastic embedding medium is usually necessary to interface with the embedding media. The standard solvent used is propylene oxide (PO). PO reduces the viscosity of the embedding media and increases its infiltration into the cells. Ethanol does not reduce the viscosity to the same degree. However, PO can severely deplete lipids from the fixed cells. As yeast cells form a loose pellet (not a dense tissue), the use of PO can be omitted from the dehydration procedure.

11. LR White is an acrylic resin preferably used for immunoelectron microscopy. Its chemical components are not fully known. The special hydrophobic nature of the LR White resin allows the penetration of aqueous solutions that contain antibodies. LR White may be polymerized by heat, addition of an accelerator, or UV light. Its polymerization is exothermic and is carried out in an anoxic atmosphere. The fixation protocol is governed by the need to preserve antigens. Ethanol dehydration is preferred.

12. Avoid acetone during dehydration because acetone acts as a radical scavenger and can interfere with polymerization.

13. LR White can penetrate and soften low-density polyethylene capsules. In addition, polyethylene is not impermeable to oxygen and may allow enough contact with atmospheric oxygen to interfere with polymerization. Both these problems can be overcome by the use of gelatin capsules.

14. Curing with LR White accelerator may also be employed but for immunolocalization the thermally cured blocks are much better, probably because when accelerator-curing the resin the exotherm produced may damage the integrity of antigens.

15. The contrast in the electron microscope depends primarily on the differences in the electron density of the organic molecules in the cell. However, these differences are usually poor in aldehyde-fixed specimens and need to be enhanced by staining. The pre-embedding treatment of (intact) cells alone may not enhance the contrast sufficiently. Therefore, the sections

are frequently "poststained" with agents that attach metals of high atomic weight (uranium and lead) to the biological structures. The most commonly used methodology consists of poststaining the grids carrying the sections with aqueous uranyl acetate followed by Reynolds' lead citrate [19]. Uranyl acetate stains membranous structures and structures containing nucleic acids. Lead binds to RNA-containing structures and hydroxyl groups of carbohydrates. Better contrast is achieved by poststaining grids immediately after the sections are produced and without exposure to the electron beam. If it is necessary to check the quality of the sections before poststaining, only one grid from a batch should be examined in the electron microscope.

16. Staining time longer than 10 min for uranyl acetate and longer than 5 min for lead citrate can result in stain precipitates (dirt) on the section.
17. Grids should be kept wet throughout the contrasting process.
18. The optimal concentration of the antibody can be determined by serial dilutions.
19. Use nickel grids, not copper. Nickel is more inert and less poisonous to immunoreactions. Copper grids can turn black due to corrosion.
20. More or less time can be used to control particle size and intensity of signal. A series of different development times should be tried, to find the optimum time for your experiment. After this time, silver may be precipitated spontaneously by self-nucleation, producing background signal.

Acknowledgment

This work was supported by the Hungarian Scientific Research Fund grant OTKA 101323

References

1. Bozzola JJ, Russel LD (1999) Electron microscopy principles and techniques for biologists. Jones and Bartlett Publishers, Sudbury, MA
2. Hayat MA (1989) Principles and techniques of electron microscopy biological applications. CRC Press, Inc., Boca Raton, FL
3. Harris JR, Adrian M (1999) Preparation of thin-film frozen-hydrated/vitrified biological specimens for cryoelectron microscopy. Methods Mol Biol 117:31–48
4. Sipiczki M, Takeo K, Yamaguchi M, Yoshida S, Miklos I (1998) Environmentally controlled dimorphic cycle in a fission yeast. Microbiology 144:1319–1330
5. Sipiczki M, Yamaguchi M, Grallert A, Takeo K, Zilahi E, Bozsik A, Miklos I (2000) Role of cell shape in determination of the division plane in *Schizosaccharomyces pombe*: random orientation of septa in spherical cells. J Bacteriol 182:1693–1701
6. Humbel BM, Konomi M, Takagi T, Kamasawa N, Ishijima SA, Osumi M (2001) In situ localization of β-glucans in the cell wall of *Schizosaccharomyces pombe*. Yeast 18:433–444

7. Glauert AM, Lewis PR (1998) Biological specimen preparation for transmission electron microscopy. In: Glauert AM (ed) Practical methods in electron microscopy, vol 17. Portland Press, London
8. Arnold WN (1991) Rapid permanganate-fixation and alcohol-dehydration of fungal specimens for transmission electron microscopy. J Microbiol Methods 13:17–22
9. Sutton JS (1968) Potassium permanganate staining of ultrathin sections for electron microscopy. J Ultrastruct Res 21:424–429
10. Sato M, Kobori K, Ishijima SA, Feng ZH, Hamada K, Shimada S, Osumi M (1996) *Schizosaccharomyces pombe* is more sensitive to pressure stress than *Saccharomyces cerevisiae*. Cell Struct Funct 21:167–174
11. Luft JH (1956) Permanganate - a new fixative for electron microscopy. J Biophys Biochem Cytol 2:799–801
12. Johnson BF, Yoo BY, Calleja GB (1973) Cell division in yeasts: movement of organelles associated with cell plate growth of *Schizosaccharomyces pombe*. J Bacteriol 115:358–366
13. Sipiczki M, Bozsik A (2000) The use of morphomutants to investigate septum formation and cell separation in *Schizosaccharomyces pombe*. Arch Microbiol 174:386–392
14. Munoz J, Cortes JCG, Sipiczki M, Ramos M, Clemente-Ramos JA, Moreno MB, Martins IM, Perez P, Ribas JC (2013) Extracellular cell wall β(1,3)glucan is required to couple septation to actomyosin ring contraction. J Cell Biol 203:265–282
15. Koch G, Schmitt U (2013) Topochemical and electron microscopic analyses on the lignification of individual cell wall layers during wood formation and secondary changes. In: Fromm J (ed) Cellular aspects of wood formation, plant cell monographs, vol 20. Springer, Berlin, pp 41–69
16. Wang H, Tang X, Liu J, Trautmann S, Balasundaram D, McCollum D, Balasubramanian MK (2002) The multiprotein exocyst complex is essential for cell separation in *Schizosaccharomyces pombe*. Mol Biol Cell 13:515–529
17. Finck H (1960) Epoxy resins in electron microscopy. J Biophys Biochem Cytol 7:27–30
18. Newman GR, Hobot JA (1987) Modern acrylics for post-embedding immunostaining techniques. J Histochem Cytochem 35:971–981
19. Reynolds ES (1963) The use of lead citrate at high pH as an electron-opaque stain in electron microscopy. J Cell Biol 17:208–212
20. Watson ML (1958) Staining of tissue sections for electron microscopy with heavy metals. J Biophys Biochem Cytol 4:475–478
21. Hagler HK (2007) Ultramicrotomy for biological electron microscopy. Methods Mol Biol 369:67–96
22. Martin-Cuadrado AB, Morrel JL, Konomi M, An H, Petit C, Osumi M, Balasubramanian M, Gould KL, del Rey F, Vazquez de Aldana CR (2005) Role of septins and the exocyst complex in the function of hydrolytic enzymes responsible for fission yeast cell separation. Mol Biol Cell 16:4867–4881

Chapter 9

Characterization of Septin Ultrastructure in Budding Yeast Using Electron Tomography

Aurélie Bertin and Eva Nogales

Abstract

Septins are essential for the completion of cytokinesis. In budding yeast, *Saccharomyces cerevisiae*, septins are located at the bud neck during mitosis and are closely connected to the inner plasma membrane. In vitro, yeast septins have been shown to self-assemble into a variety of filamentous structures, including rods, paired filaments, bundles, and rings (Bertin et al. Proc Natl Acad Sci U S A, 105(24):8274–8279, 2008; Garcia et al. J Cell Biol, 195(6):993–1004, 2011; Bertin et al. J Mol Biol, 404(4):711–731, 2010). Using electron tomography of freeze-substituted sections and cryo-electron tomography of frozen sections, we determined the three-dimensional organization of the septin cytoskeleton in dividing budding yeast with molecular resolution (Bertin et al. Mol Biol Cell, 23(3):423–432, 2012; Bertin and Nogales. Commun Integr Biol 5(5):503–505, 2012). Here, we describe the detailed procedures used for our characterization of the septin cellular ultrastructure.

Key words Septin, Budding yeast, Cytokinesis, Cryo-tomography, Image processing, Cryo-sectioning

1 Introduction

Septins were discovered through a screen for cell division cycle mutants in budding yeast more than 40 years ago [6]. Septins are indeed essential for cytokinesis, and play a variety of molecular roles, including the recruitment of proteins like myosin2 [7] or serving as a diffusion barrier for membrane-bound proteins [8]. Furthermore, the self-assembly of septin has been shown to be required for cell survival in yeast [9]. From early electron microscopy studies using standard preparation methods [10, 11], we know that septins assemble in concentric rings at the bud neck, but several studies point to a variable organization and orientation of septins through the cell cycle, likely regulated by post-translational modifications [12]. In situ FRAP experiments have shown that the assembly of septins at the bud neck is dynamic [12], while fluorescence polarization studies indicate a global reorientation of the septin filaments at the onset of cytokinesis [13]. In agreement with

Alberto Sanchez-Diaz and Pilar Perez (eds.), *Yeast Cytokinesis: Methods and Protocols*, Methods in Molecular Biology, vol. 1369, DOI 10.1007/978-1-4939-3145-3_9,

these in vivo observations, we have characterized a variety of septin structures in vitro depending on ionic strength [1], the nature of the septin subunit composition [2], presence of specific lipids [3], or the phosphorylation state of septins [2]. In high salt (above 200 mM), the mitotic septin complex made of Cdc3, Cdc10, Cdc12, and Cdc11 exists as a 32 nm long octameric, symmetric, rod-like structure [1]. At lower ionic strength, these rods self-assemble into long paired filaments resembling railroad tracks, or into bundles of filaments [1]. Remarkably, replacing Cdc11 by Shs1, a less essential and sub-stoichiometric septin, induces the formation of ring-like structure or, for a specific phosphomimetic Shs1 mutation, into gauzes of orthogonal filaments [2]. Hence, the organization of septins is highly variable and plastic. In order to get insight into the organization of septin filaments in situ, it is necessary to use advance electron microscopy methods for sample preparation and visualization that enable the quantitative description under optimized cellular preservation. Using electron tomography, we have characterized the three-dimensional organization of septin filaments in dividing budding yeasts [4, 5]. This chapter describes the methods we used for sample preparation, data collection, and computation.

2 Materials

The methods presented here require specialized equipment for sample preparation and data collection. Below we list the material we have used in our studies; electron microscopy facilities are often equipped with these or similar tool alternatives.

2.1 Preparation of Resin Embedded Samples for Sectioning and EM Analysis

1. Yeast extract peptone glucose, commonly referred to as YPD medium (1 % yeast extract, 2 % peptone, 2 % glucose), autoclaved for 20 min at 121 °C. To prevent burning the glucose, sterile, filtered glucose can be added after autoclaving. Otherwise the media will darken and cell growth will not be optimal.
2. Incubator and shaker (to be set at 30 °C) able to contain 2 L cell culture flasks.
3. Spectrophotometer.
4. Vacuum filtration device with a pump and a borosilicate glass funnel, equipped with a fritted glass of 25 mm in diameter (Millipore). 0.45 μm polycarbonate filters are used.
5. High-pressure freezing device (EMPACT2-RTS, Leica) and 100 μM deep membrane carriers (Leica). Hexadecene (Fluka) to be used to coat the membrane carriers.
6. Cryogenic vials (Nalgene) of 2 mL for sample conservation at liquid nitrogen temperature (in a nitrogen tank) or freeze substitution.

7. Freeze substitution media: 1 % osmium tetraoxide, 0.1 % uranyl acetate, 5 % water in freshly opened dry acetone. The freeze substitution medium can be prepared in advance and stored in liquid nitrogen. We used a Leica AFS2 freeze substitution apparatus.
8. Epon resin solutions in acetone at increasing concentrations of 30, 60, and 100 %. Epon polymerization molds and oven to be set at 60 °C.
9. Ultamicrotome (Ultracut E, Reichert) equipped with either a homemade glass knife (with a glass "knifemaker") or a diamond knife of 4.5 mm (Diatome). One of your own eyelash glued (with nailpolish) to a toothpick to be used to handle the sections. Dumont tweezers N7 to be used to hold the hexagonal copper grids (mesh size of 100) coated with formvar (0.5 %).
10. Solutions of uranyl acetate (2 %) and lead citrate (70 %) in methanol, and 10 nm gold beads (Aurion).
11. A carbon evaporator apparatus.

2.2 Sample Cryo-Sectioning

1. YPD medium as described above.
2. Vacuum filtration device as described above.
3. High-pressure freezing device (EMPACT2-RTS, Leica) and copper tubes (350 μm inner diameter, Leica).
4. Cryo-ultramicrotome (UC7, Leica) equipped with an anti-contamination "cryosphere" (Leica) and a CRION ionizer (Leica).
5. Trimming diamond knife (Diatome) and 3.5 mm cryo-immuno diamond knife (Diatome).
6. C-flat holey carbon grids (Protochips) coated with gold beads (Aurion) and an eyelash glued to a 20 cm long stick.

2.3 Data Collection

1. Data collection for resin-embedded samples, at room temperature: a dual-axis holder (model 2040, Fischione) enables the collection of two, perpendicular tilt series. For data collection, we use an FEG CM200 microscope (FEI) equipped with a 4K×4K CCD camera (Ultrascan 4000, Gatan). We use Digital Micrograph software (Gatan) for automated data collection.
2. Data collection for cryo-samples: we mount the sample on a 626 cryo-holder (Gatan) and visualize it using a Tecnai 12 electron microscope (FEI) equipped with a 4K×4K Eagle camera (FEI). We use the software suite Xplore 3D (FEI) for automated data collection.

2.4 Data Processing

1. We used the IMOD software [14] for both tilt series alignment and three-dimensional reconstruction. The segmentation can be performed using either Amira [15] or IMOD.
2. Our computation was carried out using workstations with 48 GB of RAM and NVDIA video cards.

3 Methods

For tomography and cryo-tomography analysis, two different strategies can be employed as described in Fig. 1. Either the harvested cells are directly sectioned after vitrification and then analyzed in a hydrated state, or the vitrified cells are gradually freeze-substituted in the presence of fixative and staining chemicals, embedded in resin, and sectioned at room temperature.

3.1 Sample Preparation for Resin-Embedded Cells

1. Grow yeast cells at 30 °C in Erlenmeyer flasks (growing 500 mL of media in 2 L flasks should provide cells with sufficient oxygenation). Grow the cells to mid-exponential phase ($OD_{600\ nm}$ = 0.4–0.6). Measure the Optical density at 600 nm with a spectrophotometer every 20 min.

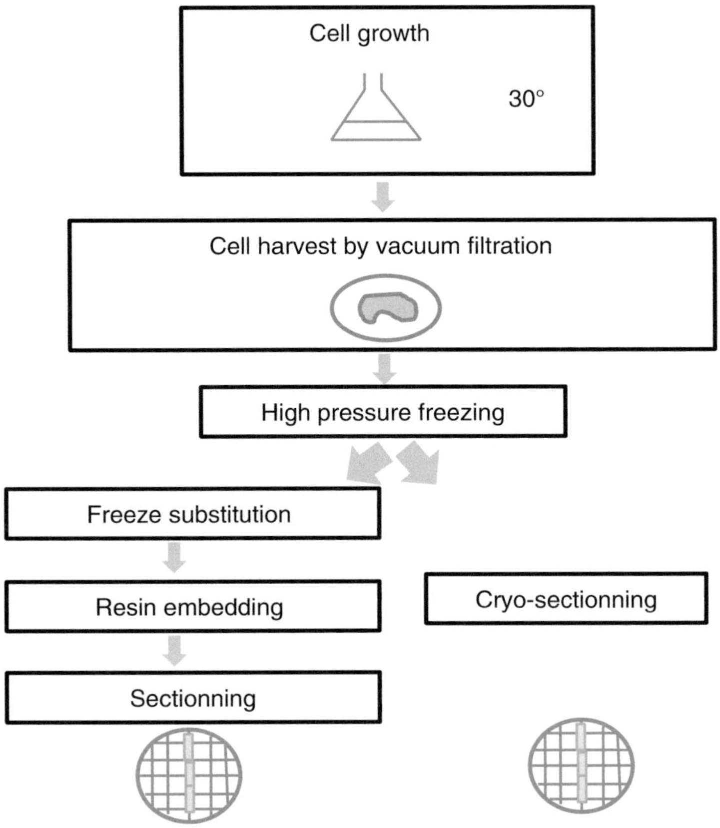

Fig. 1 Schematic of the procedure used for sample preparation for electron microscopy analysis. After cell growth, the cells are harvested by vacuum filtration and vitrified by high-pressure freezing. Two strategies can then be employed. Either the vitrified cells are cryo-sectioned right before observation in the scope (or storage) or the cells are freeze-substituted, embedded in resin at room temperature and sectioned

2. Harvest the yeast cells by vacuum filtration as described in McDonald [16] (*see* **Note 1**). After filtration, a yeast thick paste is isolated on top of the filtration membrane. Transfer the membrane in a wet Petri dish (*see* **Note 2**).
3. Coat the membrane carriers with hexadecene, which is a cryo-protectant. Use a sterile toothpick to transfer some of the yeast paste to the 100 μm deep membrane carrier already attached to its stand. This step is easier using binoculars. Then transfer the membrane carrier into the high-pressure freezing machine and vitrify the sample at once (*see* **Note 3**).
4. Transfer the vitrified samples into cryo vials containing the frozen freeze-substitution media and either store in liquid nitrogen (*see* **Note 4**) or process further for freeze substitution. During freeze substitution, steadily and accurately raise the temperature from −90 to −25 °C in increments of 2 °C per hour. Then raise the temperature to 0 °C in 5 °C per hour increments (*see* **Note 5**).
5. Rinse the samples three times for 10 min in pure acetone. The samples should be now darker due to the presence of osmium tetraoxide in the media. The samples are most likely to detach from the membrane carrier by themselves during the rinses.
6. Perform resin embedding by consecutive baths in epon/acetone solutions of increasing epon concentrations: 30, 60, and 100 % for 1 h or more each (*see* **Note 6**). During the baths, gently mix the samples on a shaker. To remove any trace of acetone, wash the samples in three consecutive solutions of pure acetone for at least 1 h each. Then put the samples into polymerization molds and label them (*see* **Note 7**) before putting them into an oven at 60 °C for 48 h.
7. Before sectioning, trim the resin-embedded sample using a homemade glass knife. Then section the resin block using a 4.5 mm diamond knife, using standard procedures. Depending on desired use, choose the proper thickness for sectioning (*see* **Note 8**). Use the eyelash to move the sections across the air/water interface. Grab the sections with the formvar-coated hexagonal grid held with tweezers and coming from below. Then let the grid dry.
8. Evaporate carefully a thin layer of carbon on top of the grid on the sample side (*see* **Note 9**). The grids are then ready to be analyzed by electron microscopy.

3.2 Sample Preparation of Vitrified Samples

1. Grow the yeast cells to mid-log phase as described above (*see* **step 1** in Subheading 3.1) in YPD media containing 20 % of sterile-filtered dextran (*see* **Note 10**).
2. Harvest the cells by vacuum filtration and keep a few milliliters in the funnel without waiting to obtain a yeast paste (*see* **Note 11**).

3. Suck some yeast solution into the copper tube mounted on its holder. Make sure the tube is entirely filled in order to ensure good vitrification. Quickly transfer the sample into the high-pressure freezing machine for vitrification and operate following standard procedures.
4. Either store the copper tubes in liquid nitrogen or transfer it into a cryo-ultramicrotome.
5. Coat some C-flat holey grids with 10 nm gold beads by incubating them upside down on a gold beads solution, followed by several rinses in PBS buffer.
6. Cool the cryo-ultramicrotome down to −150 °C. Attach the copper tube safely into the jaws of the cryo-ultramicrotome and trim one of its ends into a pyramidal shape using the trimming diamond knife (*see* **Note 12**).
7. Then use the cryo-immuno knife to section 50 nm thick ribbons. Hold the ribbon with an eyelash attached to a 20 cm long stick. When the ribbon is long enough, hold a pre-cooled grid below the ribbon and invert the charge of the "CRION Ionizer" to make the ribbon stick onto the grid (*see* **Note 13**). The cryo-sectioning procedure has been described in details by Pierson et al. [17].
8. Then transfer the grid, kept under liquid nitrogen, into the cryo-holder and either store in liquid nitrogen or transfer to the microscope.

3.3 Data Collection for Resin-Embedded Samples

1. Insert a good grid (*see* **Note 14**) into the dual axis holder.
2. Irradiate the area of interest for about 30 min to shrink the resin to its final thickness (*see* **Note 15**). Choose an area of interest in the center of a hexagonal copper hole to prevent a grid bar from coming into the field of view.
3. Make sure that the sample is at eucentric height. Tilt the sample from −70° to 70° to make sure a tilt series can be acquired within this range of angles. Collect a tilt series (*see* **Note 16**) as shown in Fig. 2a. On resin sections, tracking and focusing of the sample can be performed on the area of interest. Choose a reasonable magnification to get a resolution that will enable the visualization of filaments within the cell (*see* **Note 17**).
4. Rotate the sample by 90° to collect a second tilt series as shown in Fig. 2b (*see* **Note 18**). Once the sample has been rotated, move the grid to have the area of interest in the center of the screen. Collect the second tilt series.

3.4 Data Collection for Cryo-Sections

1. Place the frozen grid into the cryo-holder, keeping the sample always at liquid nitrogen temperature.
2. Make sure the ribbon is still vitrified (*see* **Note 19**). Using low-dose procedures, find a dividing cell in the center of a grid square.

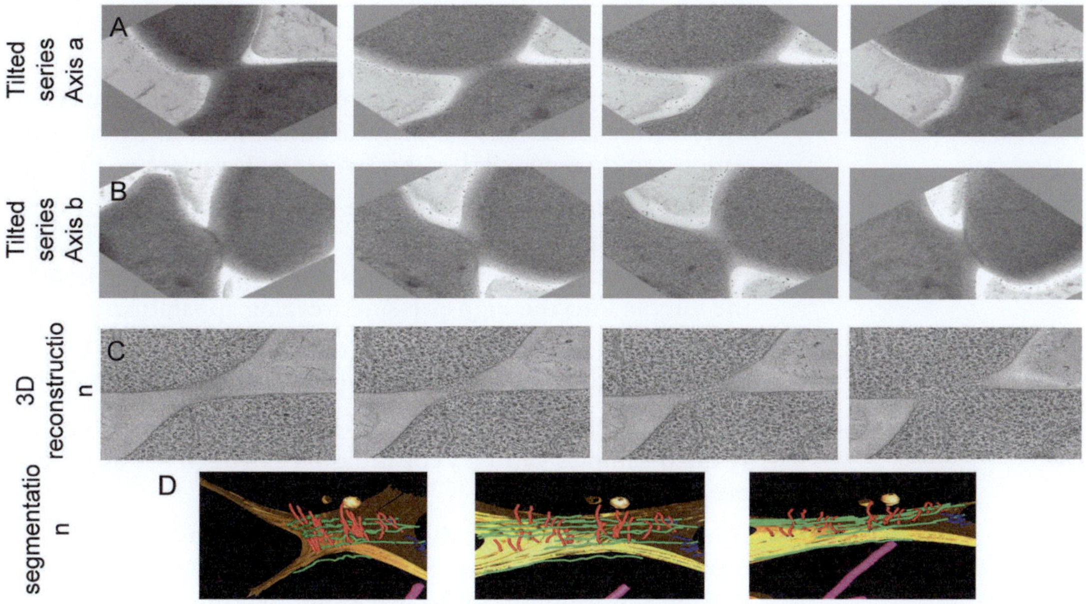

Fig. 2 Snapshots from a reconstructed tomogram and its corresponding model. (**a**) Images from a tilted series collected from a resin-embedded sample. (**b**) After the sample has been rotated by 90°, another tilted series was collected from which a few images are shown. (**c**) Images from different slices of the corresponding 3D reconstruction. (**d**) Model obtained after segmentation of the reconstructed tomogram

3. Collect a tilt series from –60° to 60° in two-degree increments (~60 images), using a total dose of about 50 electrons per Å^2. Focus and tracking have to be performed on an area a few microns away from the area of interest to prevent any irradiation and burning of the sample.

3.5 Data Processing

1. Align the tomographic dataset using the fiducial gold beads and software like IMOD [14] (*see* **Note 20**).
2. Perform the 3D reconstructions using weighted back projection (for both tilted series in the case of a dual dataset from a resin-embedded sample).
3. For resin-embedded samples, recombine the 3D reconstruction using the datasets from the two perpendicular tit series. One example is displayed in Fig. 2c.
4. For datasets from cryo-samples, enhance the signal-to-noise ratio trying different types of filtering software (*see* **Note 21**).
5. Segment the sample by tracing recognizable features (filaments, vesicles, microtubules, membranes, etc.) through sequential consecutive slices in the depth of the 3D reconstruction (*see* **Note 22**) as shown in Fig. 2d.

4 Notes

1. For optimum cell preservation, cells should not be harvested by centrifugation. The mechanical stress induced by centrifugation can disrupt cellular ultrastructural features.
2. A water-soaked filter paper can be set in the Petri dish. It prevents the cells from being dried. The yeast paste can be preserved this way for several minutes, the time needed to transfer the sample into the high-pressure freezing apparatus.
3. The transfer needs to be carried out as quickly as possible in order to prevent the cells from drying before being vitrified.
4. If the freeze substitution device is not readily available, the vitrified samples can be kept in liquid nitrogen for as long as necessary.
5. It is convenient to start the freeze substitution procedure just before the weekend. The sample will be ready, at 0 °C, Monday morning. It is important to make sure that the liquid nitrogen tank connected to the apparatus is full.
6. The epon resin can be kept at −20 °C and defrosted before use. To replace a bath one should use plastic Pasteur pipettes and make sure the samples are not disposed of with the discarded solution. The discarded and rinsing solutions need to be discarded in a dedicated chemical waste container. The preparation of solutions has to be performed under a hood for safety reasons.
7. To label the samples, you can insert small pieces of paper written with a lead pencil inside the molds.
8. 50 nm sections are fine for 2D imaging. Thicker sections (150–200 nm) can and should be used for tomography (irradiation with the electron beam will thin down the sections).
9. Carbon evaporation prevents charging and subsequent drifting of the sample during data collection. While evaporating carbon, make sure not to produce any sparks, which can damage the formvar film and tear it apart.
10. Dextran is a cryo-protectant that helps the vitrification process and also makes cryo-sectioning easier.
11. Since the media is more viscous because of the dextran, do not wait until a yeast paste is obtained. There will be enough cells when harvesting 2–3 mL from an initial 500 mL.
12. Usually the tips of the tubes are not perfectly vitrified, as seen by the presence of holes and cracks in the ice. Trim the first 50 μm of the tube before carving the pyramid. The surface of the vitrified ice, when inspected through the binoculars, should look dark and shiny.
13. This step is the most delicate and challenging part of the method. Tune the sectioning speed to make it comfortable for yourself.

The "CRION ionizer" makes the attachment of the vitrified ribbon onto the grid easier than using a mechanical "stamping" device. The ribbon has to be perfectly attached to the grid and has to lay flat on top of it. Otherwise, cryo-tomography and 3D reconstruction will be difficult to carry out.

14. It is wise to pre-screen the different grids and samples one has prepared to choose the best one for data collection. The quality of the stain, the density of cells in the sample, and the sample preservation (e.g. the quality of the cellular membranes) are important factors to keep in mind.
15. Pre-irradiation of the sample is needed to shrink the resin to its final thickness (by about 30 %). Otherwise, the resin will shrink during data collection, making it challenging to align the images and to obtain a consistent reconstruction.
16. Depending on the data collection software use, this procedure will be different (some steps may or may not be automated). The quality of the data (tracking and focusing have to be correct) is essential to be able to carry out any further processing.
17. About 4 Å per pixels should be enough. We collect our data at a magnification of 50,000 on a 4K×4K camera and bin by 2, resulting in a final resolution of about 4 Å per pixels.
18. The dual-axis holder from Fischione enables a mechanical rotation of the sample gradually so that one can still follow an area of interest (in our case a dividing cell) throughout the rotation. This way the cell of interest is kept in the center of the screen by moving the sample in the *x,y* directions. Collecting two tilt series at 90° from one another reduces the "missing wedge" in the tomographic data to a "missing pyramid" and thus minimizes the anisotropy of the tomographic reconstruction.
19. Check the diffraction pattern to make sure the ice is vitreous. Additionally, the ribbon has to lay perfectly flat onto the grid. Otherwise, this will affect negatively the quality of the reconstruction.
20. IMOD [14] is a free, downloadable software package (http://bio3d.colorado.edu/imod/). It is powerful, user-friendly, and widely used in the field. We used IMOD for almost all the processing steps. The IMOD website includes comprehensive instructions and tutorials as well as practical advice and help from the developers. The initial dataset has to be in mrc format.
21. One can use a nonlinear anisotropic diffusion filter [18] for instance, available in IMOD or TOMOAND [19] (https://sites.google.com/site/3demimageprocessing/tomoand).
22. For most of our datasets we carry out this task manually using IMOD, but Amira [15] enables some automated rendering in the process.

Acknowledgments

We thank K. McDonald for training and advice on sample preparation, Manfred Auer and Jeff Triffo for tomography advise, Jason Pierson and Peter Peters for training and hosting A.B. during the cryo-sectioning studies, and Patricia Grob for technical support. This work was funded by a Jane coffin Childs Research postdoctoral fellowship (A.B) and by the Howard Hughes Medical Institute (E.N).

References

1. Bertin A, McMurray MA, Grob P, Park SS, Garcia G 3rd, Patanwala I, Ng HL, Alber T, Thorner J, Nogales E (2008) Saccharomyces cerevisiae septins: supramolecular organization of heterooligomers and the mechanism of filament assembly. Proc Natl Acad Sci U S A 105(24):8274–8279
2. Garcia G 3rd, Bertin A, Li Z, Song Y, McMurray MA, Thorner J, Nogales E (2011) Subunit-dependent modulation of septin assembly: budding yeast septin Shs1 promotes ring and gauze formation. J Cell Biol 195(6):993–1004. doi:10.1083/jcb.201107123
3. Bertin A, McMurray MA, Thai L, Garcia G 3rd, Votin V, Grob P, Allyn T, Thorner J, Nogales E (2010) Phosphatidylinositol-4,5-bisphosphate promotes budding yeast septin filament assembly and organization. J Mol Biol 404(4):711–731. doi:10.1016/j.jmb.2010.10.002
4. Bertin A, McMurray MA, Pierson J, Thai L, McDonald KL, Zehr EA, Garcia G 3rd, Peters P, Thorner J, Nogales E (2012) Three-dimensional ultrastructure of the septin filament network in Saccharomyces cerevisiae. Mol Biol Cell 23(3):423–432. doi:10.1091/mbc.E11-10-0850
5. Bertin A, Nogales E (2012) Septin filament organization in Saccharomyces cerevisiae. Commun Integr Biol 5(5):503–505. doi:10.4161/cib.21125
6. Hartwell LH (1971) Genetic control of the cell division cycle in yeast. II. Genes controlling DNA replication and its initiation. J Mol Biol 59(1):183–194
7. Dobbelaere J, Barral Y (2004) Spatial coordination of cytokinetic events by compartmentalization of the cell cortex. Science 305(5682):393–396. doi:10.1126/science.1099892, 305/5682/393 [pii]
8. Barral Y, Mermall V, Mooseker MS, Snyder M (2000) Compartmentalization of the cell cortex by septins is required for maintenance of cell polarity in yeast. Mol Cell 5(5):841–851, S1097-2765(00)80324-X [pii]
9. McMurray MA, Bertin A, Garcia G 3rd, Lam L, Nogales E, Thorner J (2011) Septin filament formation is essential in budding yeast. Dev Cell 20(4):540–549. doi:10.1016/j.devcel.2011.02.004
10. Byers B, Goetsch L (1976) A highly ordered ring of membrane-associated filaments in budding yeast. J Cell Biol 69(3):717–721
11. Soll DR, Mitchell LH (1983) Filament ring formation in the dimorphic yeast Candida albicans. J Cell Biol 96(2):486–493
12. Dobbelaere J, Gentry MS, Hallberg RL, Barral Y (2003) Phosphorylation-dependent regulation of septin dynamics during the cell cycle. Dev Cell 4(3):345–357, S1534580703000613 [pii]
13. Vrabioiu AM, Mitchison TJ (2006) Structural insights into yeast septin organization from polarized fluorescence microscopy. Nature 443(7110):466–469. doi:10.1038/nature05109, nature05109 [pii]
14. Kremer JR, Mastronarde DN, McIntosh JR (1996) Computer visualization of three-dimensional image data using IMOD. J Struct Biol 116(1):71–76. doi:10.1006/jsbi.1996.0013
15. Stalling D, Westerhoff M, Hege H-C, Hansen C, Johnson C (2005) Amira, a highly interactive system for visual analysis. In: Hansen CD, Johnson CR (eds) The visualization handbook. Elsevier, New York, NY, pp 749–767
16. McDonald K (2007) Cryopreparation methods for electron microscopy of selected model systems. Methods Cell Biol 79:23–56. doi:10.1016/s0091-679x(06)79002-1
17. Pierson J, Fernandez JJ, Bos E, Amini S, Gnaegi H, Vos M, Bel B, Adolfsen F, Carrascosa JL, Peters PJ (2010) Improving the technique of vitreous cryo-sectioning for cryo-electron tomogra-

phy: electrostatic charging for section attachment and implementation of an anti-contamination glove box. J Struct Biol 169(2):219–225. doi:10.1016/j.jsb.2009.10.001

18. Frangakis AS, Hegerl R (2001) Noise reduction in electron tomographic reconstructions using nonlinear anisotropic diffusion. J Struct Biol 135(3):239–250. doi:10.1006/jsbi.2001.4406

19. Fernandez JJ, Li S (2003) An improved algorithm for anisotropic nonlinear diffusion for denoising cryo-tomograms. J Struct Biol 144(1–2):152–161

Chapter 10

Isolation of Cytokinetic Actomyosin Rings from *Saccharomyces cerevisiae* and *Schizosaccharomyces pombe*

Junqi Huang*, Mithilesh Mishra*, Saravanan Palani, Ting Gang Chew, and Mohan K. Balasubramanian

Abstract

Cytokinesis is the final stage of cell division, through which cellular constituents of mother cells are partitioned into two daughter cells resulting in the increase in cell number. In animal and fungal cells cytokinesis is mediated by an actomyosin contractile ring, which is attached to the overlying cell membrane. Contraction of this ring after chromosome segregation physically severs the mother cell into two daughters. Here we describe methods for the isolation and partial purification of the actomyosin ring from the fission yeast *Schizosaccharomyces pombe* and the budding yeast *Saccharomyces cerevisiae*, which can serve as in vitro systems to facilitate biochemical and ultrastructural analysis of cytokinesis in these genetically tractable model systems.

Key words Yeast, *Schizosaccharomyces pombe*, *Saccharomyces cerevisiae*, Actomyosin ring, Isolation, ATP-dependent contraction

1 Introduction

In most eukaryotes from yeast to human, cytokinesis is carried out by the regulated assembly and subsequent contraction of an actomyosin-based contractile ring, which leads to constriction of the plasma membrane at the division site. This ring is made up of a conserved set of proteins including actin, formin, myosin II, IQGAP, and F-BAR proteins [1, 2]. While genetic and cell biological analysis in the last couple of decades in yeasts, nematodes, fruit flies, and human cells have led to the identification of many conserved proteins involved in cytokinesis, the mechanism of ring formation, contraction, and disassembly remain mysterious [2–6]. For ring contraction, uncovering its mechanisms will require a

* Both the editors contributed equally.

Alberto Sanchez-Diaz and Pilar Perez (eds.), *Yeast Cytokinesis: Methods and Protocols*, Methods in Molecular Biology, vol. 1369, DOI 10.1007/978-1-4939-3145-3_10, © Springer Science+Business Media New York 2016

definitive list of factors sufficient for ring contraction, as well as experimental tools enabling more detailed biochemical and ultrastructural examination of the process. Achieving these goals will demand a system of ring contraction in vitro.

Actomyosin contractile rings are highly complex assemblies of more than 130 different proteins [2]. As an important first step toward understanding this complex macromolecular structure, intact actomyosin rings were recently isolated/partially isolated from both budding yeast and fission yeast, respectively [7, 8]. Although it is still not clear whether rings isolated from budding yeast are fully contractile, cytokinetic rings isolated from fission yeast are functional and upon addition of ATP these rings undergo complete closure in vitro [7, 8]. Fission yeast ring contraction in vitro is myosin II-dependent but surprisingly is independent of actin turnover. Though functional actomyosin rings have been isolated from newt eggs and sea urchin embryos [9, 10], the genetic intractability of these organism has limited the progress in furthering these studies. On the other hand, yeasts are easy to manipulate genetically and have a short generation time. This allows for isolation of sufficient amounts of cellular material for cytokinesis studies.

The isolated contractile ring systems allow a combination of biochemical and genetic analysis that complement cell biological studies of the cytokinetic ring. Moreover, isolated rings would also be amenable to ultrastructural analysis. In summary, these approaches to the isolation of the cytokinetic apparatus and its analysis hold great potential in increasing our understanding of the important process of cytokinesis. The subsequent sections provide detailed step-by-step protocol for isolation and analysis of yeast actomyosin rings.

2 Materials

2.1 *S. pombe* Actomyosin Ring Isolation

1. Edinburgh minimal medium + Supplements (EMM4S) + 0.5 % Glucose (filter-sterilized, room temperature storage): 3 g/l potassium hydrogen phthalate, 2.2 g/l sodium phosphate dibasic, 5 g/l ammonium chloride, 5 g/l D-(+)-glucose (anhydrous), 1× salt solution, 1× minerals solution, 1× vitamins solution, 82.5 mg/l L-histidine, 82.5 mg/l L-LEUCINE, 112.5 mg/l adenine hemisulfate, 82.5 mg/l uracil.
2. 50× salt solution stock (filter-sterilized, 4 °C storage): 0.26 M $MgCl_2 \cdot 6H_2O$, 4.99 mM $CaCl_2 \cdot 2H_2O$, 0.67 M KCl, 14.1 mM Na_2SO_4.
3. 50× minerals solution stock (filter-sterilized, 4 °C storage): 80.9 mM boric acid, 23.7 mM $MnSO_4$, 13.9 mM $ZnSO_4 \cdot 7H_2O$, 7.40 mM $FeCl_2 \cdot 6H_2O$, 2.47 mM molybdic acid, 6.02 mM KI, 1.60 mM $CuSO_4 \cdot 5H_2O$, 47.6 mM citric acid.

4. 50× vitamins solution stock (filter-sterilized, 4 °C storage): 4.20 mM pantothenic acid, 81.2 mM nicotinic acid, 55.5 mM inositol, 40.8 μM biotin.
5. E-Buffer (filter-sterilized, room temperature storage): 50 mM sodium citrate-dihydrate, 100 mM sodium phosphate.
6. E-Buffer with 1.2 M sorbitol (filter-sterilized, room temperature storage): 50 mM sodium citrate-dihydrate, 100 mM sodium phosphate, 1.2 M D-sorbitol.
7. E-Buffer with 0.6 M sorbitol (filter-sterilized, room temperature storage): 50 mM sodium citrate-dihydrate, 100 mM sodium phosphate, 0.6 M D-sorbitol.
8. EMM4S+ 0.8 M Sorbitol (filter-sterilized, room temperature storage): 3 g/l potassium hydrogen phthalate, 2.2 g/l sodium phosphate dibasic, 5 g/l ammonium chloride, 20 g/l D-(+)-glucose (anhydrous), 1× salt solution, 1× minerals solution, 1× vitamins solution, 82.5 mg/l L-histidine, 82.5 mg/l L-LEUCINE, 112.5 mg/l adenine hemisulfate, 82.5 mg/l uracil, 0.8 M D-sorbitol.
9. Wash Buffer (pH ~7.0, no need to adjust pH, filter-sterilized, 4 °C storage): 0.8 M D-sorbitol, 2 mM EGTA, 5 mM $MgCl_2$, and 20 mM PIPES–NaOH (pH 7.0).
10. Isolation Buffer (pH ~7.0, no need to adjust pH, filter-sterilized, 4 °C storage): 0.16 M sucrose, 50 mM EGTA, 5 mM $MgCl_2$, 50 mM potassium acetate, 50 mM PIPES–NaOH (pH 7.0), 0.5 % NP-40 (Nonidet P40 substitute, from Fluka, added fresh every time before experiments), 10 μg/ml leupeptin hemisulfate (stock solution: 10 mg/ml), 10 μg/ml aprotinin from bovine lung (stock solution: 10 mg/ml), 400 μg/ml benzamidine hydrochloride hydrate (stock solution: 80 mg/ml), 1 mM phenylmethylsulfonyl fluoride (PMSF, stock solution: 0.1 M), and 1 mM dithiothreitol (DTT, stock solution: 1 M).
11. Reactivation Buffer (pH adjusted to ~7.5 with NaOH, filter-sterilized, 4 °C storage): 0.16 M sucrose, 5 mM $MgCl_2$, 50 mM potassium acetate, 20 mM MOPS–NaOH (pH 7.0), 10 μg/ml leupeptin hemisulfate (stock solution: 10 mg/ml), 10 μg/ml aprotinin from bovine lung (stock solution: 10 mg/ml), 400 μg/ml benzamidine hydrochloride hydrate (stock solution: 80 mg/ml), 1 mM phenylmethylsulfonyl fluoride (PMSF, stock solution: 0.1 M), and 1 mM dithiothreitol (DTT, stock solution: 1 M). For protease inhibitors (leupeptin hemisulfate, aprotinin from bovine lung, benzamidine hydrochloride hydrate, PMSF) and DTT, aliquot the stock solutions and store at −20 °C.
12. ATP, stock solution: 0.1 M (aliquot and storage at −20 °C).

13. Lysing enzymes from *Tricoderma harzianun* (Sigma, working concentration: 0.0125 g/5 ml).
14. Zymolyase (G-Biosciences, 1.5 units/μl).
15. Glass homogenizer (Corning Potter-Elvehjem Tissue Grinders, Thomas Scientific).

2.2 S. cerevisiae Ring Isolation

1. YPDA/YP-Raffinose media.
2. S-buffer: 100 mM potassium phosphate (pH 7.0), 1.33 M D-sorbitol, 40 mM β-mercaptoethanol.
3. Ring isolation buffer: 50 mM HEPES-KOH (pH 7.5), 10 mM MgOAc (pH 7.5), 60 mM potassium acetate (pH 7.5), 1 mM EDTA, 10 % glycerol, 1 mM DTT.
4. Lyticase (Sigma-Aldrich).
5. Protease Inhibitor (Roche, EDTA-free, complete cocktail).
6. 1 mM PMSF.
7. 0.5 % NP-40.
8. Strains used for the ring isolation (kindly provided by Dr. David G. Drubin, University of California, Berkeley): *Mat* a *mob1-77 MYO1-GFP::KANMX pep4Δ::TRP1 MLC2-HPM* (*his9-precission2-myc9*)::*HIS3 ura3-52 trp1Δ1 his3Δ200 leu2-*

3 Methods

3.1 S. pombe Protoplast Formation (Fig. 1)

1. Inoculate fresh (from −80 °C freezer, 1 day on 24 °C YES plate) *cdc25-22S. pombe* cells (with fluorescent actomyosin ring marker such as Rlc1p-3GFP) into 20 ml liquid EMM4S + 0.5 % Glucose, and grow overnight at 24 °C and shaking at 200 rpm (*see* **Notes 1–3**).
2. Next day morning, measure $O.D._{595}$ (should be around 0.3; 3×10^6 cells/ml), shift up cells to 36 °C for 3.5 h, while still shaking at 200 rpm (*see* **Notes 4** and **5**).
3. Spin down cells in a 50 ml Falcon tube at 450 × *g* for 2 min at room temperature.
4. Wash cells once with E-buffer. Spin at 450 × *g* for 2 min at room temperature.
5. Resuspend cells in E-buffer supplemented with 1.2 M sorbitol up to the volume of 5 ml in the 50 ml Falcon tube.
6. Weigh 0.0125 g lysing enzymes from *Tricoderma harzianun* (Sigma, working concentration: 0.0125 g/5 ml) (*see* **Note 6**).
7. Mix lysing enzymes evenly with the cell suspension from **step 5**. Lay the Falcon tube flat on the air shaker. Digest cell walls at 36 °C, while shaking at 80 rpm for 1 h 20 min.

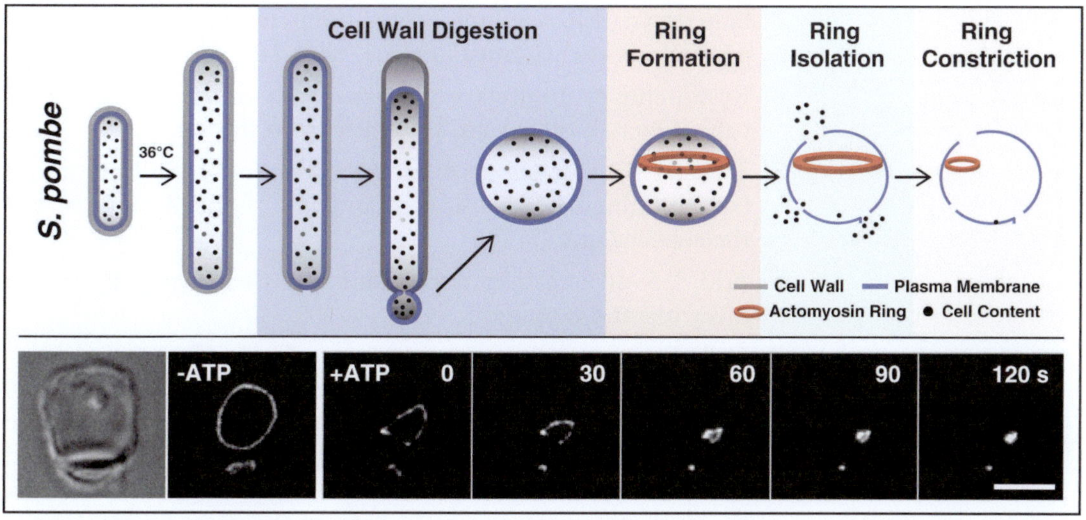

Fig. 1 *S. pombe* ring isolation. *Upper panel* shows the schematic of *S. pombe* ring isolation process. *Lower panel* shows an example of purified *S. pombe* ring constriction. Scale bar, 5 μm

8. Monitor protoplast formation by phase-contrast microscopy. Take 5 μl cells from **step** 7, gently spread on a coverslip (24 × 40 mm), and observe cells directly without another coverslip on top. After **step** 7, more than 80 % cylindrical cells should form a bud-like extrusion at the tip(s)/cell middle (*see* **Note** 7).
9. Add 10 μl of zymolyase (G-Biosciences, 1.5 units/μl) into the cell suspension from **step** 7. Cells are kept at 36 °C, while shaking at 80 rpm for another 40 min (*see* **Note 8**).
10. Spin at 450 × *g* for 2 min at room temperature.
11. Wash once with E-buffer supplemented with 0.6 M sorbitol. Spin at 450 × *g* for 2 min at room temperature.
12. Resuspend protoplasts in 10 ml EMM4S + 0.8 M Sorbitol, 80 rpm at 24 °C for 5–6 h. Check actomyosin ring formation on the coverslip (similar to **step 8** without another coverslip) under wide-field fluorescence microscopy. Isolate rings when more than 30 % of the protoplasts form a full actomyosin ring.

3.2 S. pombe Actomyosin Ring Isolation

1. Cool down the bench-top centrifuge to 4 °C for ring isolation.
2. Prepare 2 ml isolation buffer (with 0.5 % NP-40) and 6 ml reactivation buffer by adding protease inhibitors (leupeptin, aprotinin, benzamidine, PMSF) and DTT every time before experiments. To make ATP-reactivation buffer, mix 995 μl reactivation buffer from above with 5 μl ATP (stock concentration: 100 mM. Final concentration after dilution: 0.5 mM) (*see* **Note 9**).

3. Spin down (450 × *g*, 2 min, room temperature, in a 50 ml tube) protoplasts after Subheading 3.1, **step 12**.
4. Resuspend protoplasts with 1 ml wash buffer in the 50 ml tube. Operate on ice and be gentle from this step on.
5. Transfer protoplasts to a 1.5 ml eppendorf tube by pouring (no pipetting). Spin at 450 × *g* and at 4 °C for 2 min. Decant the supernatant.
6. Add 1.5 ml isolation buffer and mix thoroughly by gently inverting and tapping.
7. Cells from **step 6** are poured into a 5 ml glass homogenizer (Corning Potter-Elvehjem Tissue Grinders, Thomas Scientific) pre-chilled on ice. Leave on ice for 5 min. Homogenize the sample (no need to rotate the pestle) for six times and avoid bubble formation. Leave on ice for another 5 min. Cells after this step are called ghosts as most of them are permeabilized (*see* **Note 10**).
8. Spin at 450 × *g* in a centrifuge cooled to 4 °C for 2 min.
9. Wash twice with 1 ml reactivation buffer. Spin at 450 × *g* in a centrifuge cooled to 4 °C for 2 min.
10. Resuspend ghosts in 250–800 μl (volume depends on the specific experiment planned) reactivation buffer. Isolated rings are preserved inside the ghosts.
11. Mix ghosts with equal volume (~8 μl each) of ATP-reactivation buffer (from **step 2**) to check ring contractility on the coverslip (similar to Subheading 3.1, **step 8**). Most isolated rings should be able to finish contraction and become a dot/cluster after 6 min (*see* **Note 11**).

3.3 Poly-L-lysine-Coated Coverslips for S. pombe Ring Constriction Movies

1. Add 40 μl poly-L-lysine (Sigma) onto the centre of the coverslips (24 × 40 mm).
2. Cut parafilm (~15 × 15 mm) and place it on top of poly-lysine from **step 1**. Press the parafilm to spread the poly-lysine evenly.
3. Dry the coverslips from **step 2** in the laminar flow cabinets for 2–3 h. Peel off the parafilm.
4. Cut and paste two double-sided tapes (NICHIBAN, NW-5, 5 mm width) onto a poly-L-lysine-coated coverslip from **step 3**. The double-sided tapes should be parallel to each other and the space between them should be similar to the width (5 mm) of the NICHIBAN NW-5 double-sided tape for better perfusion efficiency (*see* **Note 12**).
5. To make sure the double-sided tapes stick tight to the coverslip, use the front end of a pipette to scratch them. Use a pair of fine point forceps to tear off the non-sticky side of the double-sided tape before perfusion experiments (Fig. 2).

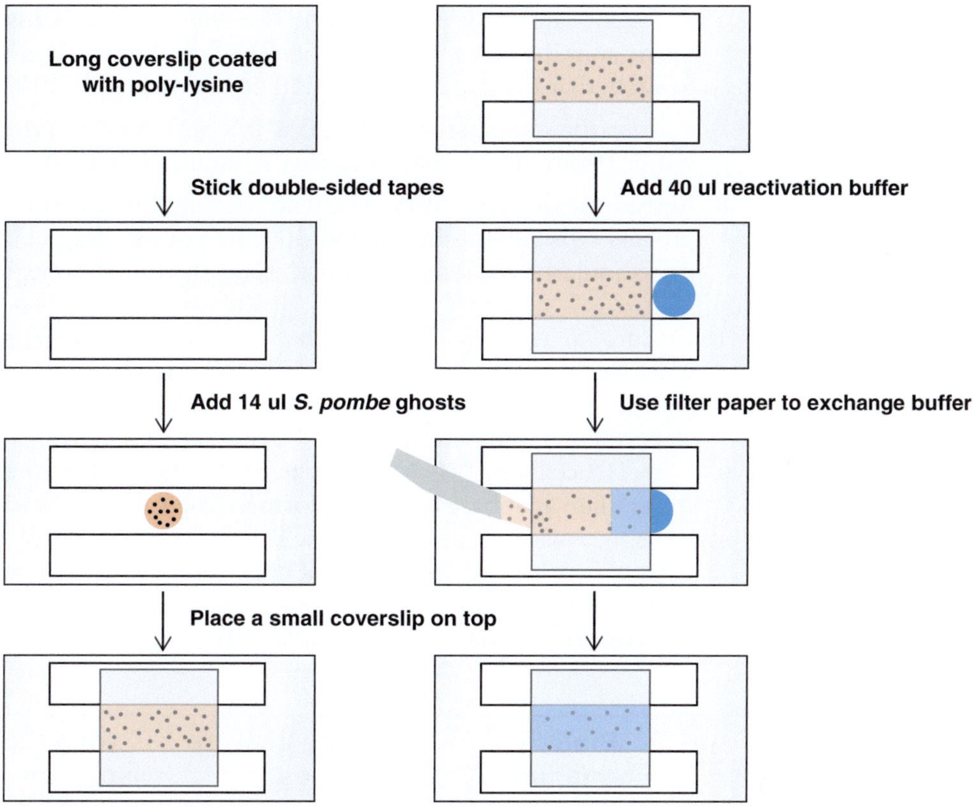

Fig. 2 Schematic of perfusion process for *S. pombe* ring constriction

3.4 ATP Perfusion for S. pombe Ring Constriction Movies

1. Add 14 μl ghosts (from Subheading 3.2, **step 10**, ghosts are kept on ice before experiments) onto a poly-L-lysine-coated coverslip from Subheading 3.3.
2. To make a chamber for perfusion, place a small coverslip (18 × 18 mm) on top of the ghosts and the double-sided tapes from **step 1**. Use the front end of a pipette to press the sides of the small coverslip to make sure it sticks tight on the double-sided tapes.
3. Make sure the chamber is stable on the microscope stage (important) (*see* **Note 13**).
4. Add 40 μl reactivation buffer (without ATP, this step is to wash away the unbound ghosts) on one side of the chamber, exchange buffer by sucking from the other side of the chamber using filter paper (Whatman, cut into a rectangular shape) (*see* **Note 14**).
5. Search for stable ghosts possessing a fully formed, unbroken ring. Optimal rings have a diameter slightly smaller than that of the ghost and are bright, even, round, and parallel to the coverslip (*see* **Notes 15–19**).

6. Start time-lapse movie acquisition (15 s interval time). Pause the movie after the first two time points (the first two time points are to show that there is no contraction before adding ATP).
7. Perfuse the chamber with 40 μl of 0.5 mM ATP-reactivation buffer (from Subheading 3.2, **step 2**) similar to **step 4**.
8. Restart movie acquisition. The whole isolated ring contraction process typically finishes within 20 time points. Ring contraction should be obvious from the third or the fourth time point.

3.5 S. cerevisiae Ring Isolation (Modified from [7]) (Fig. 3)

1. Culture *S. cerevisiae* cells [using mitotic exit temperature-sensitive (ts) alleles, e.g., *mob1-77*] with Myo1-GFP (type-II myosin) actomyosin ring marker in YPDA/YP-Raffinose media at permissive temperature (23 °C/30 °C) until $O.D._{600} \sim 0.4$ ($\sim 5 \times 10^6$ cells/ml) then shift cells to 37 °C for 4 h to arrest cells at late anaphase with fully formed rings. As an alternative to using *mob1-77* allele, galactose is added to overexpress proteins to arrest cells at specific cell cycle stage (e.g., overexpression of non-degradable version of mitotic cyclin, Clb2 to arrest the cells in late anaphase) (*see* **Note 20**).

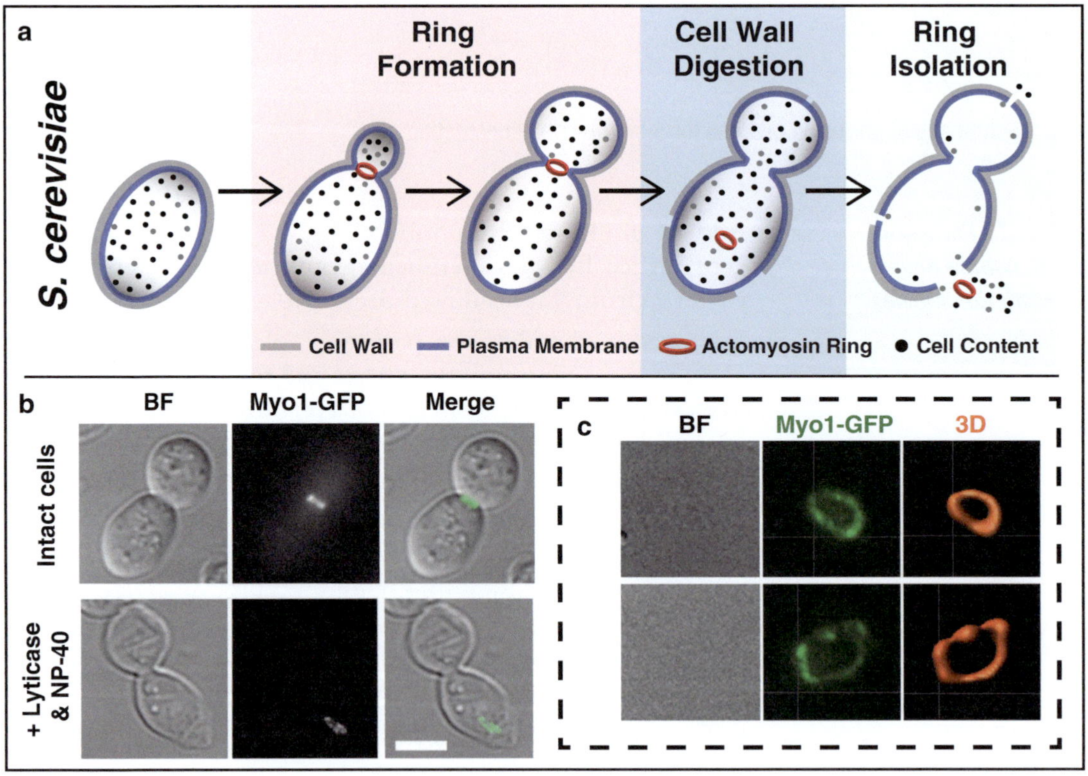

Fig. 3 *S. cerevisiae* ring isolation. (**a**) Schematic of *S. cerevisiae* ring isolation process. (**b**) Bright field and fluorescent images showing *S. cerevisiae* ring isolation process. Scale bar, 5 μm. (**c**) Bright field, fluorescent, and Imaris 3D rendering images showing the purified *S. cerevisiae* rings on poly-L-lysine-coated coverslip

2. For the purpose of simplifying the protocol, the methods described below are detailed for a 100 ml culture. Cells are pelleted at 2000 × *g* for 20 min in a rotor pre-warmed to 37 °C (in case of using ts allele) or 30 °C (*see* **Note 21**).
3. Wash pellets twice with 10 ml sterile PBS and once with 10 ml PBS + PMSF. Freeze cells using liquid nitrogen or by placing the pellets at −80 °C (*see* **Note 22**).
4. To isolate the cytokinetic rings, frozen cells (from **step 3**) are rapidly brought to 30 or 37 °C (in case of using ts alleles) by adding 0.5 ml pre-warmed sorbitol buffer (S-buffer).
5. Pre-incubate cells in a 37 °C water bath for 5–10 min before the addition of 25 μg Lyticase (recombinant *O. xanthineolytica* β-1,3-glucanase, either expressed and purified from *E. coli* (homemade, please refer to Scott and Schekman [11]) or obtained commercially (Sigma-Aldrich)). Incubate cells at 37 °C for 30 min (*see* **Note 23**).
6. Cells are centrifuged at 1000 × *g* for 3 min. The cell pellets are then washed twice with 0.5 ml sorbitol buffer and osmotically lysed using 100 μl of ice-cold ring isolation buffer supplemented with EDTA-free protease cocktail inhibitor (Roche) and 1 mM PMSF.
7. NP-40 is added to a final concentration of 0.5 % and cells are incubated on ice for approximately 10 min.
8. Lysate is cleared of debris/unbroken cells by centrifugation at 300 × *g* for 5–10 min at 4 °C (*see* **Note 24**).
9. Clarified lysate is spotted on a poly-L-lysine-coated coverslip (refer to Subheading 3.3 and Fig. 2), incubated for 10 min at room temperature and then washed twice with ring isolation buffer to clear the unattached rings and debris.
10. Rings attached on the poly-L-lysine-coated coverslips are visualized by fluorescence microscopy.
11. Actomyosin ring components from the clarified lysate are confirmed by immunoblotting using respective antibodies.

4 Notes

1. It is very important that *S. pombe* cells are fresh. It may be helpful to dilute fresh cells 2 days before the isolation experiment and to sub-culture them on the following days.
2. *cdc25*-22 background and the 36 °C shift up process are not essential for ring isolation.
3. Low glucose medium (EMM4S + 0.5 % Glucose) may be good for protoplast formation.

4. It is better to culture cells in a programmable shaker, as the whole fission yeast ring isolation process (including shift up) takes around 12–13 h. Inoculate a few flasks (each flask contains 20 ml EMM4S + 0.5 % Glucose medium, choose the best flask later for ring isolation) of fresh fission yeast cells at 9 a.m. 1 day before the experiments (24 °C, 200 rpm). At 6 p.m., dilute the culture to $O.D._{600} \sim 0.05$ (6×10^5 cells/ml), keep the volume around 20 ml. Culture cells in a programmable shaker (setting: 200 rpm, 24 °C for 10.5 h; then 200 rpm, 36 °C for 3.5 h). The next day morning at 8 a.m. check $O.D._{600}$, continue the steps from Subheading 3.1, **step 3**.
5. Do not shift cells up at 36 °C for more than 4 h.
6. Lysing enzymes activity may differ from batch to batch; one can either adjust the amount of lysing enzymes used in the experiment or adjust the lysing enzymes digestion time.
7. When observing protoplasts directly on the coverslip under bench-top phase-contrast microscopy, do not observe cells on the edge or observe for too long. Protoplasts may shrink because of water evaporation.
8. Adding Zymolyase may help improving the isolated ring constriction in the later steps.
9. For NP-40, first use double-distilled water to dilute the 100 % NP-40 into 20 %, then used at the final working concentration of 0.5 %.
10. Homogenizer in Subheading 3.2, **step 7** may not be necessary. Protoplasts can be permeabilized with isolation buffer for 10 min on ice. Invert tubes 2–3 times every 2 min.
11. It is very important to mix equal volume of ATP-reactivation buffer with ghosts to check ring contractility. This is especially important when working with rings isolated from mutants that may affect ring constriction.
12. The quality of the double-sided tape can influence the perfusion efficiency. NICHIBAN, NW-5 (5 mm width) is one of the best.
13. Scotch tapes can be used to stabilize the chamber on the microscopy stage.
14. It may be better to choose soft filter papers for perfusion experiment as the hard ones can move the chamber during buffer exchange.
15. It is necessary to make a few time-lapse movies to check ring quality before other experiments such as drug treatment. Some solvents may affect ring constriction.
16. It is very important to choose proper rings for ATP perfusion movies (refer to Subheading 3.4, **step 5**) because it is unavoidable that structures of some isolated rings are altered after permeabilization.

17. To search for proper rings under fluorescent microscopy, it is better to search by eyepiece (only if the fluorescent signal would not be bleached easily such as Rlc1p-3GFP). This is because the field of view through eyepiece is usually bigger than that through the camera.
18. After washing away the unbound ghosts with reactivation buffer (Subheading 3.4, **step 4**), the number of remaining ghosts reduces. Depending on the purpose, it may be good to concentrate the ghosts from Subheading 3.2, **step 10** or try other ways to coat the coverslips with poly-L-lysine more efficiently.
19. ATP perfusion occasionally does not work well as quality differs from chamber to chamber. It is suggested to acquire more movies for quantitative analysis.
20. *S. cerevisiae* cells can be grown in YPDA medium when no external selection is required. Appropriate selection medium (with/without supplements) can be used for growing strains containing plasmids or genes with inducible promoters.
21. Depending on the purpose of the experiment, the culture volume can be scaled up from 0.05 to 4 l.
22. Freshly frozen cells (refer to Subheading 3.5, **step 3**) yield better ring preps in comparison to ring preps from a few months old stock. Care is to be taken to minimize the storage time at permissive temperature prior to freezing.
23. Commercial lyticase may be better than homemade lyticase for cell wall digestion.
24. Isolated *S. cerevisiae* rings are stable in solution for a few hours (~2–4 h) on ice. Further lysate clarification and subcellular fractionation can be performed as described [7].

Acknowledgments

We thank Dr. David Drubin for kindly providing the *S. cerevisiae* strains. This work was supported by research funds from the Wellcome Trust, Royal Society, and the University of Warwick. M.M. was supported by funds from the Wellcome Trust India Alliance and the Tata Institute of Fundamental Research, Mumbai, India.

References

1. Wu JQ, Pollard TD (2005) Counting cytokinesis proteins globally and locally in fission yeast. Science 310(5746):310–314. doi:10.1126/science.1113230
2. Pollard TD, Wu JQ (2010) Understanding cytokinesis: lessons from fission yeast. Nat Rev Mol Cell Biol 11(2):149–155. doi:10.1038/nrm2834
3. Glotzer M (2005) The molecular requirements for cytokinesis. Science 307(5716):1735–1739. doi:10.1126/science.1096896
4. Huang J, Huang Y, Yu H, Subramanian D, Padmanabhan A, Thadani R, Tao Y, Tang X, Wedlich-Soldner R, Balasubramanian MK (2012) Nonmedially assembled F-actin cables incorporate into the actomyosin ring in fission yeast.

J Cell Biol 199(5):831–847. doi:10.1083/jcb.201209044

5. D'Avino PP (2009) How to scaffold the contractile ring for a safe cytokinesis - lessons from Anillin-related proteins. J Cell Sci 122(Pt 8):1071–1079. doi:10.1242/jcs.034785
6. van den Heuvel S (2005) The C. elegans cell cycle: overview of molecules and mechanisms. Methods Mol Biol 296:51–67
7. Young BA, Buser C, Drubin DG (2010) Isolation and partial purification of the Saccharomyces cerevisiae cytokinetic apparatus. Cytoskeleton 67(1):13–22. doi:10.1002/cm.20412
8. Mishra M, Kashiwazaki J, Takagi T, Srinivasan R, Huang Y, Balasubramanian MK, Mabuchi I (2013) In vitro contraction of cytokinetic ring depends on myosin II but not on actin dynamics. Nat Cell Biol 15(7):853–859. doi:10.1038/ncb2781
9. Walker GR, Kane R, Burgess DR (1994) Isolation and characterization of a sea urchin zygote cortex that supports in vitro contraction and reactivation of furrowing. J Cell Sci 107(Pt 8):2239–2248
10. Fujimoto H, Mabuchi I (1997) Isolation of cleavage furrows from eggs of regular sea urchins and identification of furrow-specific proteins. J Biochem 122(3):518–524
11. Scott JH, Schekman R (1980) Lyticase: endoglucanase and protease activities that act together in yeast cell lysis. J Bacteriol 142(2):414–423

Chapter 11

Measurements of Myosin-II Motor Activity During Cytokinesis in Fission Yeast

Qing Tang, Luther W. Pollard, and Matthew Lord

Abstract

Fission yeast myosin-II (Myo2p) represents the critical actin-based motor protein that drives actomyosin ring assembly and constriction during cytokinesis. We detail three different methods to measure Myo2p motor function. Actin-activated ATPases provide a readout of actomyosin ATPase motor activity in a bulk assay; actin filament motility assays reveal the speed and efficiency of myosin-driven actin filament gliding (when motors are anchored); myosin-bead motility assays reveal the speed and efficiency of myosin ensembles traveling along actin filaments (when actin is anchored). Collectively, these methods allow us to combine the standard in vivo approaches common to fission yeast with in vitro biochemical methods to learn more about the mechanistic action of myosin-II during cytokinesis.

Key words Actin-activated ATPase assay, Colorimetric assay, Actin filament motility assay, Epifluorescence microscopy, Myosin-bead motility assays, Total internal reflection fluorescence microscopy, Protein-fluorophores

1 Introduction

Robust molecular genetics, live cell imaging, and mathematical approaches have elevated fission yeast as a model cell system for studying the complex process of cytokinesis [1, 2]. However, in vivo phenomena and theoretical predictions rely on in vitro measurements of protein function to gain a clear understanding of cellular processes. This chapter focuses on three biochemical-based approaches that can be employed to study a key fission yeast cytokinesis protein: myosin-II (Myo2p).

Myosin motors use the energy derived from ATP hydrolysis to propagate a conformational change which generates force on attached actin filaments. Conventional (class II) myosin is found in all cell types in the human body and (in addition to cytokinesis) participates in many different processes, including muscle contraction, cell motility, cortical tension, and cell morphogenesis [3]. The general mechanism of myosin-II function is quite well defined

Alberto Sanchez-Diaz and Pilar Perez (eds.), *Yeast Cytokinesis: Methods and Protocols*, Methods in Molecular Biology, vol. 1369, DOI 10.1007/978-1-4939-3145-3_11,

thanks to the application of a range of in vitro approaches over the past few decades to study the readily available muscle myosin-IIs. Such approaches include kinetic methods, optical trapping, structural studies (electron microscopy and X-ray crystallography), and motility assays unique to molecular motor proteins. However, the action and regulation of non-muscle myosin-II motors in other cellular processes such as cytokinesis remains relatively poorly understood.

Cytokinesis in fission yeast relies on the assembly and constriction of an actomyosin ring, which contains actin filaments, myosin-II, and a whole host of other conserved factors [4]. Rings fail to assemble in the absence of Myo2p function, making this particular myosin essential for fission yeast cell division and growth [5–7]. Two other fission yeast myosins (myosin-II Myp2p and myosin-V Myo51p) are also found at the ring and play accessory roles in cytokinesis [8–11]. Reconstitution of motor activity using fission yeast-purified Myo2p has advanced structure-function analysis and regulatory studies, leading to a much better understanding of the mechanism of myosin-II action during cytokinesis [12–16].

Here we describe three biochemical methods that we have employed in the past to study Myo2p. Firstly, ATPase assays allow apparent actomyosin affinities and rates of actomyosin motor activity to be measured using the actin-dependent ATPase of the motor, the enzyme reaction that fuels force production. The experiment involves sampling of actomyosin reactions over time with a range of actin filament concentrations. A colorimetric assay is used to quantitate released phosphate to generate rates of P_i release [17], which are subsequently assessed in plots using Michaelis-Menten kinetics. Secondly, actin filament motility assays rely on the attachment of purified myosin molecules to a glass cover-slip surface [18]. The ability of these tethered molecules to propel the movement of actin filaments can be monitored over time by visualization of labeled actin using epifluorescence microscopy. Actin filament binding efficiency is directly visualized and the speeds of myosin-driven filament gliding can be deduced over several imaged frames. Finally, myosin-bead motility assays utilize the attachment of myosin molecules to fluorescent beads. The ability of individual beads to undergo motility along fluorescently labeled actin filaments (tethered to the cover-slip surface) can be resolved and measured using total internal reflection fluorescence (TIRF) microscopy [19]. Efficiency, speeds, and run lengths of myosin-beads are determined by analysis of the movies. Collectively, these methods offer important approaches to extend analysis and gain a greater understanding of myosin-II function and regulation during cytokinesis in fission yeast and other systems.

2 Materials

Prepare all solutions using ultrapure water and analytical grade reagents. Prepare and store all reagents at room temperature (unless otherwise specified).

2.1 Proteins

1. Fission yeast myosin-II Myo2p can be isolated from fission yeast via a two-step purification process [12] (*see* Subheading 3.1 for the detailed protocol).
2. Rabbit skeletal muscle actin is purified in a monomeric, non-polymerized form using a standard protocol [20] and stored in G Buffer (5 mM Tris–HCl, pH 8.0, 0.2 mM $CaCl_2$, 0.2 mM ATP, 0.5 mM DTT) at 4 °C (*see* **Note 1**). A standard actin purification protocol can be found at http://www.med.upenn.edu/ostaplab/documents/ActinPreparationfromAcetonePowder.pdf

2.2 Myosin Actin-Activated ATPase Assays

1. Prior to use Myo2p is dialyzed overnight into Working Buffer (10 mM imidazole, pH 7.4, 500 mM KCl, 1 mM DTT). An actin filament (F-actin) stock is polymerized by diluting KMI Buffer (100 mM imidazole, pH 6.8, 500 mM KCl, 20 mM $MgCl_2$) tenfold with G-actin. Mock KG Buffer is prepared in the same way using G Buffer lacking actin.
2. ATPase reactions are started by addition of 10× Supplements (40 mM ATP, 40 mM $MgCl_2$ in G Buffer) to actin/myosin samples.
3. Detection solution (1:1:2:2 0.572 % ammonium molybdate in 6 N HCl:0.232 % poly-vinyl alcohol:0.0812 % malachite green:double-distilled water) is used to quantify the levels of P_i in reaction samples.
4. Stop Solution (20 mM sodium citrate) is used to terminate ATPase reactions and halt color development in this colorimetric assay.
5. Phosphate Color Standard is employed to generate standard a P_i curve to derive free phosphate released in ATPase reactions. A range of P_i concentration standards are generated by back-diluting a phosphate stock (4 mM KH_2PO_4 solution) into H_2O.

2.3 Myosin-Driven Actin Filament Motility Assays

1. Myo2p and actin filament working stocks are prepared as in Subheading 2.2, **item 1** (*see* **Note 2**). Fluorescent F-actin is generated by mixing 5 μM actin filaments 1:1 with rhodamine-phalloidin (*see* **Note 3**).
2. Nitrocellulose-coated cover slips: 22 mm × 22 mm, glass #1 (*see* **Note 4**). These cover-slips are used to adhere myosin.
3. Motility chamber components: top cover glass: nitrocellulose-coated 22 mm × 22 mm cover-slips; bottom cover glass:

24 mm × 60 mm, #1; plastic shim of 0.125 mm thickness (Blue 0.005 in.; Artus Corporation, Englewood, NJ, USA); optical adhesive.

4. Standard running buffer is Motility Buffer (25 mM imidazole, pH 7.4, 50 mM KCl, 1 mM EGTA, 4 mM $MgCl_2$, 50 mM DTT) plus additional reagents as indicated at appropriate places in Subheading 3 (see below).
5. Microscopy equipment: An inverted fluorescence microscope (e.g., Nikon TE2000-E2) with a rhodamine fluorescence (543 nm excitation) filter and a Plan Apo 60× (1.45 NA) objective is used to capture epifluorescence images of actin filaments over time at room temperature. Fluorescence utilizes an EXFO X-CITE 120 illuminator. Nikon NIS Elements software is used to control the microscope, two Uniblitz shutters, and a CoolSNAP HQ2 14-bit camera.

2.4 Myosin-Bead Motility Assays

1. Working stocks of Myo2p and fluorescent actin filaments are prepared as described in Subheading 2.2, **item 1**, and Subheading 2.3, **item 1**.
2. Uncoated glass cover slips to adhere *N*-ethylmaleimide-modified skeletal muscle myosin (NEM-myosin; Cytoskeleton Inc., Denver, CO, USA). NEM-myosin is a chemically inactivated form that tightly binds to F-actin without generating any force or movement (by persisting in the "rigor" state). NEM-myosin is diluted into dilution buffer (motility buffer plus 350 mM KCl).
3. Motility chamber components (*see* Subheading 2.3, **item 3**). Unlike in the myosin-driven actin filament motility assay, nitrocellulose-coating of the top cover slip is not required to attach NEM-myosin for the myosin-bead motility assay.
4. Standard running buffer is motility buffer plus additional reagents as indicated in Subheading 3 (*see* Subheadings 3.3 and 3.4 below).
5. Quantum dots (655 nm; Invitrogen, Waltham, MA, USA) are highly fluorescent beads with which to attach multiple Myo2p molecules. These commercially available beads are obtained pre-coated with streptavidin. A Myo2p heavy chain species carrying a C-terminal biotin carrier peptide (BCP) tag is employed (*see* **Note 5**).
6. Microscopy equipment: Through-the-objective TIRF microscopy is performed at room temperature using a microscope equipped with a 100× Plan Apo objective lens (1.49 NA) and auxiliary 1.5× magnification. Still images of fluorescent actin are detected as described earlier (Subheading 2.3, **item 5**), whereas quantum dots (Qdots) are excited with a 473-nm laser line to obtain time-lapse images collected using a high-speed XR/Turbo-Z camera controlled by Piper Control software v2.3.39.

3 Methods

3.1 Myo2p Over-Expression and Purification from Fission Yeast

1. Myo2p is purified as a complex that includes the Myo2p heavy chain and its light chains (essential light chain Cdc4p and regulatory light chain Rlc1p). To this end, a Myo2p over-expression strain (*nat^R:nmt41^prom-myo2*) is co-transformed with *p-nmt3^prom-GST-rlc1* (*ura4^+*) and *p-nmt3^prom-GST-cdc4* (*LEU2*) plasmids. Transformants are isolated on EMM (Edinburgh minimal media) Leu$^-$ Ura$^-$ plates (containing 5 μg/ml of thiamine to repress the toxic over-expression of Myo2p).
2. Transformants are grown in liquid cultures to saturation in EMM Leu$^-$ Ura$^-$ containing 5 μg/ml of thiamine. Cells are harvested and washed three times in EMM Leu$^-$ Ura$^-$ medium (lacking thiamine). Over-expression of Myo2p and its light chains are induced by diluting the washed cells to an optical density at 595 nm (OD_{595}) of ~0.05 (10^6 cells/ml) in 4 l of the same medium lacking thiamine.
3. Sufficient over-expression of Myo2p, GST-Rlc1p, and GST-Cdc4p is achieved after 24–28 h of incubation at 32 °C, by which time the OD reaches ~3.0 (6×10^7 cells/ml) (*see* **Note 6**).
4. Cells are harvested and washed once in water and once in ice-cold lysis buffer (25 mM Tris–HCl, pH 7.4, 750 mM KCl, 4 mM $MgCl_2$, 20 mM $Na_4P_2O_7$, 2 mM EGTA, and 0.1 % Triton X-100). Pellets are resuspended in an equal volume of ice-cold Lysis Buffer with additives consisting of 1 mM DTT, 4 mM ATP, 2 mM PMSF, EDTA-free protease inhibitors, and diisopropyl fluorophosphate at a final concentration of 0.5 mM. From this point forward all work is conducted at 4 °C with samples stored on ice whenever possible.
5. Cells are lysed by glass bead-beating with a Fastprep-24 (MP Biochemicals, Santa Anna, CA, USA). The lysate is centrifuged at 500 × *g* for 5 min to remove unlysed cells and beads, and then further centrifuged at 100,000 × *g* for 45 min to remove insoluble matter.
6. The supernatant is batch-incubated with 2 ml of glutathione-Sepharose for 90 min, and then transferred to a 20 ml disposable column. Once the glutathione-Sepharose is packed, the lysate is then slowly flowed over the column once at a rate of ~0.5 ml/min.
7. The bound sample is washed with lysis buffer plus additives (4 × 15 ml at a flow rate of ~0.5 ml/min) before eluting in 5 ml of lysis buffer plus additives and 100 mM glutathione. Eluate contains affinity-purified GST-Rlc1p and GST-Cdc4p enriched with co-purifying Myo2p.
8. The sample is filtered with a 0.22 μm filter and dialyzed for 24 h into 1 l of A15 Buffer (10 mM imidazole, pH 7.0,

500 mM KCl, 10 mM EDTA, 1 mM DTT, and 0.3 mM NaN_3). The A15 buffer should be changed (with another liter) at the 12 h mark.

9. Thrombin is added to the sample (Amersham Biosciences; 10 units/mg GST-Cdc4p/GST-Rlc1p) and incubated at 4 °C for 24 h to detach GST from the light chains (*see* **Note 7**). The amount of GST-light chains present is estimated using the Bradford assay to measure the total protein concentration of the crude one-step purified sample. BSA is used as the protein standard here.
10. One-step purified, cleaved Myo2p is further purified by gel filtration on a 2.5 × 50 cm column of Bio-Gel A15m, 200–400 mesh (Bio-Rad Laboratories) equilibrated and run with A15 Buffer (*see* **Note 8**). The column should be run with a flow rate of 0.5 ml/min. ≥80 × 4 ml fractions are typically collected in a fraction collector.
11. Gel filtration fractions enriched in Myo2p are identified by running samples on SDS-PAGE gels. Typical of a myosin-II, Myo2p elutes in the early fractions owing to its large size and elongated molecular shape. The first two-thirds of the Myo2p-containing fractions yield ~95 % pure Myo2p. The final third of the fractions can be sacrificed (depending on purity).
12. Pooled fractions (~50 ml) are concentrated to a volume of ~1–2 ml by repeated rounds of dialysis in: (1) 500 ml A15 Buffer containing 50 % glycerol (8 h) and (2) 1 l A15 Buffer (8 h). The dialysis bag is tightened each time dialysis in A15-glycerol Buffer is completed. This process typically requires two rounds of dialysis with a final dialysis back into the A15-glycerol buffer.
13. Myo2p concentration is determined using the Bradford assay, with rabbit skeletal muscle myosin-II being used as the standard (extinction co-efficient = 0.54 at 280 nm for a 1 mg/ml solution). The two-step purification typically yields 0.5 mg Myo2p per 4 l of culture.
14. Myo2p is stored in the A15-glycerol buffer at −20 °C, where it remains stable for at least 6 months.

3.2 Myosin Actin-Activated ATPase Assays

1. Myo2p is diluted to a final concentration of 20–40 nM in both KG buffer alone (zero actin control) and KG buffer including different actin filament concentrations (5–80 μM). Sample size (myosin + actin) = 180 μl.
2. Reactions are started through addition of 20 μl 10× supplements for a total sample volume of 200 μl. A replicate set of samples lacking Myo2p are prepared in unison as blanks to account for any excess P_i from the buffer, basal Myo2p ATPase activity (i.e., in the absence of actin), or actin filaments.

Three 50 μl samples of the reaction are collected at 10, 20, and 30 min following addition of the 10× supplements.

3. Reactions (50 μl) are assayed and halted by subsequent addition of (a) 200 μl detection solution and (b) 50 μl stop solution. The 300 μl samples are transferred to a 96-well plate for detection following full color development 1 h later (*see* **Note 9**).
4. P_i levels are determined through quantification of the malachite green-dependent color change using the 595 nm setting on a spectrophotometer plate reader (Biorad, Hercules, CA, USA). Both basal Myo2p P_i release-associated color changes and color change associated with F-actin alone are subtracted from initial sample measurements to yield P_i release corresponding to the actomyosin ATPase alone (*see* **Note 10**). Molar amounts of phosphate released per Myo2p head per second (rates, s^{-1}) can be derived for each sample using the Phosphate Color Standard (working phosphate curve range for detection = 0–3 nmol).
5. Final Myo2p ATPase rates are determined and plotted versus F-actin concentration to generate an actin-activated ATPase curve (Fig. 1). Typically 3–5 experimental data sets are required to produce an accurate representative average curve. The apparent actin affinity (K_M) and ATPase rate (V_{MAX}) can be derived using Michaelis-Menten kinetics (Fig. 1). Curve fits are best obtained using graphing software such as KaleidaGraph (Synergy Software, Reading, PA, USA).

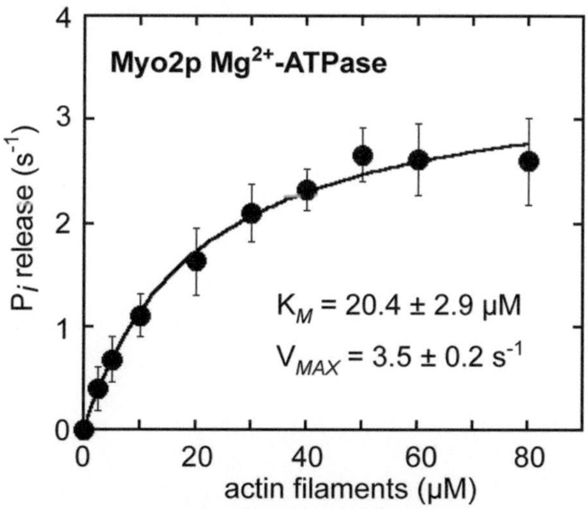

Fig. 1 Myo2p actin-activated ATPase curve. Myo2p actin-activated ATPase rates plotted versus F-actin concentration (average values from four separate experiments are shown). The plot is fit to Michaelis-Menten kinetics to derive Myo2p's apparent actin affinity (F-actin concentration yielding half-maximal activity, K_M) and ATPase rate (maximal ATPase rate, V_{MAX})

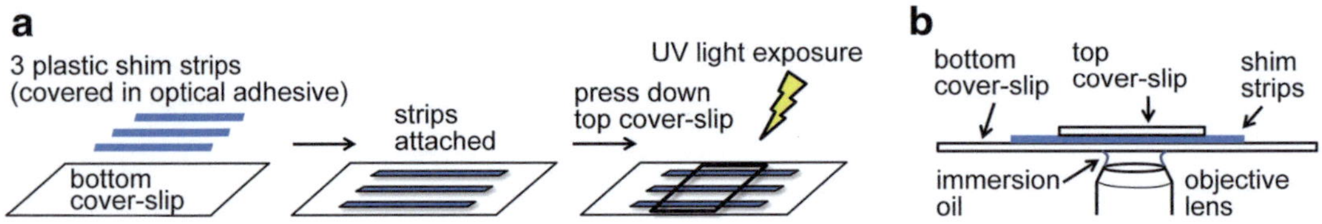

Fig. 2 Preparation and use of the motility chamber. (**a**) Summary of the individual steps used to construct a motility chamber. (**b**) Setup of the motility chamber on the microscope objective (shown here positioned on the objective for Myo2p-bead motility assays by TIRF microscopy). The chamber is inverted (positioned with the top cover slip facing the objective lens) for use in the myosin-driven actin filament motility assays

3.3 Myosin-Driven Actin Filament Motility Assays

1. Motility chambers are routinely constructed homemade using aforementioned glass cover slips (Subheading 2.3, **item 3**) (Fig. 2a). Firstly, cut the plastic shims into 3 mm × 45 mm strips and cover front and back with the optical adhesive. Place three pieces of the plastic strips on top of the 24 × 60 mm cover slip (bottom glass) to form two lanes, the width of which is usually 5–7 mm. Then, lightly press down the 22 × 22 mm cover slip (top glass) on top to the plastic strips, and make sure that there is sufficient glue to seal the contacting area between the glass cover slip and the plastic strips so that there is no leak between the two flow cells. Expose the assembled slides under UV light for 5–7 min to allow the glue to solidify (*see* **Note 11**).
2. 100 μl of 5–10 nM of Myo2p is delivered into motility chambers and adhered to the nitrocellulose-coated cover slip for 10 min.
3. The chamber is subsequently washed using filter paper and capillary action, employing 100 μl of the appropriate buffer per wash. First set of washes: 3× motility buffer plus 0.5 mg/ml BSA.
4. Three times washes with motility buffer alone.
5. Two times washes with motility buffer containing 1 μM of vortexed (30 s) unlabeled actin filaments to load non-motile, inactive myosin motors.
6. Three times washes with motility buffer plus 1 mM ATP.
7. Two times washes with motility buffer plus 25 nM rhodamine phalloidin-labeled actin filaments and oxygen scavengers (50 μg/ml catalase, 130 μg/ml glucose oxidase, and 3 mg/ml glucose) to minimize fluorescence bleaching.
8. Two times washes with motility buffer plus 20 mM DTT, 0.5 % methyl-cellulose, and oxygen scavengers.
9. Two times washes with motility buffer plus 20 mM DTT, 0.5 % methyl-cellulose, 1.5 mM ATP, and oxygen scavengers.
10. Once samples have been applied and washed, the motility chamber should be immediately transferred to the epifluorescence

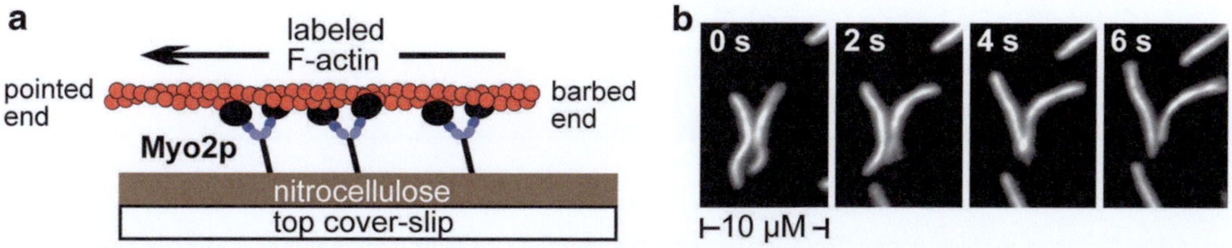

Fig. 3 Myo2p actin filament motility assay. Myo2p attached to the surface of the top glass cover slip propels rhodamine-phalloidin-labeled actin filament motility. (**a**) Illustration depicting an experiment with Myo2p attached to the nitrocellulose-coated cover slip. The orientations of the Myo2p molecules shown indicate how (ideally) these molecules will be situated to freely translocate an actin filament. The pointed and barbed ends of the actin filament are indicated for reference (given that myosin-II, like most myosins, is a barbed end-directed motor). (**b**) Time-lapse images recording Myo2p-driven actin filament motility by epifluorescence microscopy. The average speed of actin filament gliding from this experiment was 0.50 μm/s

microscope (*see* **Note 12**). We use an inverted objective where the chamber is put in place such that the top cover-slip sits on the objective lens. Note that this placement is inverse to the way in which the chamber is positioned for TIRF microscopy in Myo2p-bead motility assays, where the bottom cover slip sits on the objective (Fig. 2b).

11. Fluorescent actin filaments are observed at room temperature and their movements recorded at intervals of 1–2 s for 1 min (Fig. 3a, b). Movies are processed in Image J (http://imagej.nih.gov/ij/); plug-in MtrackJ (www.imagescience.org/meijering/software/mtrackj/) is employed to measure the speed (μm/s) of individual filaments ($n \geq 50$ filaments).

3.4 Myosin-Bead Motility Assays

1. Motility chambers are constructed as described in Subheading 3.3, **step 1** (Fig. 2a). Chamber washes are performed in a similar fashion to those described in Subheading 3.3 in that 100 μl of the indicated solutions/samples are typically used (unless indicated otherwise). Nitrocellulose-coating of the top glass cover slips is not needed in this assay.
2. 100 μl of 0.1 mg/ml NEM-myosin is delivered into motility chambers and adhered to the cover-slip surfaces for 10 min.
3. Three times washes with motility buffer plus 1 mg/ml BSA.
4. Apply 200 μl of 60 nM rhodamine-phalloidin-labeled actin filaments and incubate for 4 min. This step attaches labeled actin tracks to the adhered NEM-myosin.
5. Two times washes with motility buffer.
6. One time washes with motility buffer plus 1 mg/ml BSA and 0.15 mM ATP.
7. Flow in 20 μl of a diluted Myo2p-Qdot mixture. To achieve attachment of multiple motors per bead, Myo2p and the Qdots are mixed at a 20:1 ratio, which yields 3–5 myosin motors per

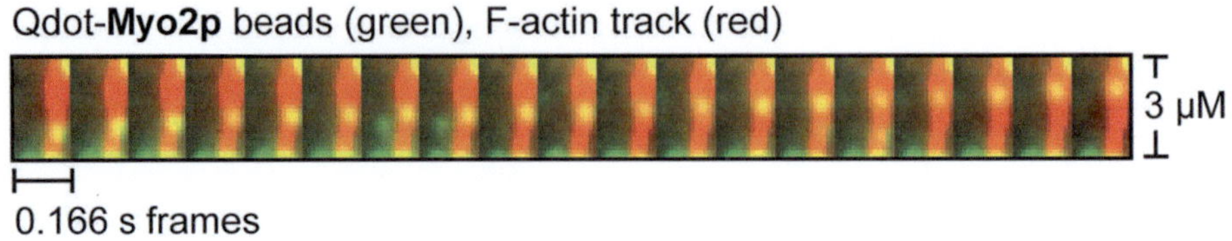

Fig. 4 Myo2p-bead motility assay. Time-lapse series of images showing movement of a Myo2p-bead (*green*) along a rhodamine phalloidin-labeled actin filament (*red*). The green Qdot (decorated with multiple Myo2p motors) moves along the actin track (from the lower portion of the image to a higher position). The experiment was performed using buffer containing 0.15 mM ATP; the bead shown is moving at a speed of 0.35 μm/s with a run length of 1.0 μm

Qdot when geometric constraints are considered [21]. Thus, 20 μl of a Myo2p stock (diluted to 0.4 μM in storage buffer) is mixed with 2 μl Qdots (0.2 μM). The mixture is stored on ice in a lighttight tube. This mixture is diluted tenfold into Motility Buffer plus 1 mg/ml BSA and 0.15 mM ATP prior to applying to the chamber (*see* **Notes 13** and **14**).

8. Once the sample has been applied, the motility chamber should be immediately transferred to the TIRF microscope (*see* **Note 12**). We use an inverted objective where the chamber is put in place such that the bottom cover slip sits on the objective lens (Fig. 2b).
9. Focus on the NEM-myosin-bound actin filaments using the rhodamine channel and capture a still image of the actin to mark the myosin tracks. Switch to the channel and use TIRF microscopy to capture Qdot movements at ~6 frames per second for 2 min (*see* Subheading 2.4, **item 6** for specifications re. laser line) (Fig. 4).
10. Movies are processed in Image J (http://imagej.nih.gov/ij/); plug-in MtrackJ (www.imagescience.org/meijering/software/mtrackj/) is employed to measure bead run lengths (μm) and speeds (μm/s). The time-lapse series shown in Fig. 4 captures movement of a Myo2p-decorated Qdot along an actin filament. ~100 runs from individual events should be sampled to derive run lengths and speeds. Average run lengths are derived from a plot of run length distributions fitted with a nonlinear regression curve $y = Ae^{(-x/\lambda)}$.

4 Notes

1. Actin stock shelf-life at 4 °C (~2 months) can be preserved by flash-freezing concentrated G-actin stocks in liquid N_2 and storing at −80 °C (for years). Avoid the need to refreeze actin stocks by preparing small, convenient-sized aliquots.

2. Inactive or 'dead' Myo2p motors can be removed from working Myo2p stocks (following dialysis into working buffer). This is achieved by a pre-clarification step using 5 μM F-actin (plus 1 mM Mg^{2+}-ATP). F-actin is sedimented using an ultracentrifuge such as the Beckman Optima MAX-E with a TLA 100 rotor (Beckman Coulter, Fullerton, CA, USA) for 1 h at 120,000 × *g* in 0.2 ml polycarbonate tubes. This will pellet most (if not all) of the 'dead' motors that assume rigor binding of F-actin (even in the weak actin-binding ATP state). Active Myo2p retained in solution is carried forward and employed in both actin filament gliding and myosin-bead motility assays. This step limits artificial reduction in motility speeds and binding of non-motile filaments.
3. Labeled rhodamine phalloidin-stabilized F-actin stocks can be stored at 4 °C and are stable for months. This avoids remaking a new stock of labeled F-actin, adding some additional consistency with experiments performed over a period of days/weeks.
4. A convenient way to produce nitrocellulose-coated cover slips involves placing one drop of high grade 1 % nitrocellulose stored in amyl acetate into ultrapure water. A Pasteur pipette is used to administer a drop of the 1 % nitrocellulose onto the surface of water housed in a 10 cm diameter container. The amyl acetate evaporates in ~1 min leaving a nitrocellulose film on the surface of the water. Cover slips are carefully placed on the film using forceps and allowed to sit for 1–2 min. A nitrocellulose-coated cover slip is then generated by holding it with forceps and pushing it down into the water where it is inverted and lifted out. This technique prevents surface action from damaging the film. Coated cover slips are allowed to air-dry and are used within a few hours of preparation.
5. Include excess biotin in fission yeast growth media when purifying Myo2p carrying the BCP tag (100-fold greater than standard media recipe). This ensures 100 % biotinylation of Myo2p tails.
6. The thiamine-repressible *nmt* promoter starts to trigger gene expression approximately 16 h following the removal of thiamine. Thus, 28 h of induction results in ~12 h of continuous Myo2p (and light chain) over-expression. Prolonged Myo2p over-expression beyond 28 h should be avoided as it becomes highly toxic leading to cell lysis and loss of yield. By 28 h, cells should appear elongated (due to cytokinesis defects associated with successful Myo2p over-expression) while largely remaining intact.
7. The efficiency of GST cleavage from the light chains should be confirmed prior to applying the sample to the gel filtration column. This can be done by simply using SDS-PAGE analysis

to compare an aliquot of the thrombin-treated sample with an untreated aliquot (i.e. a 20 μl sample removed prior to addition of thrombin). A second round of thrombin treatment may be required in some cases.

8. A small aliquot of glycerol (5 % v/v) should be added to the cleaved one-step purified sample prior to loading it on the gel filtration column. This step increases sample density and limits its diffusion when loading the top of the column. Start the column running while loading to favor immediate movement of the sample into the resin.
9. A brief spin (1 min at 800 × *g* in a microcentrifuge) is advisable for samples containing high concentrations of F-actin (which tends to precipitate in the detection solution). This clarification ensures accurate color change measurement prior to applying samples to the 96-well plate.
10. Before conducting an extensive range of ATPase reactions one should ensure the appropriate concentration of Myo2p and reactions times are being employed. This can quickly be done by performing a small number of trial experiments at low, medium, and high actin concentrations. The aim is to attain measurements that fall in the working range of your phosphate standard curve (0.2–2.0 at OD_{595}). Measurements that yield too low or too high color changes will fall outside the linear range of the standard curve leading to inaccuracies.
11. Avoid excessive glue as it will occupy chamber space when the top cover slip is pressed down.
12. Ideally microscopy should be initiated immediately following sample preparation in the motility chamber. This expeditious move limits drying of the sample before or during microscopy, which can take place within 15 min depending on the humidity of the imaging room/suite.
13. Relatively low ATP (0.15 mM) concentrations are used (cf. with 1–2 mM ATP typically employed in such assays). This adjustment facilitates capture of motility events by beads decorated with the non-processive Myo2p motors. Like most myosin-IIs, Myo2p is a low-duty-ratio motor [16], meaning it spends very little of its ATPase cycle time in the strong actin-bound ADP state. This property does not allow hand-over-hand processive motility by single dimeric Myo2p molecules (nor does it favor motility driven by multiple Myo2p molecules), in contrast to what is observed with, e.g., the high duty ratio, double-headed, processive myosin-V motors [22]. Lower ATP concentrations will introduce some degree of apo/rigor strong actin-bound Myo2p events, which provide transient tethers that help establish and propagate motility, albeit at suboptimal rates of movement. The ATP concentration

employed is ultimately determined by the type of myosin under study in such bead-based assays.

14. Myo2p is more stable when kept at high concentration. For optimal activity, the Myo2p-Qdot mixture is diluted to the appropriate concentration immediately before diffusing it into the motility chamber.

Acknowledgement

This work was supported by a National Institutes of Health RO1 grant (GM097193) to ML.

References

1. Bathe M, Chang F (2010) Cytokinesis and the contractile ring in fission yeast: towards a systems-level understanding. Trends Microbiol 18:38–45
2. Lee IJ, Coffman VC, Wu JQ (2012) Contractile-ring assembly in fission yeast cytokinesis: recent advances and new perspectives. Cytoskeleton (Hoboken) 69:751–763
3. Clark K, Langeslag M, Figdor CG, van Leeuwen FN (2007) Myosin II and mechanotransduction: a balancing act. Trends Cell Biol 17:178–186
4. Pollard TD, Wu JQ (2010) Understanding cytokinesis: lessons from fission yeast. Nat Rev Mol Cell Biol 11:149–155
5. Balasubramanian MK, McCollum D, Chang L, Wong KC, Naqvi NI, He X, Sazer S, Gould KL (1998) Isolation and characterization of new fission yeast cytokinesis mutants. Genetics 149:1265–1275
6. Kitayama C, Sugimoto A, Yamamoto M (1997) Type II myosin heavy chain encoded by the myo2 gene composes the contractile ring during cytokinesis in Schizosaccharomyces pombe. J Cell Biol 137:1309–1319
7. May KM, Watts FZ, Jones N, Hyams JS (1997) Type II myosin involved in cytokinesis in the fission yeast, Schizosaccharomyces pombe. Cell Motil Cytoskeleton 38:385–396
8. Bezanilla M, Forsburg SL, Pollard TD (1997) Identification of a second myosin-II in Schizosaccharomyces pombe: Myp2p is conditionally required for cytokinesis. Mol Biol Cell 8:2693–2705
9. Motegi F, Arai R, Mabuchi I (2001) Identification of two type V myosins in fission yeast, one of which functions in polarized cell growth and moves rapidly in the cell. Mol Biol Cell 12:1367–1380
10. Motegi F, Nakano K, Kitayama C, Yamamoto M, Mabuchi I (1997) Identification of Myo3, a second type-II myosin heavy chain in the fission yeast Schizosaccharomyces pombe. FEBS Lett 420:161–166
11. Win TZ, Gachet Y, Mulvihill DP, May KM, Hyams JS (2001) Two type V myosins with non-overlapping functions in the fission yeast Schizosaccharomyces pombe: Myo52 is concerned with growth polarity and cytokinesis, Myo51 is a component of the cytokinetic actin ring. J Cell Sci 114:69–79
12. Lord M, Pollard TD (2004) UCS protein Rng3p activates actin filament gliding by fission yeast myosin-II. J Cell Biol 167: 315–325
13. Lord M, Sladewski TE, Pollard TD (2008) Yeast UCS proteins promote actomyosin interactions and limit myosin turnover in cells. Proc Natl Acad Sci U S A 105:8014–8019
14. Sladewski TE, Previs MJ, Lord M (2009) Regulation of fission yeast myosin-II function and contractile ring dynamics by regulatory light-chain and heavy-chain phosphorylation. Mol Biol Cell 20:3941–3952
15. Stark BC, James ML, Pollard LW, Sirotkin V, Lord M (2013) UCS protein Rng3p is essential for myosin-II motor activity during cytokinesis in fission yeast. PLoS One 8, e79593
16. Stark BC, Sladewski TE, Pollard LW, Lord M (2010) Tropomyosin and myosin-II cellular levels promote actomyosin ring assembly in fission yeast. Mol Biol Cell 21:989–1000
17. Henkel RD, VandeBerg JL, Walsh RA (1988) A microassay for ATPase. Anal Biochem 169: 312–318
18. Kron SJ, Spudich JA (1986) Fluorescent actin filaments move on myosin fixed to a

glass surface. Proc Natl Acad Sci U S A 83: 6272–6276

19. Warshaw DM, Kennedy GG, Work SS, Krementsova EB, Beck S, Trybus KM (2005) Differential labeling of myosin V heads with quantum dots allows direct visualization of hand-over-hand processivity. Biophys J 88:L30–L32
20. Spudich JA, Watt S (1971) The regulation of rabbit skeletal muscle contraction. I. Biochemical studies of the interaction of the tropomyosin-troponin complex with actin and the proteolytic fragments of myosin. J Biol Chem 246:4866–4871
21. Clayton JE, Pollard LW, Sckolnick M, Bookwalter CS, Hodges AR, Trybus KM, Lord M (2014) Fission yeast tropomyosin specifies directed transport of myosin-V along actin cables. Mol Biol Cell 25:66–75
22. Mehta AD, Rock RS, Rief M, Spudich JA, Mooseker MS, Cheney RE (1999) Myosin-V is a processive actin-based motor. Nature 400: 590–593

Chapter 12

In Vitro Biochemical Characterization of Cytokinesis Actin-Binding Proteins

Dennis Zimmermann, Alisha N. Morganthaler, David R. Kovar, and Cristian Suarez

Abstract

Characterizing the biochemical and biophysical properties of purified proteins is critical to understand the underlying molecular mechanisms that facilitate complicated cellular processes such as cytokinesis. Here we outline in vitro assays to investigate the effects of cytokinesis actin-binding proteins on actin filament dynamics and organization. We describe (1) multicolor single-molecule TIRF microscopy actin assembly assays, (2) "bulk" pyrene actin assembly/disassembly assays, and (3) "bulk" sedimentation actin filament binding and bundling assays.

Key words Actin, Cytokinesis, TIRF, Microscopy, Single molecule, Micropatterning, Biomimetics

1 Introduction

Eukaryotic cells assemble diverse actin filament networks with distinct organization and dynamics to facilitate a variety of fundamental processes such as polarization, endocytosis, motility, and cytokinesis [1, 2]. Specific actin filament networks are assembled at the right time and place through the coordinated action of numerous actin-binding proteins (ABPs) with complementary properties including monomer binding, nucleating, cross-linking, and severing/disassembly [1]. Therefore, it is critical to understand how specific sets of diverse ABPs work in concert to facilitate actin filament networks for different cellular processes.

Most types of eukaryotic cells physically separate through the formation of a contractile actin filament network that is assembled, maintained, constricted, and disassembled by diverse actin binding proteins including profilin, formin, myosin, α-actinin, tropomyosin, cofilin, and capping protein [3–6]. Understanding the mechanisms by which these biochemically diverse ABPs facilitate formation of a

Alberto Sanchez-Diaz and Pilar Perez (eds.), *Yeast Cytokinesis: Methods and Protocols*, Methods in Molecular Biology, vol. 1369, DOI 10.1007/978-1-4939-3145-3_12, © Springer Science+Business Media New York 2016

contractile network depends upon our understanding of their combinatorial effects on actin filament dynamics.

Here we describe in vitro biochemical methods that can be used to measure the association of purified ABPs with actin filaments and their effects on actin filament dynamics and network organization: (1) multicolor single filament/molecule TIRF microscopy assays to directly visualize association with actin filaments and effects on actin filament dynamics and organization, (2) pyrene actin assays to measure general effects on spontaneous actin assembly, elongation and disassembly, (3) biomimetic and micropattern assays to directly observe actin filament network formation, and (4) actin filament sedimentation assays to measure actin filament affinity and actin filament cross-linking ability. Utilizing combinations of these assays will provide key mechanistic insight into cytokinesis actin filament network organization and dynamics.

2 Materials

Unless mentioned otherwise, all solutions should be prepared using cold ultrapure water (purify deionized water to attain a sensitivity of 18 MΩ cm at 25 °C) and analytical grade reagents. All reagents should be prepared fresh on ice unless otherwise indicated.

2.1 General Stock Reagents Required for Actin Assembly Assays

1. 0.1 M ATP stock: For 100 mL, dissolve 6.05 g of solid ATP (disodium-trihydrate) in 95 mL ice-cold Milli-Q water. Raise the pH to 7.4 using ~15–20 drops of 10 M NaOH. Add ice-cold water to 100 mL and prepare 0.5, 1.0 and 10 mL aliquots. Store at −20 °C.
2. 1 M DTT stock: Prepare 0.5 and 1.0 mL aliquots by dissolving 7.71 g solid DTT to a volume of 50 mL Milli-Q water. Store at −20 °C.
3. 100× Buffer G (G-actin buffer) stock: 200 mM Tris-Cl (pH 8.0), 50 mM DTT, 20 mM ATP, 100 mM Na-azide. Make 100 mL and prepare 5 mL aliquots. Store at −20 °C.
4. 1× Ca-Buffer G: Thaw frozen 100× Buffer G stock and dilute to 1× in cold Milli-Q water, and add 1 μL/10 mL of 1 M $CaCl_2$ for 0.1 mM $CaCl_2$. Store at 4 °C.
5. 10× KMEI (KCl, $MgCl_2$, EGTA, Imidazole polymerization buffer) stock: 500 mM KCl, 100 mM Imidazole (pH 7.0), 10 mM $MgCl_2$, 10 mM EGTA (pH 7.0). Prepare 10 mL and sterile-filter using a syringe-filter device (0.22 μm, Fisher). Store at room temperature.
6. 10× ME buffer (Mg exchange buffer) stock: Prepare 500 μM $MgCl_2$ and 2 mM EGTA by mixing 50 μL of 1 M $MgCl_2$ (in H_2O) and 400 μL of 0.5 M EGTA (pH 8.0, in H_2O), and

bring up to 100 mL with Milli-Q water. Sterile-filter using a syringe-filter device (0.22 μm, Fisher, Pittsburgh, PA, USA). Store at 4 °C.

7. Prepare Ca-ATP-actin from rabbit skeletal muscle acetone powder (Pel-Freez, Rogers, AR, USA) [7].

2.2 Fluorescent Labeling of Actin

1. OG (Oregon Green) label 10 mM: Dissolve 5 mg of Oregon Green 488 iodoacetamide (Invitrogen, Life Technologies Inc., Burlington, Ontario, Canada) in 907 μL of dimethylformamide (DMF).
2. TMR (Tetramethylrhodamine) label 10 mM: Dissolve 5 mg of tetramethylrhodamine-6-maleimide (Invitrogen, Life Technologies Inc.) in 964 μL DMF.
3. Pyrene label 10 mM: Dissolve 100 mg of *N*-(1-pyrene)iodoacetamide in 26 mL of DMF.

2.3 Fluorescent Labeling of SNAP-Tagged Proteins

1. pSNAP-tag® (T7)-2 *E. coli* expression plasmid (New England Biolabs, Ipswich, MA, USA).
2. SNAP-Surface™ fluorescent substrates (New England Biolabs) to label SNAP-tagged fusion proteins (*see* Table 1).

2.4 Reagents for TIRF Microscopy Assays

1. 1 M Glucose stock: Dissolve 901 mg of high-grade glucose up to 5 mL H_2O and prepare 300 μL aliquots. Store at −20 °C.
2. 2 % Methyl-cellulose (400 cP) stock: For 50 mL, dissolve 1 g methyl-cellulose in 25 mL of Milli-Q water that was preheated to 65 °C, add up to 25 mL of room temperature water, and rock/rotate overnight at 4 °C. Spin in an ultracentrifuge at 150,000 × *g* and transfer the upper 75 % of the supernatant to a fresh 50 mL tube. Store at 4 °C.
3. 2× TIRF buffer stock: 2× KMEI, 200 mM DTT, 0.4 mM ATP, 30 mM glucose, 1 % methyl-cellulose (2 %, 400 cp). Prepare

Table 1

List of SNAP-Surface substrates to label SNAP-tagged proteins

Product name	Excitation maximum (nm)	Emission maximum (nm)	Product number
SNAP-Surface® 488	506	526	S9124S
SNAP-Surface® Alexa Fluor® 546	558	574	S9132S
SNAP-Surface® 549	560	575	S9112S
SNAP-Surface® 594	606	626	S9134S
SNAP-Surface® Alexa Fluor® 647	652	670	S9136S
SNAP-Surface® 649	655	676	S9159S

10 mL and make 240 μL aliquots. Store at –20 °C. Just before use, add 10 μL of GOC mix (*see* **item 5**) and 10 μL of 5 % BSA per 240 μL 2× TIRF Buffer.

4. 1× TIRF buffer is prepared by mixing 100 μL of the above prepared 2× TIRF Buffer with 100 μL ice-cold MilliQ water.
5. 50× Glucose-oxidase/catalase (GOC) mix: Dissolve 20 mg of Catalase and 100 mg of Glucose-oxidase (Sigma) in 10 mL cold water. Clear undissolved particles by ultracentrifugation for 30 min at 100,000 ×*g* at 4 °C and prepare 50 μL aliquots from the upper 90 % of the supernatant in the cold room. Stored at –20 °C.
6. 5 % BSA stock: Dissolve 500 mg bovine serum albumin (BSA) in 10 mL Milli-Q water and prepare 0.5 mL aliquots. Store at –20 °C.
7. 5× ME (Mg-Exchange) buffer: Prepare 500 μL 5× ME buffer from a 10× ME stock with water. Sterile-filter using a syringe-filter device (0.22 μm) and store at 4 °C.
8. ABPs and associated buffers will be specific to particular interests.

2.5 Reagents for Biomimetic TIRF Microscopy

1. Microspheres: There are many different microsphere products commercially available, and as mentioned in more detail in Subheading 3.6, the best fit depends on factors such as (a) the ABP, (b) microsphere size, and (c) whether the microspheres need to be functionalized or nonfunctionalized and/or fluorescent or non-fluorescent. An overview of the microsphere products we have successfully used are listed in Table 2.

Table 2
List of a subset of fluorescent and non-fluorescent microsphere products used to specifically or nonspecifically attach actin-binding proteins

Color	Functionalization	Diameter (μm)	Excitation max.	Emission max.	Product number (Company)
Black	Carboxylate	1.0	–	–	08226 (Polysciences)
Black	Carboxylate	2.0	–	–	18327 (Polysciences)
Yellow-green	Carboxylate	1.0	505	515	F8823 (Molecular Probes)
Red	Carboxylate	1.0	580	605	F8821 (Molecular Probes)
Black	Streptavidin	2.0	–	–	24160 (Polysciences)
Black	NeutrAvidin®	1.0	–	–	F8777 (Molecular Probes)
Yellow-green	NeutrAvidin®	1.0	505	515	F8776 (Molecular Probes)
Red	NeutrAvidin®	1.0	580	605	F8775 (Molecular Probes)

2. 10× Bead buffer stock: 100 mM Hepes (pH 7.5), 1 M KCl, 10 mM $MgCl_2$, 1 mM $CaCl_2$. Prepare 5 mL and make 200 μL aliquots. Store at −20 °C.
3. 1× Bead buffers containing 0, 0.1, and 1 % BSA: Using the previously prepared 10× bead buffer stock solution, prepare a 1× solution and add 1 mM ATP as well as 0 %, 0.1 %, or 1 % BSA, respectively.

2.6 Reagents for Cover Glass Coating with PEG-Silane

1. 98.8 % Acetone.
2. 95 % Ethanol (EtOH).
3. Hellmanex III (Hellma, Southend on Sea, UK).
4. 35 % Hydrogen chloride (HCl).
5. 30 % Hydrogen peroxide (H_2O_2).
6. 96.4 % Sulfuric acid (H_2SO_4).
7. Glass Coplin jar: Slide jar with glass cover for 8 slides.
8. 24 × 40-1 mm Cover glasses.
9. 25 × 75 × 1 mm Microscope slides.
10. Acid-resistant goggles.
11. Acid-resistant coat: 100 % Tychem QC.
12. Laminate acid-resistant gloves.
13. General-purpose Oft-Tipped forceps.
14. Ceramic Tweezers (0.38 mm tip width).
15. mPEG-Silane: MW 5k (creative PEGworks, Winston Salem, NC, USA). Bring 1 L of 95 % EtOH and 0.1 % HCl to 70 °C on a hot plate while stirring. Add 1 g of mPEG-Si. Stir until dissolved. Aliquot in four 250 mL bottles and store in the dark at room temperature. Solutions can be reused up to 2 months.
16. Water bath sonicator: Branson 1510 ultrasonic cleaner (Ultrasonics Co., Danbury, Connecticut, USA) or similar.

2.7 Reagents for Micro-Patterning Assays

1. Mask creation software: AutoCAD (Autodesk, San Rafael, CA, USA).
2. Mask: Chrome mask 12.5 × 12.5 cm, containing transparent micropatterns (Applied Image Inc, Rochester, NY, USA).
3. UV machine: UVO-Cleaner 42 (Jelight, Irvine, California, USA) or similar.
4. Isopropanol: 99.9 % 2-propanol.
5. Double-sided tape: Double-coated pressure-sensitive adhesive tape with low emission values.

2.8 TIRF Microscope

1. Olympus IX-71 microscope equipped with through-the-objective TIRFM illumination.

2. Lasers: Sapphire LP 488 nm and Sapphire LP 561 nm (Coherent, Inc., Santa Clara, CA, USA) or similar.
3. Objectives: 100×/1.45 Oil, TIRFM.
4. Camera: iXon EMCCD camera (Andor Technology, Belfast, Northern Ireland) or similar.

2.9 Reagents for Sedimentation Assays

1. 1× SDS running buffer: Dissolve 30.3 g Tris, 144.1 g glycine and 10 g SDS in up to 1 L water.
2. 5× PBS: Dissolve 10 g SDS, 10 mL 2-mercaptoethanol, 100 mL glycerol and 40 mg bromophenol blue in up to 200 mL of water. Adjust the pH to 4.6.
3. Coomassie Blue stain: In a glass beaker dissolve 2 g Coomassie Brilliant Blue in 1 L MeOH, 200 mL acetic acid and add water to 2 L.
4. Gel destain: Mix 1 L MeOH with 400 mL acetic acid and water to 4 L.
5. Broad range marker: Bio-Rad, Hercules, California, USA.
6. Ultracentifuge: Optima TLX ultracentrifuge (Beckman Coulter, Brea, California, USA).
7. Table centrifuge similar to Centrifuge 5415D (Eppendorf, Hamburg, Germany).
8. Gel scanner: Odyssey Infrared Imager (LI-COR Biosciences, Lincoln, NE, USA).

2.10 Reagents for Bulk Assays with Pyrene-Actin

1. Pyrene-labeled Ca-ATP-actin: *See* Subheading 2.2, **item 3**.
2. 1× Mg-buffer G: Thaw frozen 100× stock of buffer G, dilute in appropriate volume of cold water and add 1 μL of 1 M $MgCl_2$ per 10 mL to yield a final Mg^{2+} concentration of 0.1 mM.
3. Microtiter plates: 96-Well half area, black polystyrene, flat bottom, non-binding surface assay plates (Corning, Tewksbury, MA, USA).
4. 100× Antifoam 204.
5. Fluorescence microplate reader such as Tecan Safire^2 and/or Infinite M200 Pro (Tecan Systems, Inc., San Jose, CA, USA).

3 Methods

Carry out all procedures on ice unless otherwise specified.

3.1 Fluorescent Labeling of Actin

1. Oregon green actin (OGA) labeling: Actin is labeled on Cys374 with Oregon Green iodoacetamide [8].
2. Tetramethylrhodamine (TMR) actin labeling: Actin is labeled on Cys374 with Tetramethylrhodamine-6-maleimide [9].

3. Pyrene actin labeling: Actin is labeled on Cys374 with *N*-(1-pyrenyl)iodoacetamide [10].
4. Pick the appropriate fractions of gel-filtration purified black (unlabeled) and fluorescently labeled Ca-ATP-actin (*see* **Note 1**).

3.2 Fluorescent Labeling of Actin-Binding Proteins

Before labeling a protein of interest for single-molecule TIRF microscopy imaging, it is important to consider the following factors to achieve optimal image quality:

1. For multicolor TIRF microscopy, it is important to consider how many different fluorescently labeled proteins will be imaged. Assuming actin is labeled, one (two-color TIRF) to two (three-color TIRF) other proteins can be labeled.
2. To achieve high labeling efficiency and high image quality, fluorophores with optimal photophysical properties (maximal photostability, medium to high brightness, minimal photo-blinking and narrow excitation/emission spectra) should be used to covalently label the protein of interest.
3. When working with multiple fluorophores, the excitation and emission wavelength spectra of the individual fluorophores must have as little overlap as possible. User-friendly applications are available online (Fluorescence SpectraViewer, Life Technologies) to check the excitation and emission spectra for most commercially available fluorophores along with the desired laser excitation wavelength.
4. Small-molecule fluorophores such as cyanine, AlexaFluor and SNAP-Surface dyes have enhanced photophysical properties that are superior to fluorescent proteins like GFP.
5. Even when photostable (high photon yield prior to photobleaching) fluorophores are utilized, it is important to include agents in the imaging buffer that improve the inherent problem of photobleaching. The TIRF microscopy protocol below (Subheading 3.5) employs an oxygen-scavenging system of glucose-oxidase and catalase that reduce photobleaching by depleting phototoxic free oxygen radicals.

3.3 Fluorescent Labeling of SNAP-Tagged ABPs

One possibility is to fluorescently label proteins with fluorescent dyes conjugated directly to either succinimidyl- or maleimide esters that target primary amines or exposed cysteine residues. Alternatively, commercial photostable fluorescent substrates that bind specifically to a protein fusion tag (SNAP, New England Biolabs; CLIP, New England Biolabs; Halo, Promega) are now widely used.

1. The 189 amino acid SNAP-tag (approx. 20 kDa) must be fused to either the N- or C-terminus of the protein of interest by standard cloning methods (along with any desired purification

tags such as MBP, 6×HIS, FLAG, GST). Typically, expression and purification of SNAP-tagged proteins are carried out the same way as for untagged versions. All SNAP-tagged proteins must be stored in buffers with 1 mM DTT, but without EDTA, to increase labeling efficiency (*see* **Note 2**).

2. Suspend the lyophilized fluorescent probe with the appropriate volume of DMSO to yield a 3 mM stock that can be stored at −20 °C.
3. Mix 30 μM fluorescent SNAP substrate with 5–10 μM SNAP-tagged protein, and incubate overnight at 4 °C on a rocking device under light-protected conditions (see example reaction outlined in Table 3).
4. The following day, separate unreacted fluorescent SNAP substrate from labeled protein by dialysis in the dark at 4 °C with three to four exchanges of 500 mL buffer (containing 1 mM DTT without EDTA) over a period of 6–8 h.
5. Because the SNAP-tag absorbs at 280 nm, protein concentration should be determined with a SDS-PAGE (sodium dodecyl sulfate-polyacrylamide gel electrophoresis) gel using an untagged version of the same protein (determined photometrically at 280 nm) as a standard. The fluorescence labeling efficiency can be determined by conventional photometric measurements at the appropriate wavelength (depends on the fluorophore).

3.4 Following ActinDynamics by Single-Color TIRF Microscopy

1. Prepare a stock mixture of fluorescently labeled actin at six times the concentration in the final reaction. Use 10–30 % fluorescently labeled green (Oregon-green, OG) or red (tetramethylrhodamine, TMR) actin. Example: for experiments with a final concentration of 1.5 μM of 15 % OG-actin, a mixture containing a total of 9 μM actin (1.35 μM OG-actin and 7.65 μM unlabeled (black) actin) in Ca-Buffer G is prepared (*see* **Note 3**).

Table 3
Reaction setup for the fluorescent labeling of SNAP-tagged proteins with SNAP-Surface substrate

Reagent	Final conc. (μM)	Final volume (μL)
40 μM SNAP-tagged fusion protein	10	12.5
3 mM Fluorescent SNAP substrate	30	0.5
1× Protein buffer	–	37
Total	–	50

Table 4
Reaction setup for preparing a mix of 9 μM actin containing 15 % Oregon-green actin

Reagent	Final conc. (μM)	Final volume (μL)
20 μM OG-actin [50 % labeled]	2.7	6.8
40 μM Black actin	6.3	7.9
Ca-Buffer G	–	35.3
Total	9	50

Table 5
Reaction setup for tubes A and B for following actin dynamics by single-color TIRF microscopy

Tube	6× Actin stock (μL)	5× ME (μL)	2× TIRF buffer (μL)	H_2O (μL)
A	5	1	–	–
B	–	–	15	9

Example for preparing a 9 μM OG (15 %)-labeled Ca-ATP-actin stock is outlined in Table 4:

2. 30 μL reaction setup: Prepare three reaction tubes labeled A, B, and C as outlined in Table 5. To start the Mg-exchange of Ca-ATP actin, make tube A mixture and allow to incubate for 2 min. Immediately prepare tube B mixture. Initiate the assembly reaction in tube C by thoroughly mixing 4 μL from tube A with 16 μL from tube B by pipetting gently up and down four times. Immediately add the entire reaction from tube C into a flow cell and begin imaging.
3. TIRF microscopy imaging (Fig. 1a):
   Typical settings for TIRF microscopy imaging of actin filaments are listed in Table 6 (*see* **Note 4**):

3.5 Multi-Color TIRF Microscopy with Fluorescently Labeled ABPs

1. Label the protein(s) of interest with an optimal fluorophore, such as the SNAP-tag labeling method described in Subheading 3.3 (*see* **Note 5**).
2. Prepare stock solutions of your protein(s)-of-interest by diluting in its storage buffer to concentrations that allow pipetting of volumes larger than 1 μL to the final reaction.
3. To set up the reaction follow the general steps outlined in Subheading 3.4 (Table 5), except that fluorescently labeled protein(s) of interest are included in tube B.

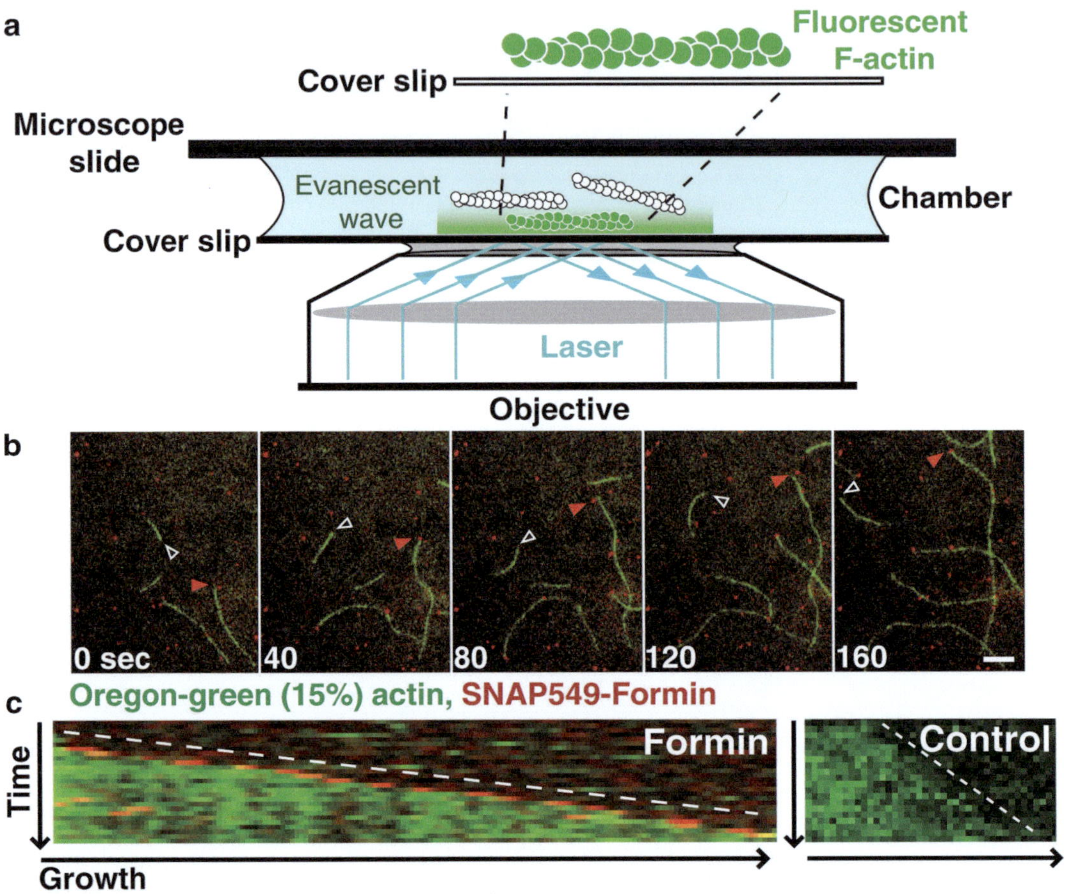

Fig. 1 Multicolor total internal reflection fluorescence (TIRF) microscopy. (**a**) Schematic of through-the-objective TIRF microscopy imaging of single fluorescently labeled actin filaments and/or molecules. At its critical angle incoming light is completely reflected. A portion of the incident light is thereby converted to an electromagnetic field forming an exponentially decaying evanescent wave at the probe surface. Only fluorophores close to the surface (i.e., 50–200 nm) are excited, which enables imaging with high signal-to-noise ratio [17]. (**b**) Time-lapse images from a dual-color TIRF microscopy experiment showing the continuous growth of control actin filaments (*white arrowheads*) and fluorescently labeled formin-associated actin filaments (*red arrowheads*). 1.5 μM Mg-ATP actin (15 % Oregon-green) was assembled with 1 nM SNAP549-labeled formin mDia2 and 2.5 μM profilin. Scale bar = 5 μm. (**c**) Kymographs reveal the change in length over time for rapidly elongating formin-associated filaments (*left*) or slower growing control filaments (*right*). *White dashed lines* indicate the rate of filament growth

Table 6
Settings for imaging individual fluorescently labeled actin filaments

Category	Setting
Exposure time [ms]	50–100
Cycle time [s]	0.1–5
Laser power [mW]	5–50

Table 7
Reaction setup for tubes A and B for following the dynamics of profilin-actin along with fluorescently labeled formin by dual-color TIRF microscopy

Tube	9 μM OG-Actin (μL)	5× ME (μL)	2× TIRF buffer (μL)	50 μM Profilin (μL)	100 nM Formin (μL)	H_2O (μL)
A	5	1	–	–	–	–
B	–	–	15	1.5	3	4.5

Table 7 provides an example for how to prepare a spontaneous assembly reaction of 1.5 μM OG-labeled actin in the presence of 2.5 μM profilin and 10 nM fluorescently labeled SNAP-tagged formin. Because the contents in tube B are diluted 1.25-fold upon mixing with tube A, 1.25-fold the desired final concentration should be added to tube B. An example of a dual-color experiment can be found in Fig. 1b along with the respective kymograph in Fig. 1c.

4. Before imaging, make sure that the microscope and laser settings are adjusted optimally. To determine optimal (a) TIRF-angle and intensity for each laser, (b) exposure time (i.e., little photo-bleaching, but good signal-to-noise ratio), and (c) frame rate (i.e., resolve the biochemical process of interest with as little photo-bleaching as possible), a control reaction containing fluorescently labeled actin and ABPs should be tested and the settings adjusted accordingly until the optimal conditions for (a)–(c) are met.

3.6 Biomimetic Assays

In biomimetic assays functionalized polystyrene microspheres are typically coated with actin assembly factors (e.g., formins) that facilitate F-actin network assembly. All proteins cannot be attached to microspheres in the same way. There are multiple ways to specifically or nonspecifically attach proteins to microspheres. Optimization of the protocol is key to success. Because of space limitations we focus on one method we have successfully established for the cytokinesis actin assembly factor formin on functionalized (carboxylated) microspheres. Consider the following before coating microspheres with your protein of interest:

1. Microsphere size. We have found microspheres of 1–2 μm in diameter work well because sufficient amounts of protein can be attached to the microspheres, and because non-fluorescent microspheres are large enough to be easily detected by bright-field illumination with a 100× TIRF objective.
2. There are diverse microspheres with different functionalized groups or proteins (e.g., amino, sulfate, antibody, streptavidin/NeutrAvidin, proteinA/B) available for specific protein attachment. Here we describe a method that uses microspheres

with carboxyl-groups through which proteins such as formins can be covalently bound to their surface.

3. Most vendors offer fluorescent versions of microspheres (Subheading 2.5, Table 2), which may be useful for identification. However, fluorescent microspheres are very bright, so the detection of single molecules on or close to fluorescent microspheres is difficult.

3.7 Coating 2 μm Carboxylated Microspheres with Formin

1. Dispense 10 μL of microspheres (Polysciences Inc., Warrington, PA, USA) into a reaction tube, centrifuge for 1 min at full-speed in a benchtop centrifuge, and aspirate the supernatant. Wash the microspheres: resuspend in 500 μL Milli-Q water, centrifuge as before, and aspirate the supernatant. Keep the pelleted microspheres on ice.
2. By gently flicking the tube, resuspend the microspheres in 30 μL of 2 μM protein (formin) solution in 1× bead buffer (0 % BSA).
3. Mount the tube(s) on a vortex-shaking device and incubate shaking (position 2.5) for 60 min at 4 °C.
4. Spin the reaction in a benchtop centrifuge for 1 min at full-speed at 4 °C. Aspirate the supernatant and wash the pellet three times by resuspending in 100 μL 1× bead buffer (with 1 % BSA) by gently flicking the tube and centrifuging as before. Resuspend the pellet in 30 μL 1× bead buffer (with 0.1 % BSA).
5. Although it is best to coat microspheres the day of the experiment, it is possible to store coated microspheres on ice for up to 2 days without experiencing significant loss in activity.

3.8 Biomimetic Assays to Follow F-ActinDynamics by Multicolor TIRF Microscopy

For TIRF microscopy imaging we typically work with final concentrations of 1.5 μM OG-labeled (10–15 %) actin and 2.5 μM profilin. For some experiments it is desirable to use flow cells made with untreated glass cover slips and objective slides because microspheres will adhere more strongly. However, nonspecific binding of proteins to the glass surface may make the use of ultra-clean and treated (hydrophobic) glasses necessary (*see* Subheading 3.15) (*see* **Note 6**).

1. Dilute 1–5 μL of the coated microspheres stock into 10 μL total 1× Bead Buffer (containing 0.1 % BSA), and perfuse into a flow cell. Incubate for 5 min. We recommend starting with a 1:1 dilution, and dilute further if the flow cell is too crowded with surface-adhered microspheres.
2. Rinse the flow cell twice with 10 μL of 1× bead buffer (with 1 % BSA) and incubate for 4 min.
3. During this incubation step, prepare the reaction containing magnesium-exchanged fluorescently labeled actin, oxygen

Table 8
Reaction setup for tubes A and B for following individual actin filaments nucleated and elongated off of formin-coated microspheres

Tube	9 μM OG-Actin mix (μL)	5× ME (μL)	2× TIRF buffer (μL)	50 μM Profilin (μL)	H_2O (μL)
A	5	1	–	–	–
B	–	–	15	1.5	7.5

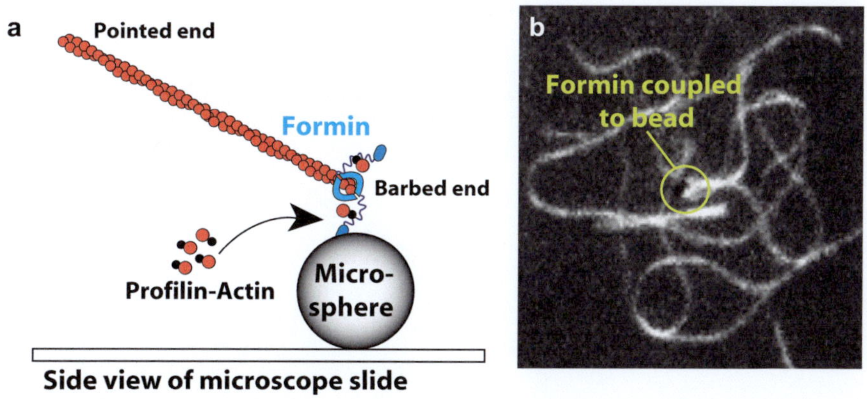

Fig. 2 Biomimetic F-actin network formation. (**a**) Schematic depicting actin filament assembly from formins adsorbed to microspheres. (**b**) TIRF microscope image of a microsphere (2 μm diameter) coated with formin molecules that nucleate and processively elongate actin filaments from 1.5 μM Mg-ATP actin (10 % Oregon-green) and 2.5 μM profilin

scavenging-system and any additional ABPs to be added (e.g., profilin). The example in Table 8 describes how tubes A and B are set up for reactions containing 1.5 μM OG-labeled (10–15 %) actin along with 2.5 μM profilin. Because the contents in tube B are diluted 1.25-fold upon mixing with tube A, 1.25-fold the desired final concentration should be added to tube B.

4. Rinse the flow cell twice with 10 μL 1× TIRF buffer (2× TIRF buffer diluted 1:1 with Milli-Q water).
5. Initiate the assembly reaction in tube C by thoroughly mixing 4 μL from tube A with 16 μL from tube B by pipetting gently up and down four times. Immediately perfuse the entire reaction from tube C into a flow cell and begin imaging (*see* Subheading 3.4). An example of a one-color TIRF experiment using formin-coated microspheres is provided in Fig. 2.

3.9 Preparation of Pyrene-Actin Assembly and Disassembly Assays

To follow the kinetics of actin assembly (Fig. 3a, b) and disassembly (Fig. 3c), pyrene-fluorescence bulk biochemistry assays are employed where the increase or decrease in signal is used to deduce the rate of polymerization (assembly) or depolymerization (disassembly). Actin is labeled at Cys374 with the thiol-specific dye

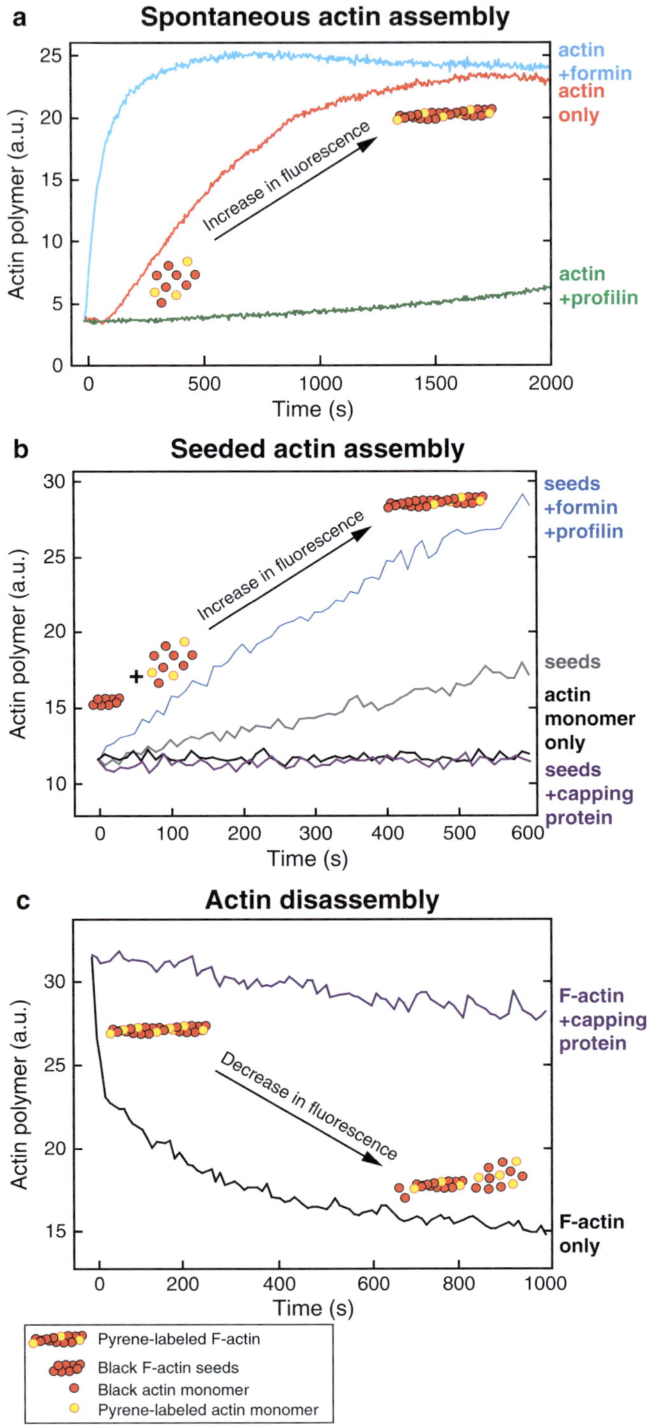

Fig. 3 Bulk pyrene actin assembly and disassembly assays. (**a**) Spontaneous actin monomer assembly: Time-course of the polymerization of 2.5 μM pyrene (15 %)-labeled Mg-ATP actin monomers. Pyrene fluorescence (F-actin) over time for actin alone (*red*), or in the presence of either formin (*blue*) or profilin (*green*). Formin nucleates actin filament assembly, whereas profilin inhibits nucleation.

N-(1-pyrenyl)iodoacetamide (pyrene). The pyrene fluorescence actin assay has proven to be very useful because the signal is proportional only to polymer weight concentration [10, 11].

1. Dialyze 1–2 mL of unlabeled (black) and pyrene-labeled Ca-ATP-actin overnight for at least 18 h at 4 °C in 250 mL of fresh 1× Ca-buffer G (Subheading 2.1). Dialysis should take place under light-protected conditions as pyrene is light-sensitive. Dialyze black actin weekly, and pyrene-labeled actin every 2 weeks.
2. Following dialysis, use an ultracentrifuge to spin both black and pyrene-labeled Ca-ATP-actin at 175,000 ×*g* for 60 min at 4 °C. Following ultracentrifugation, carefully transfer the top 80 % of the actins into fresh tubes and keep on ice until further usage.
   Determine the protein concentrations for both black and pyrene-labeled Ca-ATP-actin by using either a nanodrop or a conventional spectrophotometer (here dilute the actin solution 1:10 in 1× Ca-buffer G and use a quartz cuvette for the measurement). While unlabeled (black) actin is measured at 290 nm, measurements of the pyrene-labeled actin fraction are carried out at 344 nm. For calculating the concentrations of black as well as pyrene-labeled Ca-ATP-actin, refer to the following equations (*see* **Note 7**):
   Black Ca-ATP-actin: c (μM) = 38.5 × ($A_{290\text{ nm}}$ × dilution factor)
   Pyrene-labeled Ca-ATP-actin:
   - Pyrene-labeled actin fraction: c (μM) = 45 × ($A_{344\text{ nm}}$ × dilution factor).
   - Total Ca-ATP-actin:
     c (μM) = 38.5 × [($A_{290\text{ nm}}$ × dilution factor) − ($A_{344\text{ nm}}$ × dilution factor × 0.1277)].
3. For any other proteins that will be used in the experiment, carry out centrifugation for 30 min at 4 °C in a benchtop centrifuge to remove precipitate. Following centrifugation, transfer the top 90 % into a fresh pre-chilled reaction tube and

Fig. 3 (continued) (**b**) Seeded actin assembly: Time course of the addition of 0.5 μM pyrene (15 %)-labeled Mg-ATP actin monomers to 0.5 μM unlabeled F-actin seeds. Pre-assembled F-actin seeds (*grey*) eliminate the long nucleation lag phase (*black*) at low actin monomer concentrations. Capping protein blocks addition of monomers to F-actin seeds (*purple*), whereas formin increases the actin monomer addition rate in the presence of profilin (*blue*). (**c**) F-actin disassembly: Time course of the barbed end depolymerization of 5.0 μM pyrene (50 %)-labeled Mg-ATP preassembled F-actin upon dilution below the critical concentration for assembly (0.1 μM). Capping protein blocks barbed end depolymerization (*purple*)

store proteins on ice until further usage. Following centrifugation always determine the protein concentration photometrically at 280 nm using the respective extinction coefficient.

4. When planning the setup of reactions to be carried out, include a control reaction containing only pyrene-labeled Mg-ATP-actin without having any accessory ABP present (at best run as duplicate).
5. Prepare a mixture of pyrene-labeled Ca-ATP-actin at six times the final total actin concentration. Typically one uses an actin mixture that contains 10–20 % pyrene labeled actin. For example, for experiments where pyrene-labeled (20 %) actin at a final concentration of 2.5 μM is used, a mixture containing a total of 3 μM pyrene-labeled actin and 12 μM unlabeled (black) actin in 1× Ca-buffer G would be required. For the assay setup laid out here, calculate 25 μL of the actin stock per microtiter plate well (corresponds to one reaction), plus two extra reactions (i.e., 50 μL). Prepare the actin stock solution fresh and keep on ice until further usage (Table 9) (*see* **Note 3**).

   Table 9 shows how to prepare 350 μL (sufficient for 12 reactions) of 15 μM pyrene (10 %)-labeled Ca-ATP-actin mix:
6. Make an appropriate stock solution of additional proteins that will be included in the assay (e.g., formin and profilin) using their respective storage buffers.
7. Make sure that the plate reader is set up and ready to go before setting up the reactions. While some settings (e.g., gain, read interval) are specific to the device used, others (Table 10) work well for all pyrene-actin with final actin concentrations that range from 1.5 to 3 μM (containing 5–20 % pyrene-actin).

3.10 Spontaneous Pyrene-Actin Assembly Assays

For the actual reaction setup, carry out steps (1–7) in the order listed below. This setup is laid out for a 96-well half-area flat bottom (non-transparent black) microtiter plate with a total reaction volume of 150 μL/well, and requires two rows of the microtiter plate to prepare the reactions. A general layout of the components

Table 9
Reaction setup for preparing a mix of 15 μM actin containing 10 % pyrene-labeled actin

Reagent	Final 6× conc. (μM)	Final volume (μL)
Pyrene-actin [35 μM, 90 % labeled]	1.7	17
Black actin [50 μM]	13.3	93.1
1× Ca-buffer G	–	239.9
Total	15	350

Table 10
List of general settings used for measuring assembly and disassembly of pyrene-actin

Category	Setting
Read setting	Kinetic
Read mode	Fluorescence
Wavelengths	Excitation = 365 nm; Emission = 407 nm
Timing	Run time = 1–3 h; interval = 10–20 s
Automixing	No
Autocalibrating	Yes
Type of microtiter plate	96 half-area flat bottom (non-transparent)
Wells to read	User dependent

Table 11
A general layout of the components and their respective volumes added to either the upper or lower row of the microtiter plate used for spontaneous pyrene-actin assembly assays

Plate row	Total vol. per well (μL)	6× Pyrene-actin mix (μL)	10× ME-buffer (μL)	100× Anti-foam (μL)	Protein (μL)	Protein buffer[a] (μL)	10× KMEI (μL)	1× Mg-buffer G (μL)
Upper	29	25	2.5	1.5	–	–	–	–
Lower	145	–	–	–	x	y	18	$145-(18+x+y)$

[a]Protein buffer is added only for measurements that are carried out at different concentrations. The appropriate amount of 1× protein buffer is calculated by subtracting the amount of a protein added at that respective concentration from the amount of protein added in the condition with the highest protein concentration.

and their respective volumes added to either the upper or lower row is given in Table 11.

1. Calculate the volumes of accessory proteins to be included in the 150 μL reaction (e.g., profilin and formins). With this information the amount of 1× Mg-buffer G can be calculated, as the amount of Mg-buffer G depends on the volumes of accessory proteins. Be sure to calculate for 1.2-fold the indicated protein amount (i.e., for a total reaction of 180 μL/well).
2. Depending on the number of reactions, mix the appropriate amount of 10× KMEI (18 μL/well) with the appropriate amount of 1× Mg-buffer G (*see* Table 11) and add the appropriate amount of the KMEI-Mg-buffer G mix to each well of the lower row.

3. Add the appropriate amounts of 1× protein buffer (if required) and protein to each well of the lower row.
4. Depending on the number of reactions, mix the appropriate amount of the 6× pyrene-actin mix (25 μL/well) with the appropriate amount of 10× ME-buffer (2.5 μL/well) and let incubate for exactly 2 min for magnesium-exchange.
5. 20 s before the 2-min incubation is over, apply 27.5 μL of the 6× Mg-ATP pyrene-actin mix to each well of the upper row. Subsequently, add 1.5 μL of 100× anti-foam per well, yielding a total volume of 29 μL/well.
6. Immediately transfer the plate to the plate reader. Using a multichannel pipette, transfer 121 μL from each well of the lower row to the respective well of the upper row. This starts the reaction and marks time point zero (*see* **Note 8**).
7. After saving the data, the fluorescence intensities are plotted over time (Fig. 3a).

3.11 Preparation of Pyrene-Actin Seeds

The spontaneous pyrene-actin assembly assays described above (Fig. 3a) cannot distinguish between filament elongation and filament nucleation, because both events are reflected by an increase in fluorescence intensity. However, actin filament elongation can be exclusively analyzed by seeded pyrene-actin assembly assays (Fig. 3b). Preassembled unlabeled F-actin seeds are prepared that bypass nucleation, thereby providing barbed ends to which pyrene-labeled actin monomers can immediately add. Based on the pyrene-actin spontaneous monomer assembly assay, we describe here the key steps for seeded pyrene-actin assembly assays.

1. For picking gel-filtered black and pyrene-labeled actin, actin dialysis, determination of actin (as well as any accessory protein) concentrations, refer to **steps 1** through **5** of Subheading 3.9.
2. When planning the setup of all the reactions to be carried out, make sure to include the following control reactions: (1) pyrene-labeled monomeric actin in the absence of F-actin seeds in both the absence and presence of accessory ABPs, and (2) F-actin seeds with pyrene-labeled monomeric actin in the absence of accessory ABPs.
3. Prepare a solution containing a final concentration of 5 μM black F-actin seeds (Table 12). 15 μL of seeds will be used per reaction, but always prepare sufficient amount for an extra two reactions. Once the entire reaction of Mg-ATP-actin is gently but thoroughly mixed, deliver 15 μL/well into the upper row of a half-area flat bottom microtiter plate, seal the plate with Parafilm and let incubate for 1 h at room temperature (*see* **Note 9**).

Example for preparation of 210 μL (14 reactions) of 5 μM unlabeled (black) F-actin seeds.

Table 12
Reaction setup for preparing 5 μM unlabeled (black) F-actin seeds, used for seeded pyrene-actin assembly assays

Reagent	Final 10× conc.	Final volume (μL)
Black actin [50 μM]	5 μM	21
10× ME-buffer	1×	21
1× Ca-buffer G	–	147
2-min incubation		
10× KMEI	1×	21
Total	5 μM	210

Table 13
A general layout of the components and their respective volumes added to either the upper or lower row of the microtiter plate used for seeded pyrene-actin assembly assays

Plate row	Total vol. per well (μL)	10× F-actin (μL)	10× Pyrene-actin (μL)	1× Ca-buffer G (μL)	Protein (μL)	Protein buffer[a] (μL)	100× Anti-foam (μL)	10× KMEI (μL)	1× Mg-buffer G (μL)
Upper	60	15	–	–	x	y	1.5	15	$60-(15+15+1.5+x+y)$
Lower	108	–	18	90	–	–	–	–	–

[a]Protein buffer is added only for measurements that are carried out at different concentrations. The appropriate amount of 1× protein buffer is calculated by subtracting the amount of a protein added at that respective concentration from the amount of protein added in the condition with the highest protein concentration.

3.12 Seeded Pyrene-Actin Assembly Assay

1. Ten minutes before the 1-h incubation for generating the seeds is over, start to prepare for setting up the actual reaction containing a final concentration of 0.5 μM pyrene-labeled (5–20 %) Mg-ATP-actin monomers along with any accessory protein-of-interest. Carry out **steps 1** through **9** in the order as they are listed below. A general layout of the components and their respective volumes added to the upper and lower rows is given in Table 13.
2. Prepare an appropriate amount (18 μL/well) of a 5 μM stock solution of pyrene-labeled (5–20 %) Mg-ATPactin (Table 14). First, mix thoroughly the required amounts of 10× ME-Buffer and 1× Ca-buffer G and keep at room temperature. Approximately 2 min before the measurements are carried out, thoroughly but gently mix all components (i.e., Ca-Buffer G-ME-Buffer mix, and pyrene-labeled actin) and keep on ice under light-protected conditions (*see* **Note 3**). Table 14 shows an example of how to prepare 250 μL (sufficient for 14 reactions) of 5 μM pyrene (10 %)-labeled Mg-ATP-actin mix:

Table 14
Setup for preparing 5 μM pyrene (10 %)-labeled actin monomer, used for seeded pyrene-actin assembly assays

Reagent	Final 10× conc.	Final volume (μL)
10× ME buffer	1×	25
1× Ca-buffer G	–	198.8
Pyrene-actin [35 μM, 90 % labeled]	0.56 μM	4
Black actin [50 μM]	4.44 μM	22.2
Total	5 μM	250

3. Depending on the total number of reactions carried out, prepare an appropriate amount of master mix containing 18 μL per reaction of pyrene-labeled 5 μM Mg-ATP-actin and 90 μL per reaction of 1× Ca-buffer G. Subsequently, deliver 108 μL of the solution to each well of the lower row on the microtiter plate.
4. Depending on the number of reactions, mix the appropriate amount of 10× KMEI (15 μL/well) with the appropriate amount of 1× Mg-buffer G (Table 13) and carefully add the appropriate per-well amount of the KMEI-Mg-buffer G mix to each well of the upper row, followed by the addition of 1.5 μL 100× anti-foam.
5. Add the appropriate per-well amounts of 1× protein buffer (if required) and protein to each well of the upper row.
6. Mix components of upper wells by carefully pipetting up and down twice using a multichannel pipette.
7. Immediately transfer the plate to the plate reader. By using a multichannel pipette transfer 90 μL from each well of the lower row to the respective well of the upper row and mix gently by pipetting up and down three times. As this starts the reaction and marks time point zero, the actual measurement of the plate should be started immediately thereafter (*see* **Note 8**).
8. Due to the fact that in seeded assembly assays there is no lag phase (i.e., phase where nucleation rather than elongation takes place), reactions are relatively fast, so that depending on the applied ABP typically measurements are carried out for only 10–20 min (and often only the first few minutes are analyzed).
9. After saving the data, the fluorescence intensities are plotted over time yielding graphs where depending on the ABPs added, the slopes may be smaller (i.e., poor elongation) or larger (i.e., enhanced elongation) with respect to the actin only control (Fig. 3b).

3.13 Pyrene-Actin Depolymerization Assay Preparation

In addition to measuring actin polymerization, Pyrene-labeled actin assays can also be used to follow depolymerization (Fig. 3c). In general, actin undergoes polymerization only when the concentration of actin monomer is higher than the critical concentration (c_c). The c_c at the filament pointed end (~0.6 μM) is higher than c_c at the barbed end (~0.1 μM) [12, 13]. At steady-state F-actin depolymerization happens primarily at the pointed end, while new actin monomers are added to the barbed end. Due to the slow off-rate of actin subunits from the pointed end and the fast on-rate of subunits at free barbed ends, actin turnover in vitro is relatively slow [14, 15]. However, in vivo actin turnover is up to 100-times faster [16], which is attributed to ABPs. The depolymerization assay described below starts with 5 μM of pre-formed pyrene-labeled F-actin that is diluted to 0.1 μM (below the c_c) immediately before the measurement. Thereby, ABP-mediated effects on F-actin disassembly can be assessed quantitatively.

1. For picking gel-filtered black and pyrene-labeled actin, actin dialysis, determination of actin (as well as any accessory protein) concentrations, refer to **steps 1** through **5** of the above-described spontaneous pyrene-actin assembly assay (Subheading 3.9).
2. When planning the setup of all the reactions to be carried out, make sure to always include a reaction containing only pyrene-labeled F-actin without having any accessory ABP present.
3. Prepare a solution containing a final concentration of 5 μM pyrene-labeled (50 %) F-actin seeds as outlined in Table 15. Per reaction (i.e., well) 3 μL of F-actin will be used, however always prepare sufficient amount for an extra two reactions. Once the entire reaction of Mg-ATP-actin is gently but thoroughly mixed, deliver 3 μL/well into the upper row of a half-area flat

Table 15
Reaction setup for preparing 5 μM pyrene (50 %)-labeled F-actin seeds, used for pyrene-actin disassembly assays

Reagent	Final 10× conc.	Final volume (μL)
Black actin [50 μM]	2.2 μM	2
Pyrene-actin [35 μM, 90 % labeled]	2.8 μM	3.6
10× ME-buffer	1×	4.5
1× Ca-buffer G	–	30.4
2-min incubation		
10× KMEI	1×	4.5
Total	5 μM	45

bottom microtiter plate, seal the plate with Parafilm and incubate for 2 h at room temperature (*see* **Note 3**). Table 15 shows an example of how to prepare 45 μL (14 reactions) of 5 μM pyrene-labeled (50 %) F-actin seeds (*see* **Note 9**).

3.14 Pyrene-Actin Depolymerization Assay

1. Ten minutes before the 2-h reaction period for generating the pyrene-labeled F-actin is complete, begin setting up the reactions that will ultimately contain a final concentration of 0.1 μM pyrene-labeled (50 %) F-actin along with the accessory proteins of interest. Carry out **steps 2** through 7 in the order as they are listed below. A general layout of the components and their respective volumes added to either the upper or lower row is given below. If running all 12 reactions (entire row of wells), it is important to make sure that the fluorescence plate-reader is capable of reading 12 wells at a time. Also, use a new pipette tip for each component.
2. Calculate the exact volumes of accessory ABPs to be added to the 150 μL reaction (e.g., profilin and formins). With this information, the actual amount of 1× Mg-Buffer G can be calculated, as the amount of applied Mg-buffer G depends on the volumes of ABPs that included to the reaction. The way the reactions are setup, always calculate for 1.2 times the indicated protein amount (i.e., for a total reaction of 180 μL/well).
3. Add 1.5 μL 100× anti-foam along with 3 μL of 1× Mg-buffer G per well to the F-actin in the upper row.
4. Depending on the number of reactions, mix the appropriate amount of 10× KMEI (18 μL/well) with the appropriate amount of 1× Mg-buffer G (Table 16), and subsequently add the appropriate per-well amount of the KMEI-Mg-buffer G mix to each well of the lower row.
5. Add the appropriate amounts of 1× protein buffer (if required) and protein to each well of the lower row.

Table 16
A general layout of the components and their respective volumes added to either the upper or lower row of the microtiter plate used for pyrene-actin disassembly assays

Plate row	Total vol. per well (μL)	50× F-actin (μL)	Protein (μL)	Protein buffer[a] (μL)	100× Anti-foam (μL)	10× KMEI (μL)	1× Mg-buffer G (μL)
Upper	7.5	3	–	–	1.5	–	3
Lower	171	–	x	y	–	18	$171-(18+x+y)$

[a]Protein buffer is added only for measurements that are carried out at different concentrations. The appropriate amount of 1× protein buffer is calculated by subtracting the amount of a protein added at that respective concentration from the amount of protein added in the condition with the highest protein concentration.

6. Mix components of the lower row wells by carefully pipetting up and down twice using a multi-channel pipette.
7. Immediately transfer the plate to the plate reader. By using a multichannel pipette transfer 142.5 μL from each well of the lower row to the respective well of the upper row and mix gently by pipetting up and down three times. As this starts the reaction and marks time point zero, the actual measurement of the plate should be started immediately thereafter (Fig. 3c) (*see* **Note 8**).

3.15 PEG-Silane Coating of Piranha-Cleaned Cover Glasses

1. Place slides and cover glasses in a metal rack submerged in a 2 L acetone-filled glass beaker. Stir for 30 min in orbital shaker.
2. Transfer the rack in a 2 L ethanol-filled beaker. Stir for 15 min in orbital shaker.
3. Wash the rack in a 2 L MilliQ water-filled beaker. Stir for 5 min in orbital shaker.
4. Put the rack in a plastic container filled with Hellmanex III 2 % in a bath sonicator filled with water. Sonicate for 2 h at room temperature.
5. Wash the rack in a 2 L MilliQ water-filled beaker. Stir for 5 min in orbital shaker.
6. Put cover glasses and slides in a glass Coplin jar for piranha acid treatment (*see* **Note 10**).
7. Add first 50 mL sulfuric acid (96.4 % H_2SO_4) in the glass Coplin jar.
8. Add 25 mL peroxide (30 % H_2O_2) to fully immerse the cover glasses.
9. Stir 30 s with glass stir bar.
10. Leave the slides in piranha acid for 1 h 30 min.
11. Using acid resistant tweezers, put the slides in a dry glass Coplin jar. Let the piranha solution finish reacting overnight in the fume hood. Do not store or cover with an airtight lid as pressure will build up and an explosion can occur. Dispose piranha acid appropriately.
12. Add 75 mL of MilliQ water, and then empty the solution. Repeat ten times.
13. Add 75 mL of ethanol, and then empty the solution. Repeat five times.
14. Transfer slides in a dry glass Coplin jar.
15. Incubate cover glass with 1 mg/mL PEG-Si solution on orbital shaker for 18 h at RT. Cover glass Coplin jar to protect from light.
16. Pour the PEG-Si in glass bottle. The solution can be stored at RT in the dark and reused during up to 2 months.

17. Add 75 mL of ethanol, and then empty the solution. Repeat five times.
18. Add 75 mL of MilliQ water, and then empty the solution. Repeat ten times.
19. Dry PEG-Si coated slide with a slow air flow.
20. Store glass at 4 °C in plates sealed with Parafilm up to 2 weeks.

3.16 Creation of Micropatterned Surfaces

1. Turn on the UV machine UVO-Cleaner-42 (Jetlight Company, Irvine, CA) 5 min before use.
2. The mask has to be cleaned before each use with a MilliQ-water soaked Kimwipe, followed with a rinse of isopropanol.
3. Put a 10 μL drop of MilliQ water on the desired micropatterning area.
4. Add the PEG-Si-coated cover glass on top of the drop.
5. Put on the montage mask/cover glass in the UV machine upside down. Turn on the UV light for 1 min.
6. Wait 5 min before opening the machine and taking out the mask/cover glass.
7. Take out the cover glass. Adding 1 mL of water on top of the cover glass can help to take it out.
8. The patterned cover glass can be stored in plastic dishes sealed with Parafilm up to 1 week.

   A schematic of **steps 1** through **8** is provided in Fig. 4a.

3.17 Building Micropattern Flow Chambers

1. Cut double-sided tape at dimensions of 40 × 2 mm.
2. Stick the tapes on the cover glass to form individual flow chambers.
3. Seal the chamber sticking the slide on top of the cover glass (Fig. 4b).

3.18 Coating of Micropatterns

1. Flow purified protein (30 μL at desired concentration) into a chamber. Let the protein bind specifically on patterned areas for 1 min.
2. Wash unbound protein with protein buffer (two times 100 μL). Use a Kimwipe to absorb excess buffer solution at the opposite side of the chamber.

3.19 Experiments with Micropatterns

The actin polymerization reaction is prepared and flowed in the flow chamber as described in Subheadings 3.4 and 3.5 (Table 5).

Table 17 shows an example of how a typical micropatterning experiment is set up:

1. Flow 30 μL of 1 μM pWA into a flow chamber.
2. Wait 1 min at RT.

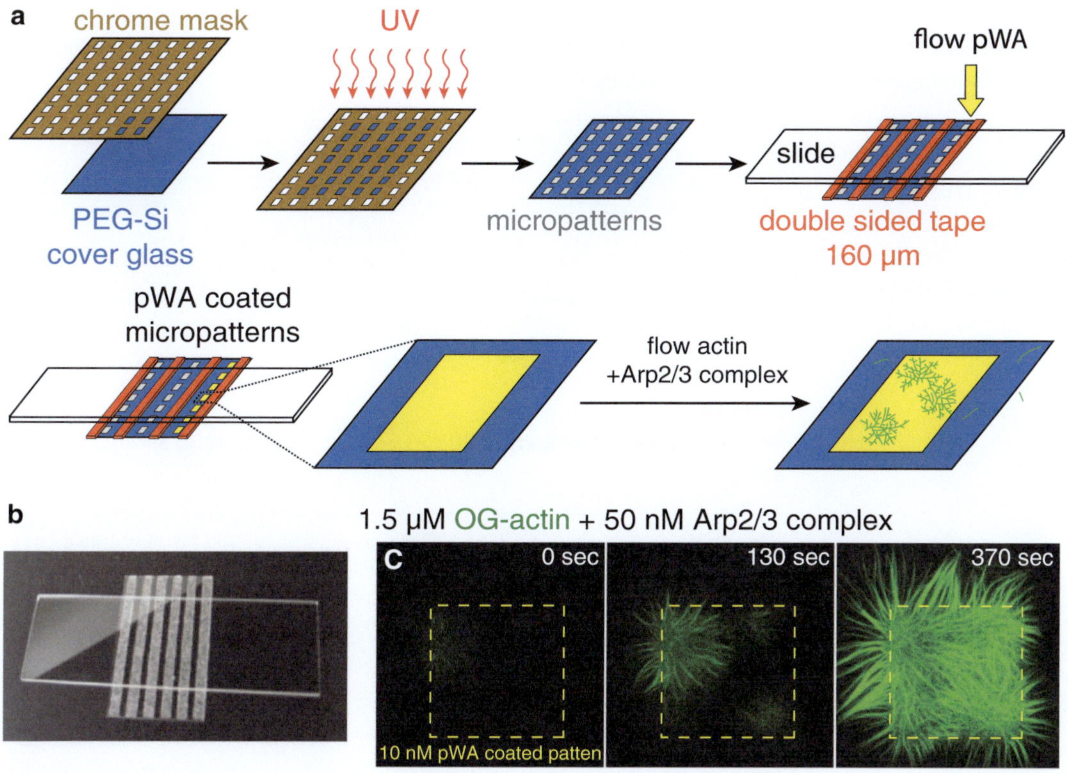

Fig. 4 F-actin network formation on micropatterns. (**a**) Schematic of the general steps for F-actin network growth on micropatterned surfaces, as described in detail in Subheadings 3.15–3.19. (**b**) Photograph of a micropatterned flow chamber. (**c**) Time-lapse fluorescent images of branched F-actin networks assembled by 50 nM Arp2/3 complex from 1.5 μM (30 % Oregon Green) Mg-ATP actin on a micropattern (*yellow box*) coated with Arp2/3 complex activator pWA

Table 17
Typical reaction setup of a micropatterning experiment using 1 μM pWA

Final concentration	5 μM pWA (μL)	pWA buffer (μL)
1 μM pWA	6	24

3. Wash two times with 100 μL pWA Buffer.
4. Prepare the TIRF polymerization reaction mix (Table 18):
5. As mentioned in Subheading 3.4 (Table 5), after a 2-min incubation of tube A, mix of 4 μL of tube A with 16 μL of tube B. Flow 16 μL of this mix into the chamber and start imaging by TIRF microscopy (Subheading 3.4, Fig. 4c).

Table 18
Reaction setup for tubes A and B for following the generation of a branched F-actin network formed by Arp2/3-complexd on a pWA-coated micropattern using TIRF microscopy

Tube	9 μM OG-Actin mix (μL)	5× ME (μL)	2× TIRF buffer (μL)	1.6 μM Arp2/3 complex (μL)	H_2O (μL)
A	5	1	–	–	–
B	–	–	15	1	8

Table 19
Reaction setup for preparing 20 μM F-actin, used in actin sedimentation assays

Reagent	Final conc.	Final volume (μL)
Black actin [40 μM]	20 μM	100
10× ME	1×	20
2-min incubation		
Ca-buffer G	–	60
10× KMEI	1×	20

3.20 Sedimentation Assays

Before beginning sedimentation experiments, spin actin and any accessory proteins at 125,000 × *g* for 30 min (*see* **Note 11**). Transfer the supernatant in a new tube and determine concentrations of all proteins.

1. Polymerize actin filaments (Table 19): Mix 100 μL of actin 40 μM with 20 μL 10× ME.
2. Wait for 2 min before adding 20 μL of 10× KMEI and 60 μL Ca-buffer G to have 200 μL of 20 μM F-actin. Let polymerize at RT for 2 h (Table 19).
3. Add the actin-binding protein of interest in a 100 μL final volume in 200 μL ultracentrifuge tubes. The F-actin has to be added last to the reaction. Incubate at room-temperature for 20 min covered with Parafilm. Two different centrifugation speeds can be used: (1) at high speed (100,000 × *g*), actin filaments will pellet with bound ABPs leaving G-actin and unbound ABPs in the supernatant (Fig. 5a, b), or (2) at low speed (10,000 × *g*) filament cross-links along with bound ABPs will pellet leaving G-actin, non-cross-linked actin filaments and remaining ABPs (bound to single filaments or unbound) in the supernatant (Fig. 5c, d).

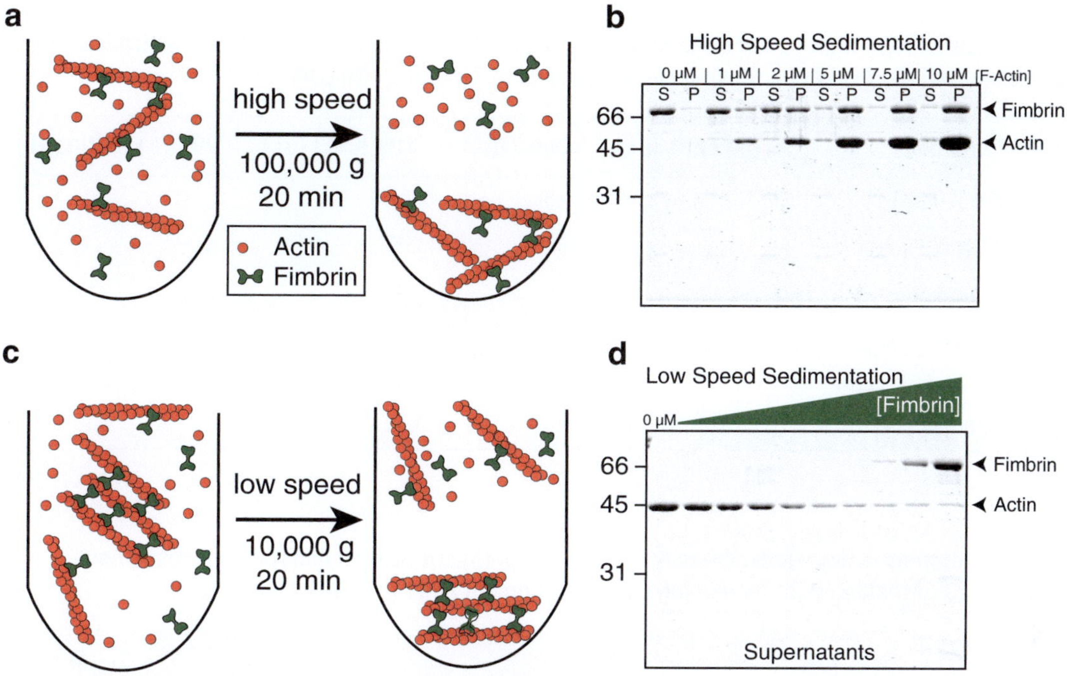

Fig. 5 F-actin sedimentation assays. (**a**, **b**) High-speed sedimentation. 3.0 μM fimbrin was incubated with a range of F-actin concentrations, centrifuged at 100,000 × *g* for 20 min, and the resulting supernatants (S) and pellets (P) were separated by 12.5 % SDS PAGE and stained with Coomassie Blue. Unbound fimbrin and G-actin remain in the supernatant, whereas fimbrin bound to F-actin pellets. (**c**, **d**) Low-speed sedimentation. 3.0 μM F-actin was incubated with a range of fimbrin concentrations, centrifuged at 10,000 × *g* for 20 min, and the resulting supernatants were separated by 12.5 % SDS PAGE and stained with Coomassie Blue. Unbundled F-actin remains in the supernatant, whereas F-actin bundled by fimbrin moves to the pellet (reproduced from [18] with permission from JBC)

An example for a high-speed co-sedimentation assay using 3 μM Fim1 and a range of F-actin concentrations is given in Table 20:

An example for a low-speed co-sedimentation assay using 3 μM F-actin and a range of Fim1 concentrations is given in Table 21:

4. Spin reactions in an ultracentrifuge at 100,000 × *g* for 20 min at 25° C for high-speed sedimentation assay and at 10,000 × *g* for 20 min at 25 °C for low-speed sedimentation assay.
5. Carefully remove 40 μL of the supernatant and place in an Eppendorf tube. Add 10 μL of 5× PSB and flick to mix.
6. Very carefully remove the remaining supernatant and discard without disturbing the pellets.
7. To resuspend the pellets, first mix 200 μL of 5× PSB with 800 μL dH_2O and boil at 95 °C.

Table 20
Reaction setup for a high-speed co-sedimentation experiment using a constant concentration of Fim1 (3 μM) along with increasing concentrations of F-actin (1–10 μM)

#	Condition	30 μM Fim1 (μL)	CaBG (μL)	10× KMEI (μL)	20 μM F-actin (μL)
1	3 μM Fim1	10	80	10	–
2	+1 μM F-actin	10	75	10	5
3	+2 μM F-actin	10	70	10	10
4	+5 μM F-actin	10	55	10	25
5	+7.5 μM F-actin	10	42.5	10	37.5
6	+10 μM F-actin	10	30	10	50

Table 21
Reaction setup for a low-speed co-sedimentation experiment using a constant concentration of F-actin (3 μM) along with increasing concentrations of Fim1 (0.025–2.5 μM)

#	Condition	Fim1	Fim1 buffer (μL)	CaBG (μL)	10× KMEI (μL)	20 μM F-actin (μL)
1	3 μM F-actin	–	5	70	10	15
2	+0.025 μM Fim1	1 μL [2.5 μM]	4	70	10	15
3	+0.05 μM Fim1	2 μL [2.5 μM]	3	70	10	15
4	+0.1 μM Fim1	4 μL [2.5 μM]	1	70	10	15
5	+0.15 μM Fim1	1 μL [15 μM]	4	70	10	15
6	+0.3 μM Fim1	2 μL [15 μM]	3	70	10	15
7	+0.5 μM Fim1	1 μL [50 μM]	4	70	10	15
8	+1 μM Fim1	2 μL [50 μM]	3	70	10	15
9	+2.5 μM Fim1	5 μL [50 μM]	–	70	10	15

8. To each sedimentation assay tube add 125 μL of the hot 1× PSB and resuspend the pellet. Load 15 μL of each supernatant on a 12.5 % SDS-PAGE gel with 5 μL broad range marker; load 15 μL of each pellet on a separate gel.
9. Run gels at 180 V for 50 min.
10. Stain gels with Coomassie blue stain for 30 min by rocking on an orbital shaker.
11. Decant the Coomassie solution and rinse with water once, followed by rocking the gel in destain solution for 16 h.
12. Pour off destain and wash with water; rock in water until gel appears rehydrated, then scan the gel to determine the optical density of the bands of interest.

13. The optical density of bands of interest can be analyzed by densitometry, such as on an Odyssey Infrared Imager (LI-COR Biosciences, Lincoln, NE, USA) (*see* **Note 12**).

Acknowledgments

Biochemical studies of cytokinesis ABPs in the Kovar lab is supported by NIH grant R01 GM079265 (to D.R.K.), ACS grant RSG-11-126-01-CSM (to D.R.K.), and DOD MURI grant W911NF-14-1-0403 (to D.R.K.). We thank Jenna Christensen, Katie Homa, and Tom Burke for helpful comments and suggestions.

References

1. Blanchoin L et al (2014) Actin dynamics, architecture, and mechanics in cell motility. Physiol Rev 94(1):235–263
2. Chhabra ES, Higgs HN (2007) The many faces of actin: matching assembly factors with cellular structures. Nat Cell Biol 9(10):1110–1121
3. Balasubramanian MK, Bi E, Glotzer M (2004) Comparative analysis of cytokinesis in budding yeast, fission yeast and animal cells. Curr Biol 14:R806–R818
4. Glotzer M (2005) The molecular requirements for cytokinesis. Science 307(5716):1735–1739
5. Pollard TD, Wu JQ (2010) Understanding cytokinesis: lessons from fission yeast. Nat Rev Mol Cell Biol 11(2):149–155
6. Lee IJ, Coffman VC, Wu JQ (2012) Contractile-ring assembly in fission yeast cytokinesis: recent advances and new perspectives. Cytoskeleton (Hoboken) 69(10):751–763
7. Spudich JA, Watt S (1971) The regulation of rabbit skeletal muscle contraction. I. Biochemical studies of the interaction of the tropomyosin-troponin complex with actin and the proteolytic fragments of myosin. J Biol Chem 246(15):4866–4871
8. Kuhn JR, Pollard TD (2005) Real-time measurements of actin filament polymerization by total internal reflection fluorescence microscopy. Biophys J 88(2):1387–1402
9. Kudryashov DS, Reisler E (2003) Solution properties of tetramethylrhodamine-modified G-actin. Biophys J 85(4):2466–2475
10. Cooper JA, Walker SB, Pollard TD (1983) Pyrene actin: documentation of the validity of a sensitive assay for actin polymerization. J Muscle Res Cell Motil 4(2):253–262
11. Kouyama T, Mihashi K (1981) Fluorimetry study of N-(1-pyrenyl)iodoacetamide-labelled F-actin. Local structural change of actin protomer both on polymerization and on binding of heavy meromyosin. Eur J Biochem 114(1):33–38
12. Cleveland DW (1982) Treadmilling of tubulin and actin. Cell 28(4):689–691
13. Neuhaus JM et al (1983) Treadmilling of actin. J Muscle Res Cell Motil 4(5):507–527
14. Pollard TD (1986) Rate constants for the reactions of ATP- and ADP-actin with the ends of actin filaments. J Cell Biol 103(6 Pt 2):2747–2754
15. Wegner A, Isenberg G (1983) 12-fold difference between the critical monomer concentrations of the two ends of actin filaments in physiological salt conditions. Proc Natl Acad Sci U S A 80(16):4922–4925
16. Theriot JA, Mitchison TJ (1991) Actin microfilament dynamics in locomoting cells. Nature 352(6331):126–131
17. Axelrod D, Thompson NL, Burghardt TP (1983) Total internal inflection fluorescent microscopy. J Microsc 129(Pt 1):19–28
18. Skau CT et al (2011) Actin filament bundling by fimbrin is important for endocytosis, cytokinesis and polarization in fission yeast. J Biol Chem 286:26964

Chapter 13

Characterization of Cytokinetic F-BARs and Other Membrane-Binding Proteins

Nathan A. McDonald and Kathleen L. Gould

Abstract

Multiple membrane-binding proteins are key players in cytokinesis in yeast and other organisms. In vivo techniques for analyzing protein–membrane interactions are currently limited. In vitro assays allow characterization of the biochemical properties of these proteins to build a mechanistic understanding of protein–membrane interactions during cytokinesis. Here, we describe two in vitro assays to characterize FCH-Bin/Amphyphysin/RVS (F-BAR) domains and other protein's interactions with membranes: liposome co-pelleting and giant unilamellar vesicle fluorescent binding.

Key words Liposomes, Membrane binding, Giant unilamellar vesicles, F-BAR domain

1 Introduction

During cytokinesis, the actomyosin-based contractile ring (CR) must associate tightly with the plasma membrane as it forms and constricts. Multiple protein components of the CR have been identified that directly bind the plasma membrane and contribute to the tight linkage. One family of proteins that has been implicated in CR–membrane linkage in yeast is the Fer/CIP4 homology-Bin/Amphyphysin/Rvs (F-BAR) family, including Cdc15, Imp2, and Rga7 in *Schizosaccharomyces pombe* and Hof1 in *Saccharomyces cerevisiae* [1–5]. F-BAR domains are crescent-shaped dimers, usually present in proteins with additional protein binding or enzymatic domains [6, 7]. In addition to binding membranes, many F-BARs are able to oligomerize and collectively bend and remodel membranes [8]. Methods to study membrane binding of F-BARs and other proteins in vivo are currently limited to fluorescent live cell imaging or biochemical fractionation techniques. It is often difficult to derive the mechanism of membrane binding and remodeling from these in vivo assays. In vitro assays offer an alternative approach to rigorously analyze how individual protein components interact with biological membranes. Here, we describe two assays

Alberto Sanchez-Diaz and Pilar Perez (eds.), *Yeast Cytokinesis: Methods and Protocols*, Methods in Molecular Biology, vol. 1369, DOI 10.1007/978-1-4939-3145-3_13, © Springer Science+Business Media New York 2016

that can be used to analyze F-BAR or other proteins' membrane interactions in vitro: (1) Liposome co-pelleting to determine membrane-binding properties and kinetics, and (2) Giant unilamellar vesicle (GUV) fluorescent binding to assay membrane bending and remodeling activities.

2 Materials

2.1 Reagents

- Recombinant or purified protein of interest (*see* **Note 1**), free of detergent (*see* **Note 2**).
- Polar lipids dissolved in $CHCl_3$ or $CHCl_3$/Methanol/water (Avanti Polar Lipids, Alabaster, AL, USA) (*see* **Note 3**). Common lipids used to create a cell-like membrane include DOPC (synthetic phosphatidylcholine), DOPE (synthetic phosphatidylethanolamine), DOPS (synthetic phosphatidylserine), phosphorylated phosphatidylinositols, and sterols [9].
- Liposome buffer: 20 mM HEPES pH 7.4, 100 mM NaCl (*see* **Note 4**).
- 5× Protein sample buffer: 300 mM Tris–HCl pH 6.8, 50 % v/v glycerol, 10 % w/v sodium dodecyl sulfate (SDS), 25 % v/v β-mercaptoethanol, 0.05 % w/v bromophenol blue. Store at −20 °C.
- Coomassie Blue stain: 0.1 % w/v Coomassie Brilliant Blue G, 50 % v/v methanol, 10 % v/v glacial acetic acid, 40 % v/v H_2O. Stir for 3–4 h before filtering through Whatman paper. Colloidal Coomassie can be substituted.
- Destain solution: 10 % v/v isopropanol, 10 % v/v glacial acetic acid, 80 % v/v water. Alternatively, Coomassie Blue stained gels can be destained in water.
- GUV buffer: 20 mM HEPES pH 7.4, 500 mM sucrose; filter-sterilize to prevent contamination. 1 mM each of EDTA and EGTA can be added to prevent clustering of certain lipids in the final GUVs.
- Glucose solution: 1 M glucose in water; filter-sterilize to prevent contamination.

2.2 Equipment

- Glass tubes.
- Vacuum chamber capable of $\sim 1 \times 10^{-3}$ torr.
- Gas-tight glass syringes (Avanti Polar Lipids, Alabaster, AL, USA) (*see* **Note 5**).
- Glass syringe mini-extruder and filters (Avanti Polar Lipids, Alabaster, AL, USA). Liposomes can also be sized with a water bath sonicator (Avanti Polar Lipids, Alabaster, AL, USA).

- Ultracentrifuge capable of 150,000 × *g* such as the Beckman Optima MAX series (Beckman Coulter, Brea, CA, USA).
- LI-COR Odyssey imager (LI-COR, Lincoln, NE, USA) or equivalent gel scanner.
- Indium tin oxide (ITO) coated glass slides, 30–60 Ω/sq (cat. 703184, Sigma-Aldrich, St. Louis, MO, USA).
- Silicon imaging chambers, press-to-seal, 2 and 0.5 mm thick (Electron Microscopy Sciences, Hatfield, PA, USA).
- Copper conductive tape (Electron Microscopy Sciences, Hatfield, PA, USA).
- Signal generator and oscilloscope capable of 10 Hz, 2.5 V sine current. USB computer modules containing both functions are available.
- Glass microscope slides and glass coverslips.
- Fluorescence microscope with 488 nm (GFP/green) and 561 nm (TRITC/red) laser lines. 60× or 100× objectives excel at visualization of 10–30 μm GUVs utilized here.

3 Methods

Carry out all procedures at room temperature unless otherwise noted. Prepare recombinant protein of interest before beginning (*see* **Note 1**).

3.1 In Vitro Liposome Preparation

1. Mix appropriate amount of lipid stocks at desired ratios in a glass tube using glass syringes (*see* **Note 5**). A concentration of 1 mg/mL lipids in the final liposome sample is common and recommended (*see* **Note 6**).
2. Evaporate solvent away from lipids under a stream of N_2 while gently swirling to form a thin film of lipid on the bottom of the glass tube.
3. Remove residual solvent by high vacuum for 1 h (*see* **Note 7**).
4. Add appropriate amount of liposome buffer to yield a 1 mg/mL sample (*see* **Note 6**). Rehydrate in liposome buffer for 30 min without disturbance to swell and separate the lipid films into bilayers.
5. Vortex vigorously for ~1 min to form liposomes. This sample is now composed of large multilamellar vesicles (LMVs).
6. Freeze and thaw the liposome sample 5–10× by moving between a dry ice-ethanol bath and a warm water bath (~36 °C). This step forces the LMVs to fracture and reform into smaller unilamellar liposomes.
7. Heat liposome sample above the transition temperature of lipids within the mixture (*see* **Note 8**) and extrude liposomes to

the desired size (100, 200, 400, 800 nm diameters are common) by passing 10× through a filter with the mini-extruder. Alternatively, sonicate lipids in a water sonication bath for 1 min to form small (<50 nm) vesicles.

8. Use vesicles immediately or store at 4 °C for up to 24 h.

3.2 Liposome Co-pelleting

The following protocol describes a simple equilibrium binding assay between liposomes and protein. Many conditions can be experimentally altered to investigate different aspects of membrane binding, including but not limited to: liposome size, liposome lipid composition, protein concentration, or buffer compositions.

1. Mix 100 μL of a 1 mg/mL liposome sample with 100 μL recombinant protein and incubate at room temperature for 15 min (*see* **Note 9**). Also include liposome-free samples (100 μL liposome buffer + 100 μL recombinant protein) for all experimental conditions (*see* **Note 10**).
2. Spin reactions at 150,000 × *g* for 15 min at 25 °C.
3. Carefully remove supernatant from small liposome pellet to another tube (*see* **Note 11**). This is the unbound fraction.
4. Resuspend the small liposome pellet with 200 μL liposome buffer. This is the bound fraction.
5. Add 50 μL 5× protein sample buffer to each fraction, boil for 5 min, and run a sample (~15 μL) of pellet and supernatant fractions on SDS-PAGE.
6. Place gel in ~10 mL Coomassie Blue stain for 10 min. Destain in destain solution for 1 h.
7. Quantify band intensity of pellet and supernatant fractions with LI-COR Odyssey imager or equivalent gel scanner for each experimental condition. Subtract background pelleting from the no liposome control to determine final % binding (Fig. 1) (*see* **Note 12**).

3.3 Giant Vesicle Preparation

1. Prepare a mixture of $CHCl_3$ lipid stocks at 10 mg/mL with desired lipid ratios using glass syringes (*see* **Note 5**). Include ~0.5 % fluorescently labeled lipid for visualization of membrane (*see* **Note 13**).
2. Evaporate 10 μL of prepared lipid stock into a film on the Indium Tin Oxide (ITO)-coated side of each glass slide under N_2 stream.
3. Evaporate residual $CHCl_3$ under high vacuum for 1 h (*see* **Note 7**). Protect from light to prevent fluorophore bleaching.
4. Attach conductive copper tape to the edge of each ITO glass slide (Fig. 2). Remove plastic from both sides of silicon imaging chamber and press firmly to one ITO glass slide to attach.

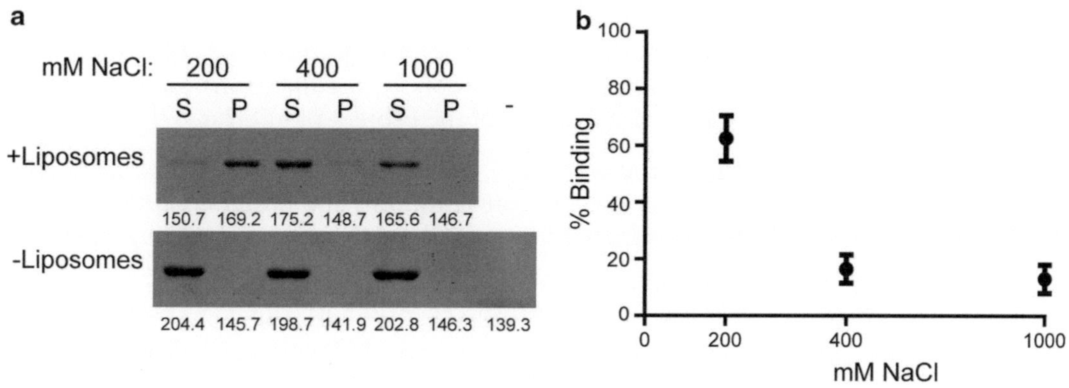

Fig. 1 Liposome co-pelleting. (**a**) Example liposome binding experiment between recombinant *S. pombe* Cdc15 F-BAR (aa19-312) and liposomes composed of 70 % DOPC, 15 % DOPE, 10 % DOPS, 5 % PI4P at the indicated concentration of NaCl. *S* supernatant, unbound fraction; *P* pellet, bound fraction. Gels were scanned with a LI-COR Odyssey infrared imager at 700 nm and lane intensities were quantified in ImageJ. (**b**) Average binding from three liposome co-pelleting experiments at different NaCl concentrations. Error bars indicate SEM

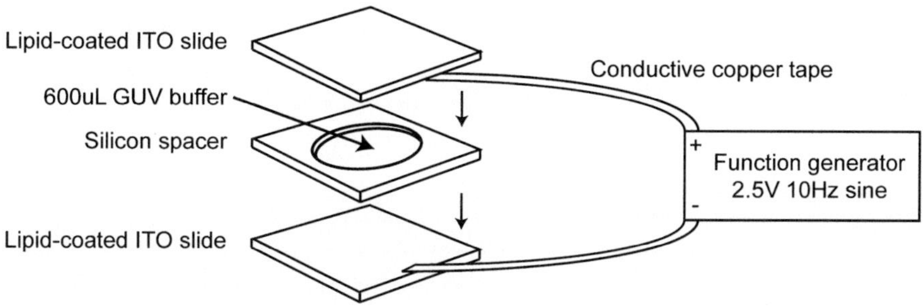

Fig. 2 The GUV formation chamber. A 2 mm thick silicon spacer is sandwiched between two indium tin oxide glass slides coated with evaporated lipids. 600 μL GUV buffer is added, the chamber is sealed, and a 2.5 V, 10 Hz sine current is passed through the chamber

5. Fill the silicon spacer chamber with 600 μL GUV formation buffer and form a seal with the remaining ITO glass slide, the lipid-coated side facing inward (*see* **Note 14**). Attach one copper tape strand to the anode and one to the cathode of a signal generator (Fig. 2).
6. Pass a 10 Hz, 2.5 V sine current across the chamber for 2 h or up to overnight to form the GUVs.
7. Carefully open chamber and remove the 600 μL GUV solution to a microcentrifuge tube. Add 300 μL of a 1 M glucose solution to form an isosmolar solution (*see* **Note 15**). This GUV sample can be used as-is, or can be concentrated by sedimentation overnight at 4 °C, after which ~50–75 % of buffer from the top is removed. Use GUVs immediately or store at 4 °C for up to 24 h.

3.4 Giant Vesicle Fluorescent Binding

Prepare fluorescently labeled protein of interest before proceeding. GFP fusions, covalent epitope modifications [10, 11], or covalent thiol, amine, or ester fluorophore modifications can be used for fluorescent visualization of the protein of interest.

1. Mix 2.5 μL of GUV sample with 2.5 μL of protein sample. Different protein concentrations and buffer conditions should be investigated for optimal binding and remodeling activity.
2. Assemble a chamber on a glass microscope slide with a 0.5 mm press-to-seal silicon spacer. Add 5 μL of GUV/protein sample and seal with a glass coverslip, suspending the sample between slide and coverslip (*see* **Note 16**).
3. Mount the imaging chamber on a fluorescent microscope. GUVs will settle onto the lower coverslip. A small amount of thermal motion in the GUVs is unavoidable (*see* **Note 17**).
4. Image GUV fluorescence and protein fluorescence channels to assay membrane binding, indicated by colocalization of both fluorescent signals. Membrane deformations and remodeling can be visualized through colocalization of both signals on GUVs bent, deformed, invaginated, or tubulated from circular (*see* **Note 18**). *Z*-stacks can be taken to build a 3D image of membrane binding and bending. Time-lapse imaging may also be employed to visualize membrane binding and deformation over time (Fig. 3).

4 Notes

1. Proteins and protein domains can be produced recombinantly in *E. coli* and purified with affinity tags (see [12] for general overview). Alternatively, proteins can be purified to isolation from yeast using an epitope tag of choice. We find that the

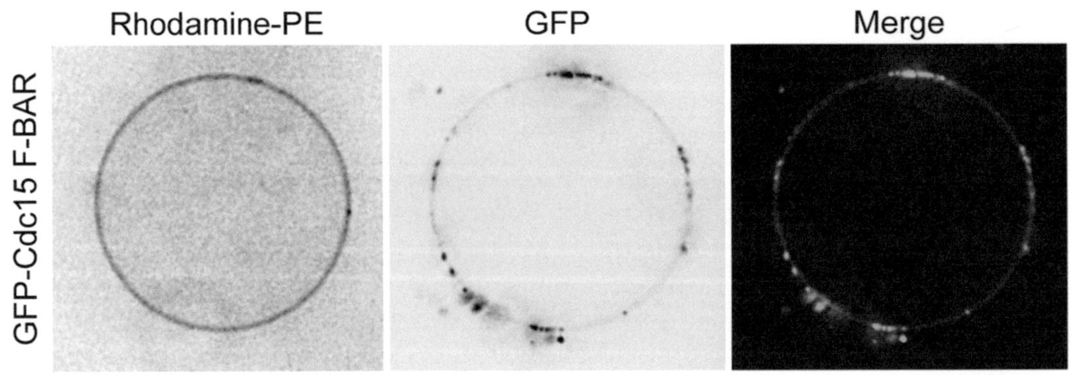

Fig. 3 Fluorescent binding to giant unilamellar vesicles. Recombinant *S. pombe*GFP-Cdc15 F-BAR (aa19-312) binds to GUVs composed of 69 % DOPC, 15 % DOPE, 10 % DOPS, 5 % PI4P, 1 % Rhodamin-PE

affinity tags must often be removed from the protein in order to avoid pleiotropic binding effects.

2. If affinity tag purification includes detergents, they must be removed from proteins of interest to prevent liposome "popping." A second ion exchange or gel filtration purification step can accomplish this.
3. Solvent-dissolved lipids should be stored in glass vials at −20 °C for no longer than 6 months. Sealing lipid stock containers in the absence of oxygen (e.g. under N_2 stream) can prolong storage life by slowing oxidation.
4. This generic buffer is recommended to start, but liposome formation is robust in multiple buffers and conditions.
5. Before liposomes are formed, lipid stocks should only touch glass or metal vials and syringes, as the hydrophobic tails can bind to plastic tips and containers.
6. For example, with 10 mg/mL lipid stocks, 1 mL of 1 mg/mL liposomes composed of 65 % DOPC, 25 % DOPE, and 10 % DOPS will require 65 μL DOPC, 25 μL DOPE, and 10 μL DOPS $CHCl_3$ stocks, respectively. Alternatively, molar percentages can be calculated based on individual lipid mass.
7. High vacuum (>10^{-3} torr) is required to remove all residual $CHCl_3$; bench-top vacuum lines are usually insufficient.
8. The transition temperatures for synthetic lipids are available from Avanti Polar Lipids. For unsaturated lipids such as DOPC, this temperature is lower than 0 °C. However, saturated lipids regularly have higher transition temperatures and require heating.
9. The amount of protein in each sample depends on the experiment being performed. Initially, 10 μg of protein is a reasonable amount per reaction to yield easy quantification with SDS-PAGE.
10. This sample is crucial to control for background sedimentation of the protein of interest. If background sedimentation is problematic, a pre-sedimentation spin of the protein sample can be carried out at 175,000 × *g* before the binding reaction.
11. It is crucial that the liposome pellet remains intact and unmixed with the supernatant during this step. Use a thin pipet tip on the opposite side of the tube as the pellet for supernatant removal.
12. Ensure the software used subtracts the background gel intensity from each band. Alternatively, measure a background region and manually subtract from each intensity value. For example, to calculate % binding of the 200 mM NaCl condition in Fig. 1, first subtract the background intensity of the gel

from each value and determine the fraction of intensity present in the pellet lane: (169.2–139.3)/[(150.7–139.3) + (169.2–139.3)] = 0.724. Next calculate the amount of background pelleting in the no liposome control: (145.7–139.3)/[(145.7–139.3) + (204.4–139.3)] = 0.090. Subtract the background pelleting value from the pellet fraction value to yield % binding: 0.724–0.090 = 0.634; 63.4 % binding.

13. Lissamine rhodamine (red emission) or 7-nitro-2-1,3-benzoxadiazol-4-yl (NBD; green emission) are excellent choices.
14. Small air bubbles are usually unavoidable, but attempt to maximize the amount of lipid-coated glass slide in contact with GUV buffer in the chamber.
15. The difference in refraction from sucrose inside GUVs and glucose outside allows visualization with differential interfering contrast (DIC) optics as well.
16. An imaging chamber is necessary, as GUV samples suspended between a glass slide and glass coverslips will deform and easily "pop."
17. If GUVs fail to settle or move too quickly to image, a biotin–streptavidin immobilization technique can be used to tether GUVs to coverslips [13]. We find in the highly viscous sucrose/glucose solution that GUVs remain stable enough to image without this technique.
18. For instance, certain F-BAR proteins bend GUVs into thin tubules.

Acknowledgements

This work was supported by NIH grant GM101035 to K.L.G and AHA fellowship 15PRE21780003 to N.A.M.

References

1. Roberts-Galbraith RH, Ohi MD, Ballif BA et al (2010) Dephosphorylation of F-BAR protein Cdc15 modulates its conformation and stimulates its scaffolding activity at the cell division site. Mol Cell 39:86–99
2. Fankhauser C, Reymond A, Cerutti L et al (1995) The S. pombe cdc15 gene is a key element in the reorganization of F-actin at mitosis. Cell 82:435–444
3. Nishihama R, Schreiter JH, Onishi M et al (2009) Role of Inn1 and its interactions with Hof1 and Cyk3 in promoting cleavage furrow and septum formation in S. cerevisiae. J Cell Biol 185:995–1012
4. Martín-García R, Coll PM, Pérez P (2014) F-BAR domain protein Rga7 collaborates with Cdc15 and Imp2 to ensure proper cytokinesis in fission yeast. J Cell Sci 127:4146–4158
5. Lippincott J, Li R (1998) Dual function of Cyk2, a cdc15/PSTPIP family protein, in regulating actomyosin ring dynamics and septin distribution. J Cell Biol 143:1947–1960
6. Roberts-Galbraith RH, Gould KL (2010) Setting the F-BAR: functions and regulation of the F-BAR protein family. Cell Cycle 9:4091–4097
7. Frost A, Unger VM, De Camilli P (2009) The BAR domain superfamily: membrane-molding macromolecules. Cell 137:191–196

8. Frost A, Perera R, Roux A et al (2008) Structural basis of membrane invagination by F-BAR domains. Cell 132:807–817
9. Ejsing CS, Sampaio JL, Surendranath V et al (2009) Global analysis of the yeast lipidome by quantitative shotgun mass spectrometry. Proc Natl Acad Sci 106:2136–2141
10. Keppler A, Gendreizig S, Gronemeyer T et al (2003) A general method for the covalent labeling of fusion proteins with small molecules in vivo. Nat Biotechnol 21:86–89
11. Kindermann M, George N, Johnsson N, Johnsson K (2003) Covalent and selective immobilization of fusion proteins. J Am Chem Soc 125:7810–7811
12. Structural Genomics Consortium et al (2008) Protein production and purification. Nat Methods 5:135–145
13. Stamou D, Duschl C, Delamarche E, Vogel H (2003) Self-assembled microarrays of attoliter molecular vessels. Angew Chemie Int Ed 42:5580–5583

Chapter 14

Analysis of Three-Dimensional Structures of Exocyst Components

Johannes Lesigang and Gang Dong

Abstract

The exocyst is an octameric protein complex implicated in tethering secretory vesicles to the plasma membrane during exocytosis. To provide a mechanistic understanding of how it functions, it is of critical importance to elucidate its three-dimensional structure. This chapter briefly describes the protocols used in our structure determination of Exo70p and Exo84p, two subunits of the exocyst from *Saccharomyces cerevisiae*. Folding and domain arrangements of both proteins are predicted using bioinformatics tools. Limited proteolysis is carried out to define the boundaries of folded structures, which guides the design of suitable constructs for protein crystallization. The solved structures of both proteins validate the strategy and suggest it might be also used for structural studies of other proteins alike.

Key words Crystallization, Exo70p, Exo84p, Exocyst, Limited proteolysis, Protein structure

1 Introduction

Intracellular vesicular transport is a major transport system for all eukaryotic cells to deliver thousands of internally generated molecules [1–3]. The last step in vesicular transport is to target and fuse vesicles to their destination compartments, which is a stepwise process including tethering by tethering factors, docking by *s*oluble *N*-ethylmaleimide-sensitive factor *a*ttachment protein *re*ceptor (SNARE) proteins, and membrane fusion [4–6]. The exocyst is an evolutionarily conserved protein complex initially identified for tethering vesicles at the plasma membrane during exocytosis [7–10]. Recent studies have shown that the exocyst also participates in cell migration and many other cellular processes [8, 11, 12].

Previous electron microscopic studies of purified exocyst show irregular flower-like structures with multiple relatively rigid petals, which suggests that although each subunit forms a folded structure, the whole complex might have a dynamic structure [13]. It is believed that side-to-side interactions of the rod-shaped exocyst subunits might serve to bring opposing membranes together for

Alberto Sanchez-Diaz and Pilar Perez (eds.), *Yeast Cytokinesis: Methods and Protocols*, Methods in Molecular Biology, vol. 1369, DOI 10.1007/978-1-4939-3145-3_14, © Springer Science+Business Media New York 2016

SNARE-mediated homotypic membrane fusion [14]. Although high-resolution structures of the intact exocyst complex or any of its subcomplexes are unavailable, likely owing to the dynamic structure of the complex, crystal structures of several of its subunits and their domains (either alone or in complex with binding partners) have been reported [15–22]. In this chapter we describe the protocols we used to determine the crystal structures of Exo70p and Exo84p, which are two subunits of the *S. cerevisiae* exocyst complex.

Selection of these two exocyst subunits is based on our bioinformatic analysis, which predicts that Exo70p is overall well-folded and Exo84p contains a distinct C-terminal segment connected to the preceding sequences by a long loop. Limited proteolysis of purified proteins overexpressed in *E. coli* reveals that Exo70p bears a flexible N-terminus, whereas the Exo84p C-terminal region forms a protease-resistant compact structure. Both an N-terminal truncation of Exo70p and the originally designed Exo84p C-terminal construct are successfully crystallized; their structures are determined using selenomethionine (SeMet)-substituted proteins for phase determination. Our results and reported structures of other exocyst subunits together demonstrate that most of the exocyst subunits crystallize by longitudinal packing of the rod-like molecules in the crystal lattice, which leaves no big cavities to accommodate large flexible parts of the molecules. Therefore, bioinformatic analysis, together with limited proteolysis by multiple proteases, proves to be critical in precisely defining domain boundaries of exocyst components that can significantly facilitate the crystallization of purified proteins.

2 Materials

2.1 Bioinformatics Tools

1. A computer with basic setups and an Internet connection. Software for general photo/file processing includes Adobe Photoshop and Acrobat Pro, for crystallographic structure determination is CCP4 Program Suite [23], and for structural analysis is PyMOL (the PyMOL Molecular Graphics System, Version 1.5.0.4 Schrödinger, LLC).
2. All bioinformatic analyses are carried out online. Links of the used servers are listed here: the IUPred server for the prediction of intrinsically unstructured regions of proteins—http://iupred.enzim.hu/, and the PSIPRED server for protein secondary structure predictions—http://bioinf.cs.ucl.ac.uk/psipred/.

2.2 Molecular Cloning and Protein Expression

1. Plasmid vector pET15b (Novagen).
2. Restriction enzyme *Nde*I.
3. Restriction enzyme *Bam*HI.

4. PCR kit with DNA polymerase, dNTP mixture, and stock buffer.
5. PCR thermal cycler.
6. DNA purification kit such as the Qiagen gel extraction kit.
7. T4 ligase and 10× ligation buffer.
8. Agarose.
9. SYBR® Safe DNA Gel Stain (Life technologies).
10. 1× Tris-acetate-EDTA (TAE) buffer: 40 mM Tris, 20 mM acetic acid, and 1 mM EDTA.
11. Sterile luria broth (LB) medium: 10 g tryptone, 5 g yeast extract, and 10 g NaCl (for 1 l).
12. LB-agar plates: 10 g tryptone, 5 g yeast extract, 10 g NaCl, and 15 g agar (for 1 l).
13. Filtered sterile ampicillin stock solution (50 mg/ml).
14. Water bath at 42 °C.
15. Competent cells of DH5α and BL21(DE3).
16. 6× DNA loading dye: 30 % (v/v) glycerol, and 0.25 % (w/v) bromophenol blue.

2.3 Protein Expression and Purification

1. Sterile luria broth (LB) medium.
2. FPLC.
3. Size-exclusion column (Superdex 200, 16/60; GE Healthcare).
4. 5-ml prepacked Ni-HiTrap column (GE Healthcare).
5. French Press: high-pressure homogenizer (EmulsiFlex-C5, Avestin).
6. Bacteria lysis buffer: 20 mM Tris–HCl (pH 8.0), 0.3 M NaCl, 15 mM 2-mercaptoethanol, 20 mM imidazole, 5 % (v/v) glycerol, and 2 μg/ml DNase I.
7. Ni-HiTrap elution buffer: 20 mM Tris–HCl (pH 8.0), 0.3 M NaCl, 15 mM 2-mercaptoethanol, and 200 mM imidazole.
8. Gel filtration buffer: 20 mM Tris–HCl (pH 8.0), 100 mM NaCl, and 5 % (v/v) glycerol.
9. Thrombin (Sigma-Aldrich).
10. Shaking incubator with temperature control.
11. IPTG (isopropyl-β-D-thiogalactopyranoside).
12. Bench-top centrifuge.
13. Photometer.
14. Liquid nitrogen.
15. Standard centrifuges and set of rotors.
16. Amicon Ultra-15 Centrifugal Filter Units with MWCO of 10 and 30 kDa.

17. 4× Tris/SDS (pH 6.8) buffer: dissolve 60.5 g Tris-base in 400 ml distilled water, adjust pH to 6.8 with 1 N HCl, add 4 g SDS, and finally add distilled water to bring the total volume to 1000 ml.
18. 6× SDS-PAGE sample buffer: 7 ml 4× Tris–HCl/SDS buffer (pH 6.8), 3 ml glycerol, 1 g SDS, 0.93 g DTT, 1.2 mg bromophenol blue, and finally bring to 10 ml with distilled water.

2.4 SDS Polyacrylamide Gel Components

1. Stock 30 % acrylamide/bis-acrylamide solution (Serva).
2. *N*, *N*, *N*, *N*′-tetramethyl-ethylenediamine (TEMED) (Sigma-Aldrich).
3. 10 % (w/v) ammonium persulfate (APS).
4. 4× Stacking gel buffer: 0.5 M Tris–HCl (pH 6.8), and 0.4 % (w/v) SDS (*see* Subheading 2.3, **step 17**).
5. 4× Resolving gel buffer: 1.5 M Tris–HCl (pH 8.8), and 0.4 % (w/v) SDS.
6. 1× SDS-PAGE electrophoresis running buffer: 25 mM Tris–HCl (pH 8.3), 192 mM glycine, and 0.1 % (w/v) SDS.

2.5 Limited Proteolysis

1. Proteases: trypsin, chymotrypsin, thermolysis, and elastase (Sigma-Aldrich).
2. Prepare protease stock solutions (1 mg/ml or 10 mg/ml) by dissolving the powder of each selected protease into a buffer containing 10 mM Tris–HCl (pH 8.0) and 100 mM NaCl (*see* **Note 1**).
3. 1× Transfer buffer without SDS: 25 mM Tris–HCl (pH 8.3) and 192 mM glycine.
4. PVDF membrane.
5. Semi-Dry Western blot transfer apparatus.

2.6 Crystallization and Structure Determination

1. Crystallization screen kits (Hampton Research).
2. Crystallization plates.
3. Coverslips.
4. Crystal handling tools.
5. Crystallization incubators (4 and 22 °C).
6. Access to X-ray synchrotron beamlines with tunable wavelengths.
7. Computer with software for structure determination.

3 Methods

The following methods outline the sequential steps in our structure determination of Exo70p and Exo84p (*see* Fig. 1). Details in buffer usage and protein purification procedures have been

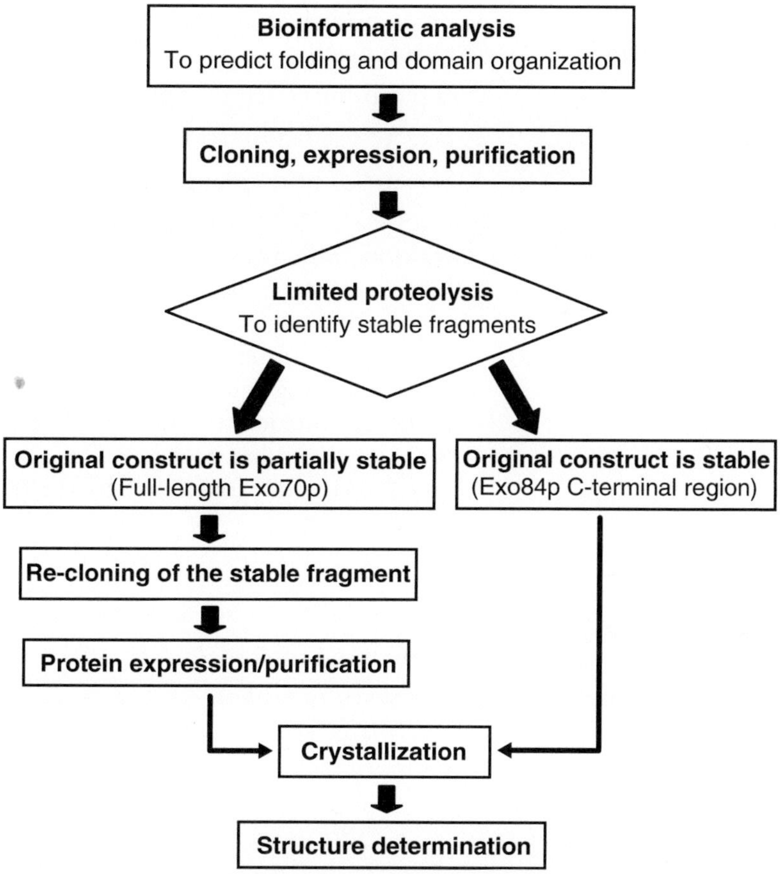

Fig. 1 Flowchart of the structure determination procedure of Exo70p and Exo84p

reported previously [22]. Both proteins contain an N-terminal 6× His-tag and are overexpressed in *E. coli*. Proteins are purified using a two step protocol on a Ni-HiTrap column followed by size exclusion chromatography (SEC). Limited proteolysis is carried out to check whether the proteins contain any long flexible regions that often hinder protein crystallization. It turns out that the N-terminal 60–65 residues of Exo70p are sensitive to proteolysis, whereas the Exo84p C-terminal region (residues 523–753) is proteolysis-resistant. Based on these results, a truncation of Exo70p (residues 63–623) is recloned. Purified protein is subjected to crystallization trials, which yield crystals diffracting X-rays to a resolution of 2.0 Å. The stably folded Exo80p C-terminal region is subjected to crystallization trials directly, which yield crystals diffracting X-rays to a resolution of 2.85 Å. Both structures were solved using SeMet-substituted proteins exploiting the anomalous signal at the K absorption edge of incorporated selenium atoms in the proteins [22].

3.1 Folding Prediction and Domain Boundary Identification by Bioinformatic Analysis

1. Open webpage of the IUPred Server [24] (http://iupred.enzim.hu/).
2. Input primary sequences of Exo70p and Exo84p separately.
3. Select "Prediction type: structured regions" (*see* **Note 2**).
4. Select "Output type: generate plot" (*see* **Note 3**).
5. Click "SUBMIT." Prediction results will come out in a new window, which contains a list of globular domains (disordered loops are highlighted in red in the sequence) and a graphic report displaying the disorder tendency of each residue (on the *Y*-axis with numerical scores between 0 and 1.0) and residue numbers along the *X*-axis. Regions that are predicted to form globular domains (the default disorder tendency cutoff is 0.5) are shown as blue bars (*see* Figs. 2a and 3a). Exo70p is predicted to be fully folded, whereas Exo84p contains three structured regions.

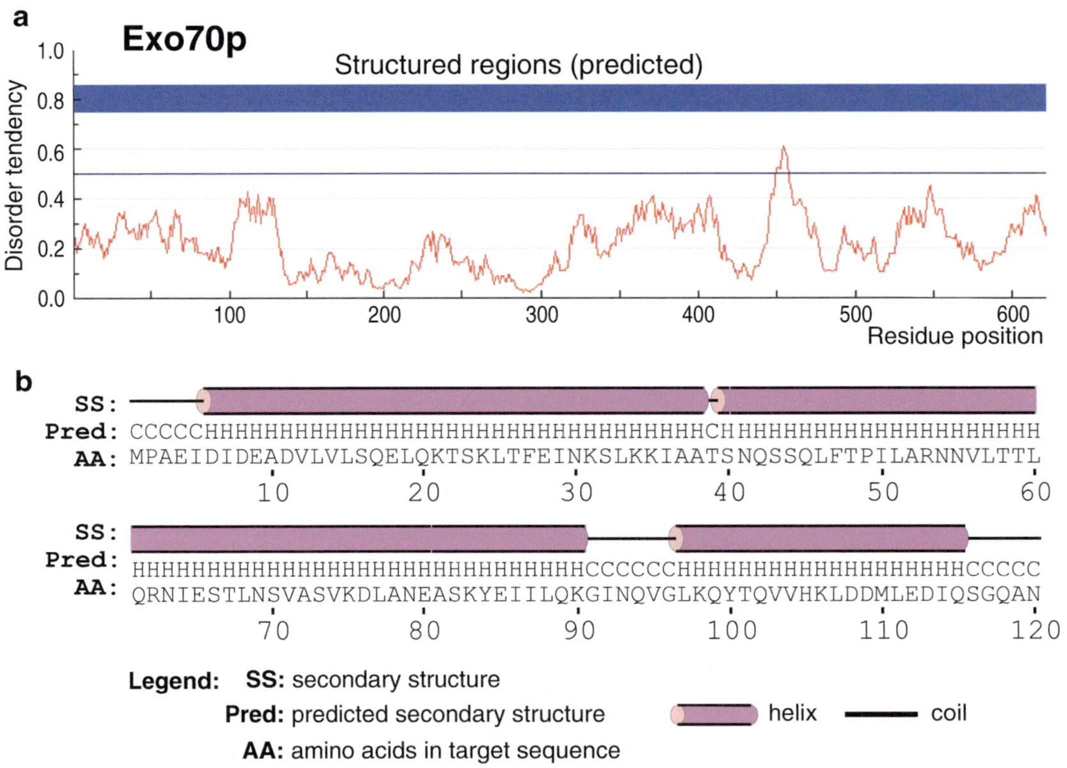

Fig. 2 Folding prediction of Exo70p. (**a**) The whole protein is predicted to be folded based on the prediction carried out on the IUPred Server (http://iupred.enzim.hu/). The *Y*-axis represents the disorder tendency scored between 0 and 1.0 (lower values suggest higher likelihood of folded structures). The region that is predicted to be structured is shown as a horizontal bar. (**b**) Secondary structures of Exo70p predicted by the PSIPRED server using the position-specific scoring matrices (http://bioinf.cs.ucl.ac.uk/psipred/). For clarity, only the N-terminal region of Exo70p is shown for a direct comparison with the limited proteolysis results

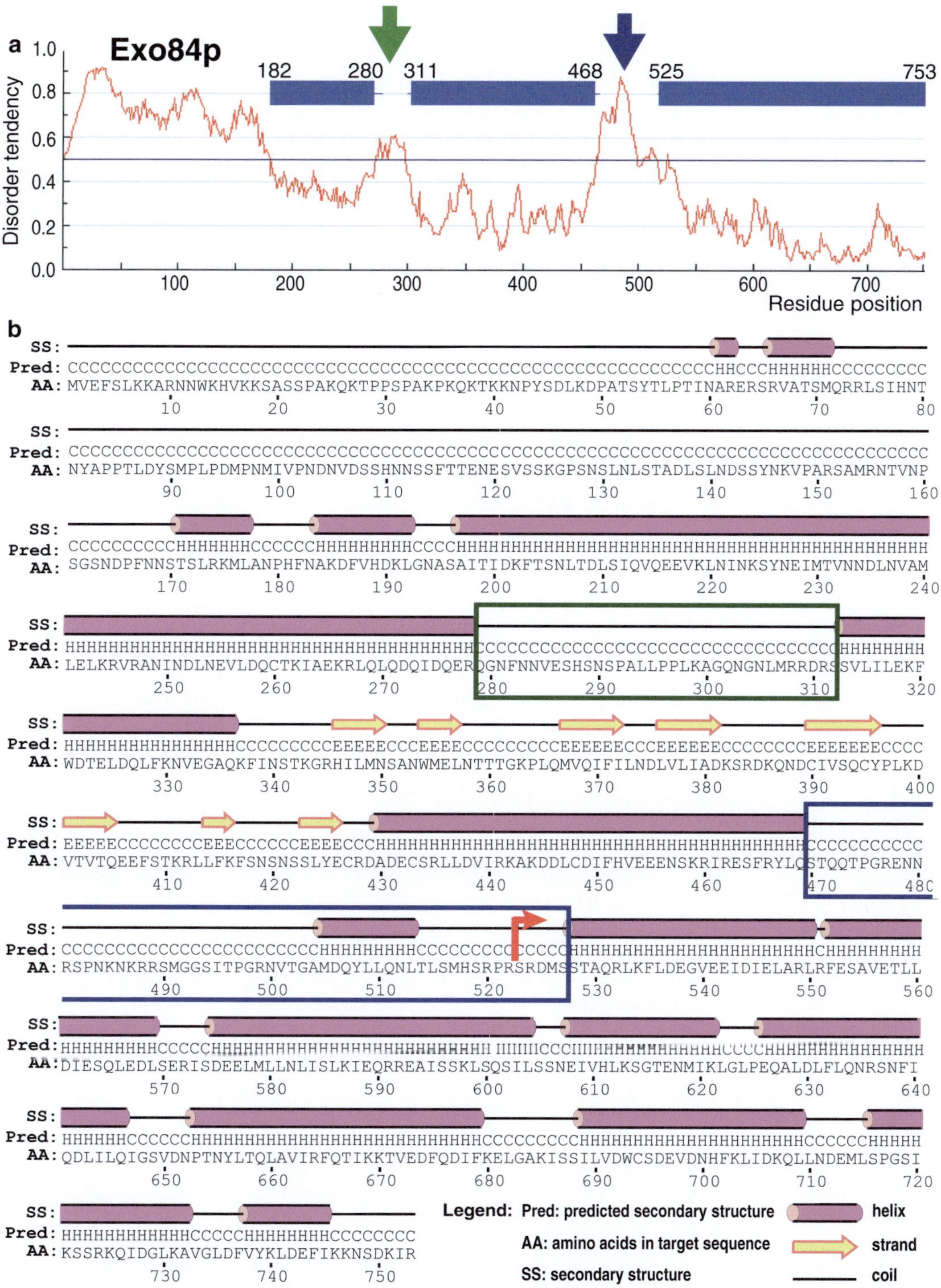

Fig. 3 Folding and secondary structure predictions of Exo84p. (**a**) Predictions carried out on the IUPred Server suggest that Exo84p likely contains a disordered N-terminus plus three structured regions which are shown as horizontal bars. Between the structured regions are long loops (marked by vertical *arrows*) that have high tendencies to be intrinsically disordered. (**b**) Secondary structures of Exo84p predicted by the PSIPRED server. Consistent with the folding prediction results in (**a**), Exo84p has a flexible N-terminus that contains very few folded secondary structures. In contrast, the rest of the protein is mostly folded except for the two long disordered loops (marked by *boxes* with the same color scheme as the *arrows* in (**a**)). The *red arrow* indicates the starting point of our selected cloning fragment, which spans residues 523–753

6. Perform secondary structure predictions using the web server of the PSIPRED Protein Sequence Analysis Workbench (http://bioinf.cs.ucl.ac.uk/psipred/) [25] (*see* **Note 4**).
7. Exo70p is suggested to consist exclusively of helices. Although the N-terminal part of Exo70p is predicted to form condensed helical structures (*see* Fig. 2b), this region of the protein turns out to be sensitive to proteolysis (*see* Fig. 4). In the case of Exo84p, the predicted disordered/folded regions are in good agreement with the secondary structure predictions (*see* Fig. 3b).

3.2 Molecular Cloning

1. Based on the bioinformatic analysis results obtained above, full-length Exo70p (residues 1–623) and the C-terminal region of Exo84p (residues 523–753) are cloned into the plasmid pET15b, which provides an N-terminal 6× His tag cleavable by thrombin. In both cases, the 5′- and 3′-primers contain a *Nde*I and *Bam*HI endonuclease cleavage sites, respectively (*see* **Note 5**).
2. Perform PCR reactions using the *S. cerevisiae* cDNA library as the template (*see* **Note 6**).
3. Extract DNA following the protocols of the Qiagen Gel Extraction Kit. Eluted DNA are double digested by *Nde*I/*Bam*HI and then ligated into pET15b which is precut by the same endonucleases (*see* **Note 7**).

3.3 Protein Expression and Protein Purification

1. Protein expression is carried out in the *E. coli* strain BL21(DE3). Constructs confirmed by DNA sequencing are transformed into BL21(DE3) competent cells. Cells are grown in a 37 °C shaker till OD_{600} reaches 0.6–0.8 (*see* **Note 8**).
2. Add IPTG to the cell culture to a final concentration of 1 mM (*see* **Note 9**).
3. Induced cells are harvested after 3–4 h of incubation at 37 °C (*see* **Note 10**).
4. Harvest cells by centrifugation (4000 × *g*, 12 min). Harvested cells are broken using a high-pressure homogenizer (*see* **Note 11**).
5. Purify proteins on a Ni-HiTrap column. Examine fractions at elution peaks on an SDS-PAGE gel and pool those containing target proteins. To remove the N-terminal 6× His tag, add approximately 1 % (w/w) of thrombin to the each protein sample and leave it at 4 °C overnight (*see* **Note 12**).
6. Further purify untagged Exo70p and Exo84p by SEC (Superdex 200, 16/60). Fractions containing target proteins are pooled for subsequent experiments (*see* **Note 13**).

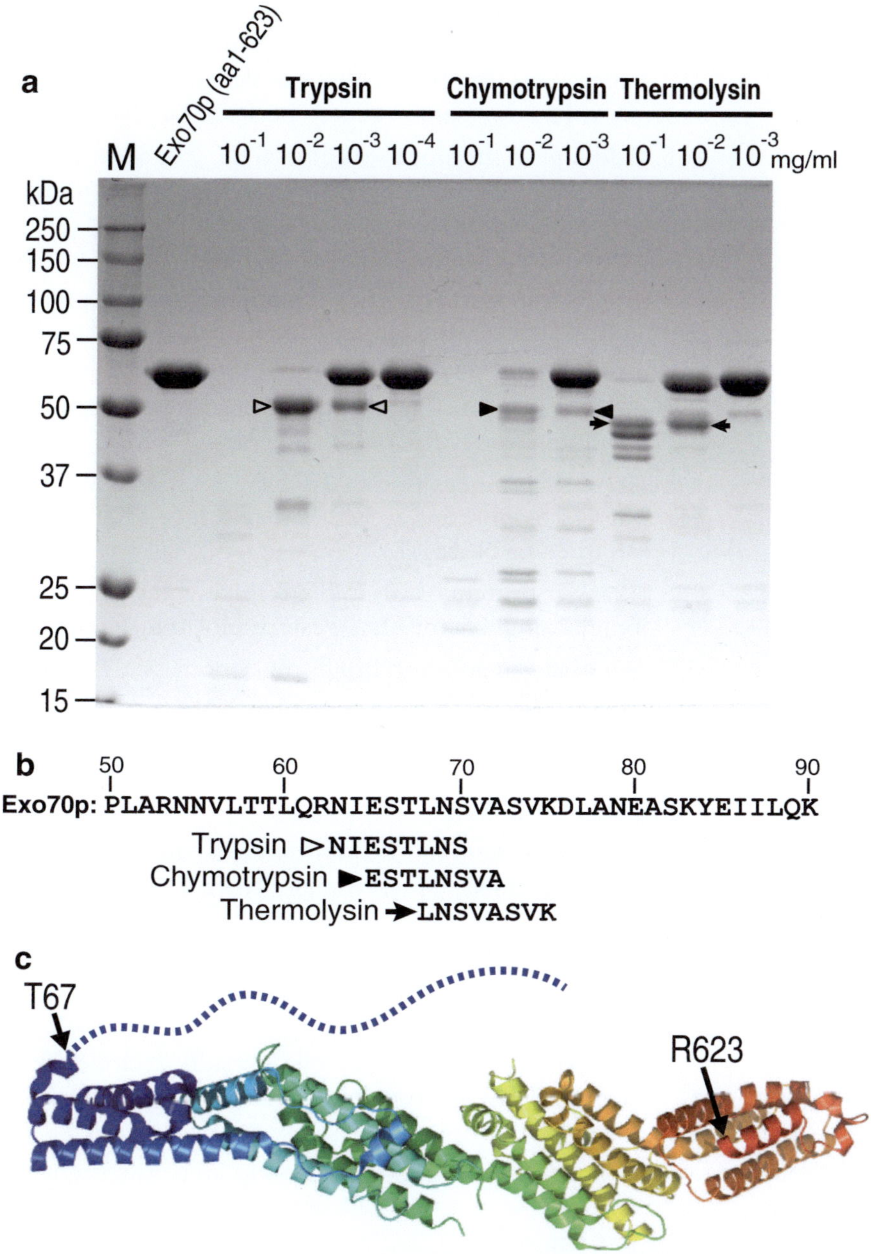

Fig. 4 Limited proteolysis analysis of Exo70p. (**a**) Digestions of Exo70p by a serial dilution of three proteases: trypsin—cutting after K/R, chymotrypsin with low specificity—cutting after F/Y/W/M/L but not before P, and thermolysin—cutting before A/I/L/V but not after D/E. Proteolytic bands that were sent for N-terminal sequencing are marked by *empty triangles* (trypsin digestion), *arrowheads* (chymotrypsin digestion), and *arrows* (thermolysis digestion). (**b**) N-terminal sequencing results of the proteolytic products of Exo70p. Shown are the primary sequence of Exo70p (residues 50–90) and the first eight residues of each digestion bands determined by the N-terminal sequencing. (**c**) The crystal structure of the truncated Exo70p (residues 63–623; 2B1E.pdb), which is color ramped from *blue* at the N-terminus to *red* at the C-terminus. Positions of the first (T67) and last residues (R623) in the structure are indicated by *arrows*. The presumably flexible N-terminal region is shown as a *dashed line*

3.4 Limited Proteolysis

1. Dilute purified Exo70p and Exo84p to ~1 mg/ml, which is then divided to 10-μl aliquots in 1.5-ml sterile Eppendorf tubes. The number of aliquots depends on how many proteolysis reactions one wants to perform (*see* **Note 14**).
2. Add 1 μl of the stock solution to the first aliquot of either protein. Mix it by tapping the tube a few times, and then transfer 1 μl from the mixture to the next aliquot. Keep doing the same thing till the final planned dilution is achieved (*see* **Note 15**).
3. Briefly spin the tubes (now containing both the target proteins and the proteases) and let the incubation go for 30–60 min at room temperature.
4. Add 2 μl of 6× SDS loading buffer to each sample and boil it for 3 min (*see* **Note 16**).
5. Load the samples onto a 15 % (w/v) SDS-PAGE gel and run the electrophoresis at 120 V for ~1 h or at 300 V for roughly 25 min (*see* **Note 17**).
6. Transfer protein bands in the resolving gel onto a PVDF membrane using a Semi-Dry Western blot transfer apparatus.
7. Stain the membrane briefly by Coomassie Brilliant Blue (*see* **Note 18**).
8. Rinse the stained membrane first by distaining solution and then by distilled water. Let the membrane dry on a clean tissue paper for 3–5 min.
9. Identify and cut out major digestion bands. Send the gel pieces for N-terminal sequencing (*see* **Note 19**).
10. Map the sequencing results onto the primary sequence of the target protein (Exo70p) to identify the proteolysis sites of each protease (*see* Fig. 4b).

3.5 Recloning of a Stable Fragment of Exo70p Identified by Limited Proteolysis

1. Based on the limited proteolysis results, reclone the stable fragment of Exo70p containing residues 63–623 into pET15b (*see* **Note 20**).
2. No recloning is done for Exo84p since the original construct does not seem to contain any protease-accessible flexible parts (*see* Fig. 5a).

3.6 Crystallization and Structure Determination

1. Carry out protein expression and purification using the same protocols as described in Subheading 3.3 (*see* **Note 21**).
2. Purified proteins are subjected to crystallization trials using the commercial screening kit (Crystal Screens 1 & 2) from Hampton Research (*see* **Note 22**).
3. In both cases, the target proteins (Exo70p and Exo84p) are able to form crystals in numerous crystallization conditions. The structures are solved eventually using SeMet-substituted

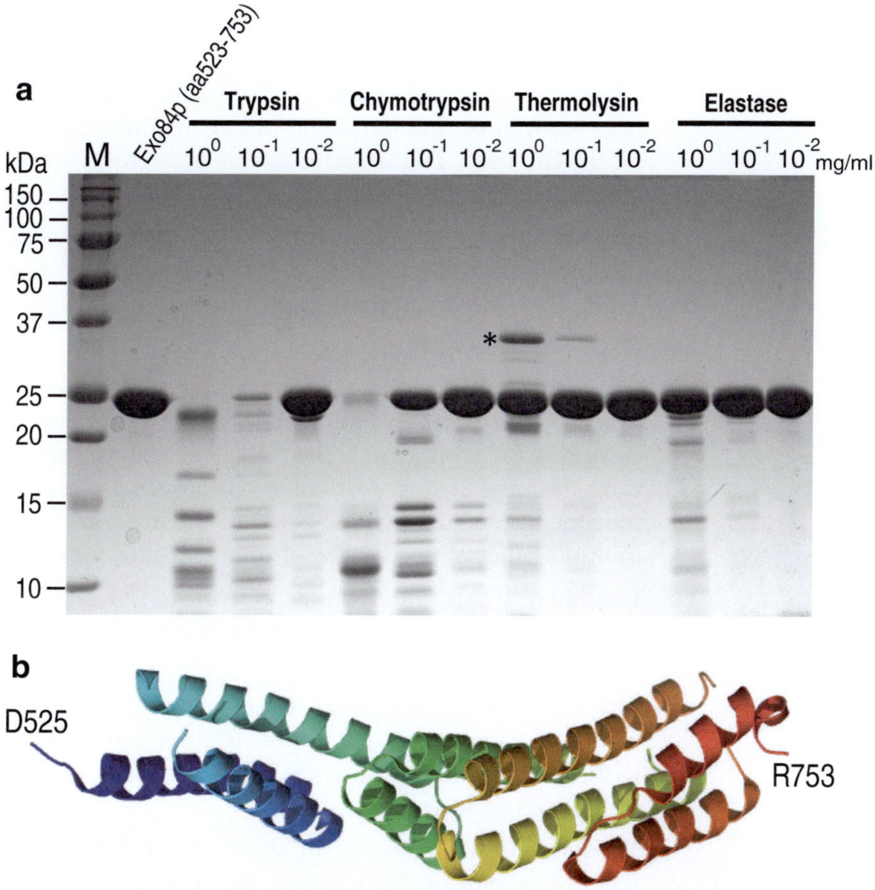

Fig. 5 Limited proteolysis analysis of Exo84p. (**a**) Digestions of the Exo84p C-terminal region (residues 523–753) by a serial dilution of four proteases: trypsin, chymotrypsin, thermolysin, and elastase—cutting preferentially after A/G/V. Marked by an *asterisk* is a protein from the protease solution—likely thermolysin (Mr ~35 kDa). (**b**) The crystal structure of Exo84p (residues 523–753; 2D2S.pdb), which is color ramped from *blue* at the N-terminus to *red* at the C-terminus. The first (D525) and last residues (R753) in the final structure are indicated. Only two residues at the N-terminus of the protein are invisible in the final electron density map used for model building

proteins employing the single-wavelength anomalous dispersion (SAD) techniques [22].

4. The solved structures turn out to be in good agreement with the initial proteolysis results. In the case of Exo70p, residues from position 67 all the way to the C-terminus (R623) are structurally ordered (except for several short loops within the structure) (*see* Fig. 4c). For Exo84p, residues 525–753 form a compact structure (*see* Fig. 5b). These results validate the strategies employed in the structure determination of the two exocyst components.

4 Notes

1. Divide the protease stock solutions into small aliquots (5–10 μl), snap-freeze them in liquid nitrogen, and store them at –80 °C for late uses. The stocks may stay good for a few years as we have found out in our laboratory.
2. Use the other two options on the screen if you want to check the disordered distributions along your target protein.
3. Adjust the "plot window size" to allow the whole sequence displayed in a continuous graph.
4. Select "PSIPRED v3.3 (Predict Secondary Structure)" in the "Input" window if you want to do just secondary structure prediction. Check options in the "Sequence Filter" window if you want to mask some particular regions (e.g. low complexity, transmembrane, and coiled coils) in the protein. There are also multiple other options in the "Input" window for you to choose.
5. Include 6–8 extra nucleotides in front of the *Nde*I cutting site when designing the 5′-primer to achieve accurate and efficient cleavage.
6. Use a precloned construct of your gene if you have it, which would be more efficient and give fewer nonspecific bands. Play with different PCR strategies if no products are obtained or yields are too low.
7. To get a higher recovery of the target DNA, (a) use a large volume of elution buffer (e.g. 80–100 μl), (b) incubate for 2–3 min before centrifugation, and (c) reapply the eluent back onto the same binding membrane and centrifuge again.
8. Try induction at a higher cell density (OD_{600} 0.8–1.2) if you want to get a better yield of the target protein.
9. Adjust IPTG concentration to control protein production. Usage of as low as 10–20 μM of IPTG for induction has been reported.
10. Adjust the post-induction incubation time if necessary. Most target proteins are expressed within 1–2 h after adding IPTG.
11. Avoid extensive tumbling of the cell resuspension because Exo70p aggregates upon agitation.
12. This protocol is usually sufficient to completely cut off the 6× His tag. You are advised to adjust the amount of thrombin in your digestion depending on the efficiency of your thrombin.
13. Avoid unnecessary pipetting and any extensive agitation because the proteins (particularly Exo70p) tend to form aggregates upon mixing.
14. Try 10^{-1}–10^{-5} mg/ml of proteases in your proteolysis reactions first in case you are not sure about how much protease you should use.

15. Keep all samples on ice before starting the next step.
16. Do not forget to include a control which is the protein sample alone containing no proteases.
17. Make sure the buffer tank is safely sealed, particularly if you run the electrophoresis with a high voltage (e.g. 200–300 V).
18. Keep the staining time as short as possible (usually 30–60 s is sufficient).
19. Cut off protein bands of interest using sterile scissors; do not touch the membrane with bare hands.
20. Cautiously leave a few extra residues at the N-terminus of the protein for protein stability.
21. Yields of both Exo70p and Exo84p are very high (10–20 mg per L of cells).
22. Crystallization droplets can be set up either by hands or using a robot. Try not to mix the protein sample with reservoir solution if you manually set up the droplets; over-mixing operations tend to cause irreversible aggregates of the proteins.

Acknowledgements

The original crystallization and structure determination were carried out in the group of Karin M. Reinisch at the Yale School of Medicine. Proteolysis analysis was repeated using the same protocols to generate Figs. 4a and 5a. The experiments were carried out in our group at the Max F. Perutz Laboratories in the Medical University of Vienna under the support of grant P24383-B21 from the Austrian Science Fund (FWF) to GD.

References

1. Benarroch EE (2012) Membrane trafficking and transport: overview and neurologic implications. Neurology 79(12):1288–1295
2. Cheung AY, de Vries SC (2008) Membrane trafficking: intracellular highways and country roads. Plant Physiol 147(4):1451–1453
3. Stow JL (2013) Nobel Prize discovery paves the way for immunological traffic. Nat Rev Immunol 13(12):839–841
4. Chia PZ, Gleeson PA (2014) Membrane tethering. F1000Prime Rep 6:74
5. Brown FC, Pfeffer SR (2010) An update on transport vesicle tethering. Mol Membr Biol 27(8):457–461
6. Whyte JR, Munro S (2002) Vesicle tethering complexes in membrane traffic. J Cell Sci 115(Pt 13):2627–2637
7. Liu J, Guo W (2012) The exocyst complex in exocytosis and cell migration. Protoplasma 249(3):587–597
8. Heider MR, Munson M (2012) Exorcising the exocyst complex. Traffic 13(7):898–907
9. Zarsky V et al (2013) Exocyst complexes multiple functions in plant cells secretory pathways. Curr Opin Plant Biol 16(6):726–733
10. Lipschutz JH, Mostov KE (2002) Exocytosis: the many masters of the exocyst. Curr Biol 12(6):R212–R214
11. Wang S et al (2004) The mammalian exocyst, a complex required for exocytosis, inhibits tubulin polymerization. J Biol Chem 279(34):35958–35966
12. Zuo X, Guo W, Lipschutz JH (2009) The exocyst protein Sec10 is necessary for primary

ciliogenesis and cystogenesis in vitro. Mol Biol Cell 20(10):2522–2529
13. Hsu SC et al (1998) Subunit composition, protein interactions, and structures of the mammalian brain sec6/8 complex and septin filaments. Neuron 20(6):1111–1122
14. Munson M, Novick P (2006) The exocyst defrocked, a framework of rods revealed. Nat Struct Mol Biol 13(7):577–581
15. Yamashita M et al (2010) Structural basis for the Rho- and phosphoinositide-dependent localization of the exocyst subunit Sec3. Nat Struct Mol Biol 17(2):180–186
16. Moore BA, Robinson HH, Xu Z (2007) The crystal structure of mouse Exo70 reveals unique features of the mammalian exocyst. J Mol Biol 371(2):410–421
17. Hamburger ZA et al (2006) Crystal structure of the S.cerevisiae exocyst component Exo70p. J Mol Biol 356(1):9–21
18. Wu S et al (2005) Sec15 interacts with Rab11 via a novel domain and affects Rab11 localization in vivo. Nat Struct Mol Biol 12(10):879–885
19. Fukai S et al (2003) Structural basis of the interaction between RalA and Sec5, a subunit of the sec6/8 complex. EMBO J 22(13): 3267–3278
20. Jin R et al (2005) Exo84 and Sec5 are competitive regulatory Sec6/8 effectors to the RalA GTPase. EMBO J 24(12):2064–2074
21. Sivaram MV et al (2006) The structure of the exocyst subunit Sec6p defines a conserved architecture with diverse roles. Nat Struct Mol Biol 13(6):555–556
22. Dong G et al (2005) The structures of exocyst subunit Exo70p and the Exo84p C-terminal domains reveal a common motif. Nat Struct Mol Biol 12(12):1094–1100
23. Winn MD et al (2011) Overview of the CCP4 suite and current developments. Acta Crystallogr D Biol Crystallogr 67(Pt 4): 235–242
24. Dosztanyi Z et al (2005) IUPred: web server for the prediction of intrinsically unstructured regions of proteins based on estimated energy content. Bioinformatics 21(16):3433–3434
25. Buchan DW et al (2013) Scalable web services for the PSIPRED Protein Analysis Workbench. Nucleic Acids Res 41(Web Server issue):W349–W357

Chapter 15

Analysis of Rho-GTPase Activity During Budding Yeast Cytokinesis

Masayuki Onishi and John R. Pringle

Abstract

Rho-type small GTPases are involved in cytokinesis in various organisms, but their precise roles and regulation remain unclear. Rho proteins function as molecular switches by cycling between the active GTP-bound and inactive GDP-bound states; the GTP-bound proteins in turn interact with their downstream effectors to transmit the signal. Biochemical assays using Rho-binding domains of effector proteins have been used to specifically pull down GTP-bound Rho proteins from cell extracts. Here, we describe the application of such a method in combination with cell-cycle synchronization in the budding yeast *Saccharomyces cerevisiae*; this approach allows dissection of the activity of Rho1 at different stages of cytokinesis. We also present data showing the importance of caution in interpreting such biochemical data and of comparing to the results obtained with other approaches where possible. The principle of this protocol is also applicable to analyses of other Rho-type GTPases and cell-cycle events.

Key words *Saccharomyces cerevisiae*, Abscission, RhoA, Rho1, Cdc42, Rhotekin, Pkc1, Rho effector

1 Introduction

Rho-type small GTPases have been found at the division site in many eukaryotes; examples include Rho and Cdc42 (animals and yeasts), Rac (animals and slime molds), and ROPs (plants) [1–4]. These observations suggest a deep evolutionary root for the roles of these proteins in cytokinesis. However, the precise roles and regulation of each Rho GTPase in each organism are not fully elucidated, in part due to the complexity of their functions throughout the multiple processes involved in cytokinesis. For example, in animals and yeasts, the RhoA/Rho1 proteins play roles during at least two distinct stages in cytokinesis, namely at the onset, to promote assembly and contraction of the actomyosin ring [5–7], and at the end, for final resolution of the plasma membrane during abscission [8–11]. However, some data also suggest that RhoA/Rho1 activity is dispensable for the intervening cleavage-furrow ingression in many mammalian cells [12, 13] and budding yeast [11].

Alberto Sanchez-Diaz and Pilar Perez (eds.), *Yeast Cytokinesis: Methods and Protocols*, Methods in Molecular Biology, vol. 1369, DOI 10.1007/978-1-4939-3145-3_15,

Whether RhoA/Rho1 remains active throughout these processes or is switched on and off at specific times and locations remains unclear.

Rho GTPases operate as signaling switches that alternate between the active GTP-bound and the inactive GDP-bound forms; this alternation is highly regulated by guanine-nucleotide-exchange factors (GEFs) and GTPase-activating factors (GAPs). In the budding yeast *Saccharomyces cerevisiae*, Rho1 can be activated by three GEFs (Rom1, Rom2, and Tus1) [14, 15] and is inactivated mainly by two major GAPs (Sac7 and Lrg1) [10, 16]. In addition, activation of Rho1can be inhibited by the guanine-nucleotide dissociation inhibitor Rdi1 [17]. The active Rho1-GTP forms complexes with effectors including the protein kinase Pkc1, the formin Bni1, the exocyst subunit Sec3, the glucan synthase Fks1 (and its paralog Fks2) [18–23], and there are probably other effectors that are not yet known. The highly selective binding by such effectors has been utilized to specifically detect GTP-bound Rho proteins by biochemistry and microscopy in both yeasts and other organisms. The original method was developed using the Rho-binding domain (RBD) of human Rhotekin: bacterially expressed GST-Rhotekin RBD was used to specifically pull down human RhoA-GTP, which was then detected and quantified by Western blotting [24, 25]. In yeast, the same Rhotekin RBD and other RBDs from the native effectors Pkc1 (C2-HR1 and C1 domains) and Bni1 have been used in similar biochemical assays [6, 11, 26, 27], with Rhotekin RBD and the Pkc1 C1 domain (amino acids 377–640, hereafter Pkc1 RBD) proving most efficient for pulling down Rho1-GTP in vitro [27] and in vivo (our unpublished results), although the other domains have been used for microscopic detection of Rho1-GTP in specific biological contexts [28]. The availability of these probes and the ease and the variety of methods for cell-cycle synchronization and genetic manipulation in yeast have allowed detection of Rho1 activity at specific times and places during the cell cycle [11, 26].

Here we describe a method in which Pkc1-RBD is used to monitor Rho1 activity through cytokinesis after cell-cycle synchronization using a *cdc15* mutation. Upon *cdc15* inactivation, cells complete chromosome segregation but fail to exit from mitosis. In this situation, the Polo-kinase Cdc5 is active at the bud neck, which in turn activates Rho1 to promote actin-ring formation [6]. However, because the machinery for primary-septum synthesis requires an active mitotic-exit network (MEN), in which Cdc15 plays a central role [29, 30], actual cleavage-furrow ingression does not take place until after release from the *cdc15*-mediated arrest. Thus, actomyosin-ring formation and subsequent processes in cytokinesis are now separated by 35–40 min as compared to the normal 10–15 min, allowing for a temporal dissection of the Rho1 activities involved in these processes. In such experiments,

Rho1 activity is initially high for formation of the actomyosin ring (and perhaps its maintenance) but is then quickly reduced before cleavage-furrow ingression, consistent with the apparent dispensability of Rho1 activity for this process. Rho1 is then slightly activated concomitant with secondary-septum formation/abscission. We also describe how Pkc1-RBD and Rhotekin-RBD detect different pools of Rho1-GTP during cytokinesis, to demonstrate how choice of the probe could affect the results. These observations highlight the importance of performing independent and parallel experiments, such as imaging-based visualization of Rho1 activity or genetic assays, to verify and support the biochemical data. The principle of this protocol can also be applied in similar experiments using other methods for cell-cycle synchronization and/or using probes for other Rho-type GTPases, such as Cdc42 [31], to gain a better understanding of the overall regulation and crosstalk among GTPases during cytokinesis [32].

2 Materials

2.1 Preparation of GST-Pkc1-RBD Beads

1. *E. coli* strain: BL21 (DE3) pLysS (*see* **Note 1**).
2. Plasmids: pGEX-PKC1RBD (also known as pB2535). This plasmid expresses amino-acid residues 377–640 of yeast Pkc1 (*see* **Note 2**) as a GST-fusion protein.
3. LB medium: (per liter) 10 g Bacto Trypton, 5 g Bacto Yeast Extract, 7.5 g NaCl, 300 μl 10 N NaOH, autoclaved. Add ampicillin (from 100 mg/ml stock in water) and/or chloramphenicol (from 30 mg/ml stock in ethanol) before use as described below.
4. IPTG stock solution: 100 mM in water; store at −20 °C.
5. STE Buffer: 10 mM Tris–HCl, pH 8.0, 150 mM NaCl, and 1 mM EDTA; store at 4 °C.
6. STE-15 % sarkosyl: STE Buffer containing 15 % (w/v) Sodium lauroyl sarcosinate; store at 4 °C.
7. STE-20%TX100: STE Buffer containing 20 % (v/v) Triton X-100; store at 4 °C.
8. STE-2%TX100: STE Buffer containing 2 % (v/v) Triton X-100; store at 4 °C.
9. DTT stock solution: 1 M in water; store at −20 °C.
10. Glutathione Sepharose 4 Fast Flow (GE Healthcare) or equivalent beads.
11. RPD Buffer: 50 mM Tris–HCl, pH 7.5, 150 mM NaCl, 12 mM $MgCl_2$, 1 mM EDTA (*see* **Note 3**), 1 mM DTT; store at 4 °C.
12. Lysozyme: 100 μg/ml in water.

Table 1
Yeast strains used in this protocol

Strain	Genotype	Source
YEF473A	*MAT*a *his3 leu2 lys2 trp1 ura3*	[37]
YEF473B	*MAT*α *his3 leu2 lys2 trp1 ura3*	[37]
MOY522	As YEF473A except *rho1Δ::his3MX6 URA3:3HA-RHO1*	[11]
MOY553	As YEF473B except *rho1Δ::his3MX6 URA3:3HA-RHO1 cdc15-2*	[11]
MOY686	As YEF473A except *lrg1Δ::His3MX6 rho1Δ::His3MX6 URA3:3HA-RHO1*	Our laboratory[a]
MOY718	As YEF473A except *sac7Δ::His3MX6 rho1Δ::His3MX6 URA3:3HA-RHO1*	Our laboratory[a]
MOY801	As YEF473A except [YCp-P_{GAL}-3HA-RHO1]	[11]
MOY803	As YEF473A except [YCp-P_{GAL}-3HA-RHO1^{Q68L}]	[11]
MOY824	As YEF473A except [YCp-P_{GAL}-3HA-RHO1^{T24N}]	[11]

[a]Constructed by crossing MOY522 to MOY407 (as YEF473B except *lrg1Δ::His3MX6*) or MOY408 (as YEF473B except *sac7Δ::His3MX6*) [11]

2.2 Preparation of Control and Cell-Cycle-Synchronized Yeast Samples

1. Yeast strains: *see* Table 1 for the strains that we have used with this protocol. Strains expressing N-terminally tagged 3HA-Rho1 were used for sensitive detection of Rho1 by a commercially available anti-3HA antibody (*see* **Note 4**).
2. SC-Leu (3 % raffinose) media: Dissolve 6.7 g Difco Yeast Nitrogen Base without amino acids and 2 g leucine-dropout powder (mix of amino acids, purines, and pyrimidines without leucine) in 900 ml water and autoclave. Separately filter-sterilize 30 g raffinose in 100 ml water and combine the two.
3. 40 % Galactose: 40 g galactose in 100 ml water; filter-sterilize.
4. YM-P medium (*see* **Note 5**): Dissolve 6.3 g Difco Yeast Nitrogen Base without amino acids, 4.5 g Yeast Extract, 9 g Difco Peptone, 9 g succinic acid, and 5.4 g NaOH in 900 ml water and autoclave. Separately autoclave 20 g glucose in 100 ml water and combine the two.
5. 37 % Formaldehyde: molecular-biology grade solution containing a low concentration of methanol is adequate.
6. Liquid nitrogen.

2.3 Preparation of Whole-Cell Extracts for Rho1-GTP Pull-Down Assays

1. Protease-inhibitor cocktail: EDTA-free cocktail purchased from Roche or Pierce. A tablet is dissolved in water at 100× manufacturer's recommendation, used at 2× concentration. The stock solution can be stored at −20 °C in small aliquots.

2. Glass beads: 0.4–0.6 mm diameter, purchased sterilized or autoclaved (*see* **Note 6**).
3. RPD-1.2 % NP40: RPD (Subheading 2.1) containing 1.2 % (w/v) Nonidet P40 (United States Biochemical Corp.); store at 4 °C.
4. RPD-1 % NP40: RPD containing 1 % (w/v) Nonidet P40 (*see* **Note 7**); store at 4 °C.

2.4 Rho1-GTP Pull-Down Assay Using GST-Pkc1-RBD

1. SDS sample buffer: 63 mM Tris–HCl, pH 6.8, 2 % SDS, 1 % beta-mercaptoethanol, 0.01 % bromophenolblue.
2. Antibody: Peroxidase-conjugated mouse anti-HA antibody (e.g. 3F10, Roche).

2.5 Rho1-GTP Pull-Down Assay Using GST-Rhotekin- RBD

1. pGEX-RBD (Rhotekin): This plasmid expresses amino-acid residues 7–89 of human Rhotekin as a GST fusion (Addgene plasmid #15247) [24].

3 Methods

3.1 Preparation of GST-Pkc1-RBD Beads

1. Preculture: Transform *E. coli* strain BL21 (DE3) pLysS with plasmid pGEX-PKC1RBD. Inoculate 2 ml of LB medium containing 50 μg/ml ampicillin and 30 μg/ml chloramphenicol with a single colony of the transformant. Shake or roll the culture overnight at 37 °C.
2. Expression: Inoculate 100 ml of fresh LB medium containing 50 μg/ml ampicillin (*see* **Note 8**) with the preculture to a final OD_{600} of ~0.1 (~1.6×10^6 cells/ml) and shake at 150 rpm at 37 °C. At $OD_{600} \approx 0.5$ (~8×10^6 cells/ml), add IPTG to a final concentration of 1 mM to induce expression of GST-Pkc1-RBD. Shake at 37 °C for 4 h.
3. Harvest the cells by centrifugation (3000 × *g*, 15 min, 4 °C) in a 50 ml conical tube. Wash the cells with 50 ml of ice-cold STE, centrifuge, and resuspend in 5.4 ml of STE containing 100 μg/ml lysozyme. After incubation on ice for 15 min, add 30 μl of 1 M DTT (5 mM final concentration) and 600 μl of 15 % STE-sarkosyl (1.5 % final concentration) (*see* **Note 9**). Lyse the cells by 6–12 × 15-s bursts of sonication with a 10-s cooling period between bursts. Clear the lysate by centrifugation (12,000 × *g*, 5 min, 4 °C) in 2-ml microcentrifuge tubes and transfer the supernatants to a 15-ml conical tube. Measure the volume, and add STE-20 % TX100 to achieve a final Triton X-100 concentration of 2 %.
4. Equilibrate 200 μl (packed volume) of Glutathione Sepharose 4 Fast Flow beads by washing three times with 1 ml STE-2 %

TX100. Resuspend the beads with the lysate from **step 3**, and incubate at 4 °C for 1 h on a rotator. Wash beads three times with 1 ml of RPD buffer, then resuspend in 1.4 ml RPD + 10 % glycerol (final volume including the beads is 1.6 ml). Store the beads in 200-μl aliquots (each containing 25 μl of beads) at −80 °C (*see* **Note 10**).

3.2 Preparation of Control and Cell-Cycle-Synchronized Cell Samples

1. Control samples: Grow wild-type cells transformed with galactose-inducible *3HA-RHO1*, *3HA-RHO1T24N*, or *3HA-RHO1Q68L* constructs (strains MOY801, MOY803, and MOY824) at ~24 °C in 47.5 ml of SC-Leu (3 % raffinose) medium to an $OD_{600} \approx 0.5$ (~8×10^6 cells/ml). Add 2.5 ml of 40 % galactose to induce expression of the transformed genes and incubate for 1 h (*see* **Note 11**). Grow wild-type, *lrg1Δ*, and *sac7Δ* cells harboring *3HA-RHO1* as the only endogenous copy of *RHO1* (strains MOY522, MOY686, and MOY718) in 50 ml of YM-P medium to an $OD_{600} \approx 0.5$ (~8×10^6 cells/ml). Collect the cells from each culture by centrifugation (3000 × *g*, ~24 °C, 2 min), freeze the pellets in liquid nitrogen, and store at −80 °C.
2. Synchronized samples: Grow *3HA-RHO1 cdc15-2* cells (strain MOY553) to exponential phase in 5 ml of YM-P medium at 24 °C. Dilute the cells to a final $OD_{600} \approx 0.05$ (~0.8×10^6 cells/ml) in 600 ml of fresh YM-P in a 2 L baffled Erlenmeyer flask and shake the culture until an $OD_{600} \approx 0.25$ (~4×10^6 cells/ml) is reached.
3. Transfer the flask to a 37 °C air (not water) incubator in order to gradually increase the medium temperature to 37 °C (*see* **Note 12**).
4. Start monitoring the progress of cell-cycle arrest at ~2 h by taking small samples of cells and examining by phase-contrast microscopy. Typically by 3 h, more than 95 % of cells are large-budded, and the OD_{600} reaches ~0.5 (~8×10^6 cells/ml).
5. Rapidly cool the culture to 24 °C by shaking the flask in ice-water bath. Use a thermometer (not necessary to sterilize it) to monitor the temperature. This process should be completed within 5 min for best time resolution. Start the timer.
6. Take samples at 15-min intervals, starting at the time of temperature shift to 24 °C, for 120 min. At each sampling, (a) take a 1-ml sample, fix the cells by adding 100 μl 37 % formaldehyde, and store at room temperature until all samples are collected and (b) collect the cells from a 50-ml sample by centrifugation (3000 × *g*, ~24 °C, 30 s) in a 50-ml conical tube, discard most of the medium by decantation, resuspend the cell pellet in the residual medium and transfer to a 1.7-ml microcentrifuge tube, and collect the cells by centrifugation

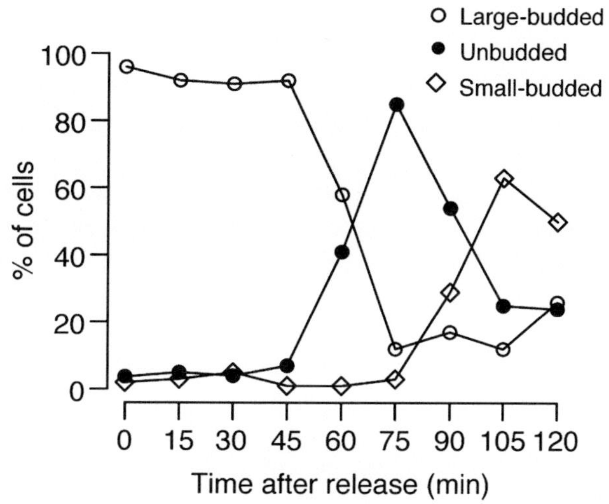

Fig. 1 Representative results of cell-cycle synchronization using *cdc15-2*. *3HA-RHO1 cdc15-2* cells (strain MOY553) were treated as described in Subheading 3.2, and >200 cells were counted for their morphologies at each time point

(12,000 × *g*, 24 °C, 10 s) in a benchtop microcentrifuge. Carefully remove the supernatant, snap-freeze the cell pellet in liquid nitrogen, and store at −80 °C.

7. To monitor the progress of cytokinesis and cell separation, the fixed cells should be briefly sonicated (<5 s) to separate post-cytokinesis cell pairs (*see* **Note 13**), and examined for budding indices by phase-contrast microscopy (Fig. 1).

3.3 Preparation of Whole-Cell Extracts for Rho1-GTP Pull-Down Assays

1. Thaw the cell pellets (~50 μl) from Subheading 3.2 on ice and resuspend in 100 μl RPD buffer supplemented with the protease-inhibitor cocktail at 2× concentration (see above).
2. Add 150–200 μl of pre-chilled glass beads to the resuspended cells. The top of the beads should be near the meniscus of the cell suspension. Vortex three to five times for 1 min at max speed with 30 s cooling periods. Confirm that 95–100 % of the cells have been disrupted by the dark appearance of disrupted cells in phase-contrast microscopy (*see* **Note 14**).
3. Remove the supernatants using a 200 μl pipet and transfer to fresh microcentrifuge tubes. Wash the glass beads with 400 μl RPD-1.2 % NP40 (with the protease-inhibitor cocktail at 2× concentration) buffer, and combine the supernatant with the previous sample. Each sample should have a final volume of ~500 μl and final NP40 concentration of 1.0 %. Incubate the samples on ice for 15 min.
4. Clear the samples of unbroken cells and cell debris by centrifugation (12,000 × *g*, 4 °C, 15 min). Determine the total protein

concentrations by Bradford assay and adjust it to 6 mg/ml in each sample with RPD-1 % NP40 (with the protease-inhibitor cocktail at 2× concentration). Use the extracts immediately for the Rho1-GTP pull-down assays.

3.4 Rho1-GTP Pull-Down Assay Using GST-Pkc1-RBD

1. Thaw tubes of GST-Pkc1-RBD beads from Subheading 3.1 and equilibrate the beads with RPD-1%NP40 by washing twice in this buffer. After the last wash, aliquot the suspended beads into microcentrifuge tubes so that 5 μl packed volume of beads are distributed to each tube. Centrifuge and carefully remove the supernatants using a 30G needle connected to a vacuum manifold.
2. In each tube, add a 500 μl sample of extract (3 mg of protein) from Subheading 3.3 to the beads, incubate at 4 °C on a rotator for 1 h, and wash the beads three times with 500 μl per wash of RPD-1 % NP40 buffer. After the last wash, spin the tube briefly, and remove the residual buffer from above and in-between the beads using a 30G needle connected to a vacuum manifold.
3. Elute the bound Rho1-GTP by adding 40 μl of SDS sample buffer and boiling for 5 min. For input controls, boil 15 μg (5 μl) of the extracts from Subheading 3.1 in SDS sample buffer. The samples can then be stored at −20 °C for at least a week or at −80 °C indefinitely.
4. Analyze the samples by SDS-PAGE (14 % Tris-glycine gel) and Western blotting using appropriate antibodies. We use a peroxidase-conjugated mouse anti-HA antibody. Quantify band intensities by densitometry and normalize Rho1-GTP values against input values. With control samples (Fig. 2), strong bands should be observed with *3HA-RHO1Q68L* and *sac7Δ 3HA-RHO1* but not with the other samples (*see* **Note 15**). With *cdc15-2* synchronization (Fig. 3), Rho1 activity is initially high for formation and maintenance of filamentous actin in the actomyosin ring but quickly decreases after release from the block and before the onset of cleavage-furrow ingression at ~30 min. At ~45 min, consistent and transient re-activation is observed, which coincides with the initiation of secondary-septum formation. Rho1 is then activated more strongly at later time points, presumably reflecting its involvement in bud growth in the new cell cycle (*see* **Note 16**).

3.5 Rho1-GTP Pull-Down Assay Using GST-Rhotekin RBD

Here, we briefly describe a similar assay using GST-Rhotekin-RBD to demonstrate how the choice of RBD can affect the results.

1. Preparation of GST-Rhotekin-RBD beads. Follow Subheading 3.1 except using plasmid pGST-RBD (Rhotekin) in place of

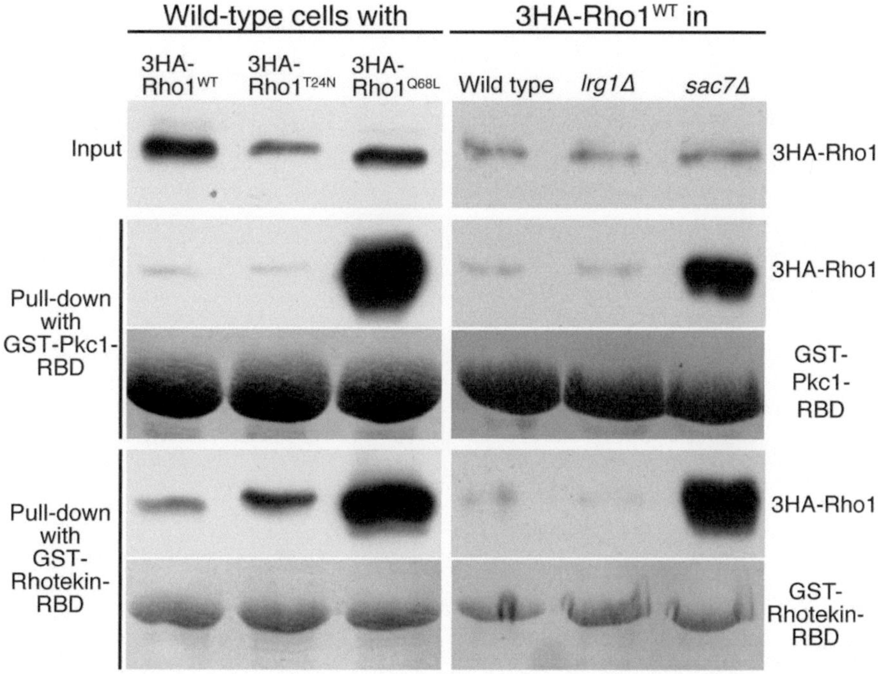

Fig. 2 Results of a control experiment to verify the specificity and capacity of GST-RBD beads to pull down active Rho1. *See* Subheadings 3.4 and 3.5 for details. Rho1-GTP was pulled down using GST-Pkc1-RBD and GST-Rhotekin-RBD in parallel from extracts of wild-type cells transiently overexpressing the indicated 3HA-Rho1 constructs (*left* three lanes; strains MOY801, MOY803, MOY824) or from extracts of wild-type and GAP-mutant cells expressing 3HA-Rho1 as the sole copy of Rho1 (*right* three lanes; strains MOY522, MOY686, and MOY718). 3HA-Rho1 in the input and pull-down samples was analyzed by SDS-PAGE and Western blotting using an anti-HA antibody. The membrane was subsequently stained with Ponceau S to reveal the GST-tagged proteins

pGEX-PKC1RBD. Alternatively, pre-bound beads can be purchased from commercial sources (*see* **Note 17**).

2. Follow Subheadings 3.2–3.4 using the GST-Rhotekin-RBD beads. Figure 2 (bottom) shows that GST-Rhotekin-RBD behaves similarly to GST-Pkc1-RBD in control experiments (*see* **Note 18**). In contrast, Fig. 3 shows an example of pull-down experiments carried out in parallel using the same yeast cell extracts and either GST-Pkc1-RBD or GST-Rhotekin-RBD. Note that Rhotekin-RBD fails to recognize Rho1-GTP at 0 min although it does detect the subsequent activation of Rho1. Moreover, the temporary inactivation of Rho1 after secondary-septum formation was less obvious with GST-Rhotekin-RBD than with GST-Pkc1-RBD. These results suggest that the two RBDs recognize different pools of Rho1-GTP in the cell, emphasizing the importance of choosing the right probe for the experiment and of using more than one experimental approach (*see* **Note 19**).

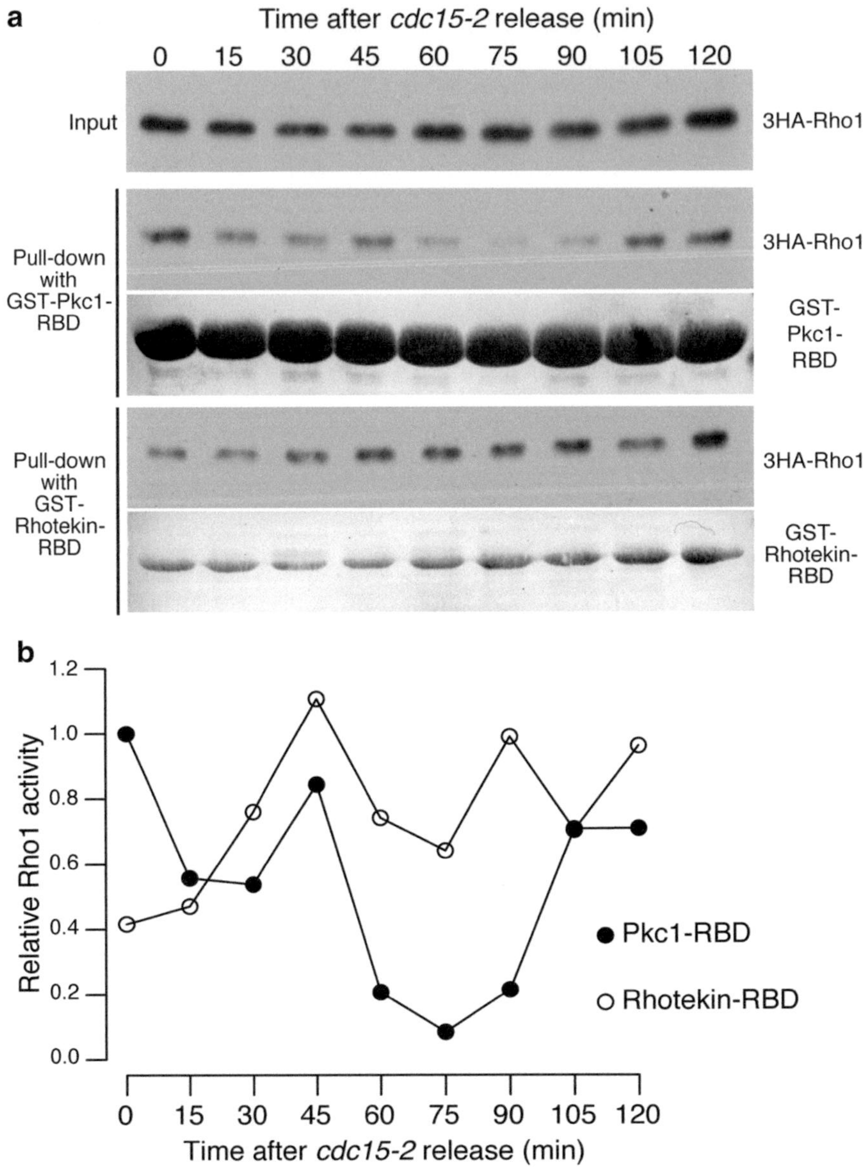

Fig. 3 A representative analysis of Rho1 activity during cytokinesis using two different probes for active Rho1 and synchronized *3HA-RHO1 cdc15-2* cells (strain MOY553). Analyses were as in Fig. 2. (**b**) The band intensities in (**a**) were quantified, and the relative intensities of active ("pull-down") relative to total ("input") Rho1 were plotted. The value for GST-Pkc1-RBD at time 0 was set to 1.0

4 Notes

1. Several other commonly available strains including BL21 and Rosetta (DE3) pLysS, and even the non-protease-deficient DH5α, have also been used successfully in our own and other labs.

2. The construction of this plasmid has been described previously [26], and a similar plasmid has also been used [27]. Note that fragments of Pkc1 (amino-acid residues 87–243 [27] or 87–391 [28]) containing another Rho1-binding domain (HR1 [33]) cannot be used for the pull-down assay described in this protocol ([27], our unpublished data).
3. We have found that inclusion of 1–2 mM EDTA in this buffer is important for reproducible detection of Rho1-GTP, especially with wild-type (i.e., not GTP or GDP-locked) Rho1.
4. It is important to tag Rho1 at its N terminus, because any alteration to the C terminus will render the protein nonfunctional by interfering with its prenylation. The functionality of the tagged gene should be tested by constructing a strain in which only the tagged allele is expressed and culturing the strain at 37 °C, where Rho1 activity is more important than at lower temperatures. Alternatively, untagged *RHO1* strains can be used if an anti-Rho1 antibody is available [18, 34]. However, our attempts to utilize a commercially available anti-Rho1 antibody (Santa Cruz Biotechnology sc-25818) were unsuccessful.
5. Conventional unbuffered YPD medium also works well for the assay described in this protocol.
6. Lower-grade glass beads with a wide diameter range appear to be most efficient in disrupting yeast cells. The same effect can also be achieved by mixing glass beads of different nominal diameters.
7. Such assays have sometimes used other non-denaturing detergents (or their combinations), such as 0.6 % CHAPS; 1 % Triton X100; or a combination of 1 % NP-40, 0.1 % SDS, and 1 % sodium deoxycholate (or "RIPA"). In our own experiments using control samples expressing 3HA-Rho1, 3HA-Rho1^{T24N}, or 3HA-Rho1^{Q68L}, replacing 1 % NP-40 with these detergents did not have any discernible effect on the results. We have not tested the effects of these other detergents on cell-cycle-synchronized samples.
8. Chloramphenicol is omitted in this step, because it is our experience that this translation inhibitor can interfere with the expression of induced proteins. pLysS (conveying chloramphenicol resistance) is not lost during a short incubation without selection.
9. For more details about this method for solubilization and purification of GST-tagged proteins using sarkosyl and Triton X100, *see* ref. [35].
10. We have successfully used up to 12-month-old beads with no signs of deterioration.

11. Presumably due to the "leaky" nature of the *GAL* promoter, *RHO1*^{T24N}-expressing cells may grow slowly even in raffinose-containing media. However, we prefer not to add glucose to suppress the *GAL* promoter during preculture, in order to ensure a rapid induction upon galactose addition. Upon induction, *RHO1*^{T24N} is lethal to the cells, so that the duration of expression should be kept short. *RHO1*^{G68L}, on the other hand, is not lethal to wild-type cells although it is not functional enough to support the growth of *rho1*Δ cells.
12. It is important not to increase the temperature too rapidly [36], and an increase of 2–3 °C per 10 min or slower gives optimal results. With a volume of 600 ml, simply shake the culture in a 37 °C air incubator. For a culture of less volume, it is best to first place the culture in a 30 °C air incubator for ~30 min, then transfer it to a 37 °C air incubator.
13. In a population of *cdc15-2*-synchronized cells, the timing of cell separation is delayed and heterogeneous in relation to cleavage-furrow ingression and abscission [36], and many cells remain connected by undigested primary septum after cytokinesis is complete. Such cell pairs can be separated after fixation by brief sonication.
14. In our experience, having no detergent present during vortexing with glass beads and vortexing at a high glass-beads-to-cell ratio appear to significantly improve the success of disruption.
15. This control experiment should be done every time a new batch of GST-Pkc1-RBD beads is prepared to confirm their specificity and sufficient capacity to pull-down all Rho1-GTP from cell extracts in subsequent experiments. In our strain background, deletion of *LRG1* seems to have little effect on the global Rho1-GTP level. Similarly, deletions of Rho1 GEFs have little effects on Rho1-GTP levels [10].
16. For quantitative assays, it is imperative not to saturate the GST-Pkc1 RBD beads with Rho1-GTP. Lack of saturation can be confirmed by comparing the maximum band intensities from the control samples and the cell-cycle-synchronized samples.
17. For example, Cytoskeleton, Inc. (http://www.cytoskeleton.com).
18. The slightly increased affinity of Rho1^{T24N} for the beads is reproducible in our hands. We do not know the explanation for this observation.
19. These results suggest that binding of Rho1 to RBDs is not governed exclusively by the nucleotide-binding state of Rho1. Other possible factors include binding of Rho1 to GEFs and GAPs that may compete with RBDs, thereby potentially defining a specificity of signaling to downstream effectors. These data also raise the possibility that RBDs from other Rho1

effectors might provide different results, although so far only Pkc1-RBD and Rhotekin-RBD have been used successfully in this assay. Thus, any conclusions based on this type of biochemical assay should also be supported by other independent data, such as microscopic assays for Rho1-GTP [28] and/or genetic evidence.

Acknowledgments

We thank Keiko Kono and Satoshi Yoshida for their helpful advice in developing this protocol. This work was supported in part by grants to JRP from the National Institutes of Health (GM31006) and RJEG Foundation and a postdoctoral fellowship to MO from the Uehara Memorial Foundation.

References

1. Fededa JP, Gerlich DW (2012) Molecular control of animal cell cytokinesis. Nat Cell Biol 14(5):440–447. doi:10.1038/ncb2482
2. Hong Z, Delauney AJ, Verma DP (2001) A cell plate-specific callose synthase and its interaction with phragmoplastin. Plant Cell 13(4): 755–768
3. Miller AL, Bement WM (2008) Regulation of cytokinesis by Rho GTPase flux. Nat Cell Biol 11(1):71–77
4. Piekny A, Werner M, Glotzer M (2005) Cytokinesis: welcome to the Rho zone. Trends Cell Biol 15(12):651–658
5. Tolliday N, VerPlank L, Li R (2002) Rho1 directs formin-mediated actin ring assembly during budding yeast cytokinesis. Curr Biol 12(21):1864 1870
6. Yoshida S, Kono K, Lowery DM, Bartolini S, Yaffe MB, Ohya Y, Pellman D (2006) Polo-like kinase Cdc5 controls the local activation of Rho1 to promote cytokinesis. Science 313(5783):108–111
7. Matsumura F (2005) Regulation of myosin II during cytokinesis in higher eukaryotes. Trends Cell Biol 15(7):371–377
8. Gai M, Camera P, Dema A, Bianchi F, Berto G, Scarpa E, Germena G, Di Cunto F (2011) Citron kinase controls abscission through RhoA and anillin. Mol Biol Cell 22(20):3768–3778. doi:10.1091/mbc.E10-12-0952
9. Hu CK, Coughlin M, Mitchison TJ (2012) Midbody assembly and its regulation during cytokinesis. Mol Biol Cell 23(6):1024–1034. doi:10.1091/mbc.E11-08-0721
10. Yoshida S, Bartolini S, Pellman D (2009) Mechanisms for concentrating Rho1 during cytokinesis. Genes Dev 23(7):810–823
11. Onishi M, Ko N, Nishihama R, Pringle JR (2013) Distinct roles of Rho1, Cdc42, and Cyk3 in septum formation and abscission during yeast cytokinesis. J Cell Biol 202(2): 311–329. doi:10.1083/jcb.201302001
12. O'Connell CB, Wheatley SP, Ahmed S, Wang YL (1999) The small GTP-binding protein rho regulates cortical activities in cultured cells during division. J Cell Biol 144(2):305–313
13. Chalamalasetty RB, Hummer S, Nigg EA, Sillje HH (2006) Influence of human Ect2 depletion and overexpression on cleavage furrow formation and abscission. J Cell Sci 119(Pt 14):3008–3019. doi:10.1242/jcs.03032
14. Ozaki K, Tanaka K, Imamura H, Hihara T, Kameyama T, Nonaka H, Hirano H, Matsuura Y, Takai Y (1996) Rom1p and Rom2p are GDP/GTP exchange proteins (GEPs) for the Rho1p small GTP binding protein in Saccharomyces cerevisiae. EMBO J 15(9): 2196–2207
15. Schmelzle T, Helliwell SB, Hall MN (2002) Yeast protein kinases and the RHO1 exchange factor TUS1 are novel components of the cell integrity pathway in yeast. Mol Cell Biol 22(5):1329–1339
16. Schmidt M, Bowers B, Varma A, Roh DH, Cabib E (2002) In budding yeast, contraction of the actomyosin ring and formation of the primary septum at cytokinesis depend on each other. J Cell Sci 115(Pt 2):293–302

17. Masuda T, Tanaka K, Nonaka H, Yamochi W, Maeda A, Takai Y (1994) Molecular cloning and characterization of yeast rho GDP dissociation inhibitor. J Biol Chem 269(31): 19713–19718
18. Qadota H, Python CP, Inoue SB, Arisawa M, Anraku Y, Zheng Y, Watanabe T, Levin DE, Ohya Y (1996) Identification of yeast Rho1p GTPase as a regulatory subunit of 1,3-beta-glucan synthase. Science 272(5259):279–281
19. Guo W, Tamanoi F, Novick P (2001) Spatial regulation of the exocyst complex by Rho1 GTPase. Nat Cell Biol 3(4):353–360
20. Nonaka H, Tanaka K, Hirano H, Fujiwara T, Kohno H, Umikawa M, Mino A, Takai Y (1995) A downstream target of *RHO1* small GTP-binding protein is *PKC1*, a homolog of protein kinase C, which leads to activation of the MAP kinase cascade in Saccharomyces cerevisiae. EMBO J 14(23):5931–5938
21. Drgonova J, Drgon T, Tanaka K, Kollar R, Chen GC, Ford RA, Chan CS, Takai Y, Cabib E (1996) Rho1p, a yeast protein at the interface between cell polarization and morphogenesis. Science 272(5259):277–279
22. Kamada Y, Qadota H, Python CP, Anraku Y, Ohya Y, Levin DE (1996) Activation of yeast protein kinase C by Rho1 GTPase. J Biol Chem 271(16):9193–9196
23. Kohno H, Tanaka K, Mino A, Umikawa M, Imamura H, Fujiwara T, Fujita Y, Hotta K, Qadota H, Watanabe T, Ohya Y, Takai Y (1996) Bni1p implicated in cytoskeletal control is a putative target of Rho1p small GTP binding protein in *Saccharomyces cerevisiae*. EMBO J 15(22):6060–6068
24. Ren XD, Kiosses WB, Schwartz MA (1999) Regulation of the small GTP-binding protein Rho by cell adhesion and the cytoskeleton. EMBO J 18(3):578–585. doi:10.1093/emboj/18.3.578
25. Kimura K, Tsuji T, Takada Y, Miki T, Narumiya S (2000) Accumulation of GTP-bound RhoA during cytokinesis and a critical role of ECT2 in this accumulation. J Biol Chem 275(23):17233–17236. doi:10.1074/jbc.C000212200
26. Kono K, Nogami S, Abe M, Nishizawa M, Morishita S, Pellman D, Ohya Y (2008) G1/S cyclin-dependent kinase regulates small GTPase Rho1p through phosphorylation of RhoGEF Tus1p in *Saccharomyces cerevisiae*. Mol Biol Cell 19(4):1763–1771
27. Logan MR, Jones L, Eitzen G (2010) Cdc42p and Rho1p are sequentially activated and mechanistically linked to vacuole membrane fusion. Biochem Biophys Res Commun 394(1):64–69. doi:10.1016/j.bbrc.2010.02.102
28. Kono K, Saeki Y, Yoshida S, Tanaka K, Pellman D (2012) Proteasomal degradation resolves competition between cell polarization and cellular wound healing. Cell 150(1):151–164. doi:10.1016/j.cell.2012.05.030
29. Weiss EL (2012) Mitotic exit and separation of mother and daughter cells. Genetics 192(4):1165–1202. doi:10.1534/genetics.112.145516
30. Wloka C, Bi E (2012) Mechanisms of cytokinesis in budding yeast. Cytoskeleton (Hoboken). doi:10.1002/cm.21046
31. Atkins BD, Yoshida S, Saito K, Wu CF, Lew DJ, Pellman D (2013) Inhibition of Cdc42 during mitotic exit is required for cytokinesis. J Cell Biol 202(2):231–240. doi:10.1083/jcb.201301090
32. Balasubramanian MK, Tao EY (2013) Timing it right: precise ON/OFF switches for Rho1 and Cdc42 GTPases in cytokinesis. J Cell Biol 202(2):187–189. doi:10.1083/jcb.201306152
33. Denis V, Cyert MS (2005) Molecular analysis reveals localization of *Saccharomyces cerevisiae* protein kinase C to sites of polarized growth and Pkc1p targeting to the nucleus and mitotic spindle. Eukaryot Cell 4(1):36–45. doi:10.1128/ec.4.1.36-45.2005
34. Drgonova J, Drgon T, Roh DH, Cabib E (1999) The GTP-binding protein Rho1p is required for cell cycle progression and polarization of the yeast cell. J Cell Biol 146(2):373–387
35. Frangioni JV, Neel BG (1993) Solubilization and purification of enzymatically active glutathione S-transferase (pGEX) fusion proteins. Anal Biochem 210(1):179–187
36. Futcher B (1999) Cell cycle synchronization. Methods Cell Sci 21(2-3):79–86
37. Bi E, Pringle JR (1996) *ZDS1* and *ZDS2*, genes whose products may regulate Cdc42p in *Saccharomyces cerevisiae*. Mol Cell Biol 16(10):5264–5275

Chapter 16

Detection of Phosphorylation Status of Cytokinetic Components

Franz Meitinger, Saravanan Palani, and Gislene Pereira

Abstract

Yeast cells can be easily cultured, synchronized, and genetically modified making them a convenient model system to study molecular mechanisms underlying cytokinesis. Here, we describe simple methods that allow the analysis of the phosphorylation profile of cytokinetic proteins, both in vivo and in vitro, using standard laboratory equipment. In addition, we compare the ability of three different, standard, and optimized acrylamide gel conditions to separate phosphorylated forms, using the protein Inn1 as an example.

Key words Cytokinesis, Phosphorylation, Cell cycle synchronization, Kinase, Phosphatase, Mitotic exit network (MEN), Dbf2, Cdc5, Cdk1, SDS-PAGE, Yeast

1 Introduction

Protein phosphorylation is one of the most important post-translational modifications that regulates the function and the activity of target proteins, either through a conformational change in the protein structure or by changing the charges of the protein surface thereby influencing protein–protein interactions [1–3]. Cytokinesis is controlled by spatio-temporal phosphorylation events, which connect the cell cycle machinery with structural rearrangements at the cell division site [4, 5]. Monitoring and characterizing phospho-regulation events during cytokinesis is key to understand the molecular basis of cell division.

The differential mobility of phosphorylated and non-phosphorylated forms of proteins on sodium dodecyl sulfate polyacrylamide gel electrophoresis (SDS-PAGE) has been vastly used as a measure for the analysis of phosphorylation events. In most of the studies, cells were stopped at a specific cell cycle stage (e.g. by addition of drugs) and allowed to resume the cell cycle after drug removal. This procedure of cell cycle synchronization allows one to

Alberto Sanchez-Diaz and Pilar Perez (eds.), *Yeast Cytokinesis: Methods and Protocols*, Methods in Molecular Biology, vol. 1369, DOI 10.1007/978-1-4939-3145-3_16, © Springer Science+Business Media New York 2016

follow protein mobility shifts by SDS-PAGE and Western blotting techniques in a cell cycle-dependent manner [6].

In yeast, cytokinesis initiates after the down-regulation of mitotic cyclin-dependent kinase activity that is triggered by the mitotic-exit-network (MEN) signaling pathway at the end of mitosis [4]. After its onset, cytokinesis takes only a few minutes to complete. Therefore, due to this short duration, one major challenge in the analysis of phosphorylation events during cytokinesis by SDS-PAGE/Western blotting is to obtain a well-synchronized cell culture, in which most of the cells undergo cytokinesis at the same time. The second major challenge is the ability to visualize phosphorylation shifts by SDS-PAGE and Western blotting. One should also keep in mind that not all phosphorylations cause a mobility shift, and the degree of the shift varies depending on proteins and phosphorylation sites. Also, mobility shifts may be caused by post-translational modifications other than phosphorylation. Here, we describe the synchronization strategy and SDS-PAGE gel conditions that we have used to study the phosphorylation of cytokinesis components by MEN kinases. Furthermore, we describe protocols to study phosphorylation of the kinase and substrate of interest in vitro.

2 Materials

Prepare all solutions with ultrapure distilled water. Make sure that hazardous materials are handled and disposed according to local safety regulations. Unless otherwise specified, solutions should be stored at room temperature (23 °C).

2.1 Sample Preparation for In Vivo Phosphorylation Detection

2.1.1 Yeast Growth Media, Synchronization, and Monitoring

1. YPDA medium: dissolve 10 g Bacto yeast extract (BD Biosciences), 20 g Bacto peptone (BD Biosciences), 20 g glucose and 0.1 g adenine (Sigma) in 1 l distilled water. Autoclave or filter-sterilize using 0.22 μm filter [7].
2. Ammonium persulfate: 10 % solution in water.
3. Nocodazole (Sigma): 1.5 mg/ml stock solution (100×) in dimethyl sulfoxide (DMSO), store in aliquots at −20 °C.
4. α-factor (Sigma): 1 mg/ml stock solution in DMSO, store in aliquots at −20 °C. Use at 1:1000 dilution for *bar1Δ* strains and 1:100 dilution for *BAR1* strains.
5. 4′,6-Diamidino-2-phenylindole (DAPI, Sigma): 1 mg/ml stock solution (10,000×) in PBS. Store in small aliquots at −20 °C. Prepare 0.1 μg/ml DAPI working solution by diluting stock 10,000 times in PBS, which can be kept protected from light at 4 °C for a couple of months.

2.1.2 Protein Precipitation and Sample Preparation

1. Trichloroacetic acid (TCA): prepare a 55 % (w/w) solution in water. Store solution at 4 °C and protected from light. Use protection gloves when working with TCA.
2. HU buffer: 8 M urea, 5 % sodium dodecyl sulfate, 200 mM $NaPO_4$ pH 6.8, 0.1 mM ethylenediaminetetraacetic acid (EDTA), 0.05 % bromophenol blue. Make 10 ml aliquots and store at −20 °C. To prepare HU-DTT, add 100 mM dithiothreitol (DTT) before use.
3. 5× SDS sample buffer: 4.5 ml glycerol, 0.5 g SDS, 1.75 ml 0.5 M Tris–HCl pH 6.8, 600 mM DTT (0.93 g), 5 mg bromophenol blue, adjust to a final volume of 10 ml with water. Store the buffer either at room temperature without DTT or at −20 °C with DTT.

2.2 Western Blots

1. SDS-PAGE equipment for casting, running, and blotting protein gels: Casting system, electrophoresis chambers, power supply, semi-dry blotter, Whatman® blotting paper and nitrocellulose transfer membrane.
2. 10 % SDS solution in water, filter-sterilized using a 0.22 μm bottle top filter (solution should not be autoclaved as it may precipitate irreversibly). Store at room temperature.
3. 99 % Tetramethylethylenediamine (TEMED).
4. 2-Amino-2-(hydroxymethyl)-1,3-propanediol (Tris base, Sigma): prepare 1.5 M (for Laemmli and Phospho gels) and 3 M (for Kornberg gels) stock solutions pH 8.8 and 0.5 M solution pH 6.8 (for stacking gels). Adjust the pH with HCl. Filter-sterilize using a 0.22 μm filter or autoclave. Store at room temperature.
5. Ammonium persulfate: 10 % solution in water. Make aliquots and freeze at −20 °C. Alternatively, store at 4 °C for up to 4 weeks.
6. Isopropanol.
7. SDS-PAGE protein standard markers.
8. Blotting buffer: 25 mM Tris, 192 mM glycine, 0.25 % SDS, 20 % methanol.
9. Ponceau S: 0.2 % (w/w) Ponceau S, 3 % (w/w) TCA in water.
10. Blocking solution: 3 % skimmed milk powder (electrophoresis grade, SERVA) in PBS-T.
11. PBS-T: 137 mM NaCl, 2.7 mM KCl, 10 mM Na_2HPO_4, 1.8 mM KH_2PO_4, 1 mM $CaCl_2$, 0.5 mM $MgCl_2$, pH 7.2, 0.2 % Tween 20.
12. Running buffer for Laemmli and Phospho gels [8]: 25 mM Tris base pH 8.3, 192 mM glycine, 0.1 % SDS to a final volume of 1 l in water. Autoclave and store at room temperature.

Table 1
Ingredients for 100 ml 30 % acrylamide stock solutions

	Acrylamide in g	Bisacrylamide in g	Ratio
Laemmli gel [8]	29.3	0.781	37.5:1
Phospho gel	29.7	0.266	111.7:1
Kornberg gel [9]	30.0	0.149	200:1

Running buffer for Kornberg gels [9]: 50 mM Tris base, 0.38 M glycine, 0.1 % SDS. Autoclave and store at room temperature.

13. 30 % Acrylamide stock solutions: dissolve acrylamide (Sigma, electrophoresis grade) and *N*,*N*′-methylene-bisacrylamide powder (hereafter referred to as bisacrylamide, GE Healthcare) in 60 ml water (for amounts *see* Table 1 and **Note 1**). Adjust the volume to 100 ml. Filter the solution using a 0.22 μm bottle top filter. Store the solution at 4 °C protected from light (e.g. in a dark-colored glass bottle or in a glass bottle wrapped in aluminum foil). The solution can be kept refrigerated for at least 6 months.

2.3 Dephosphorylation Assays

1. Screw cap tubes (1.5 ml tubes, screw caps with integral O-ring, Sarstedt).
2. Lysis buffer: 50 mM Tris–HCl pH 7.5, 100 mM NaCl, 1 mM EDTA pH 8.0, 10 % glycerol.
3. Triton X100: prepare 10 % stock solution and store it at 4 °C.
4. Glass beads (acid-washed, 250–500 μm, Sigma).
5. Ribolyser (e.g. Thermo Savant BIO101 Fastprep FP120).
6. Protease inhibitors:

Roche complete, EDTA-free protease inhibitor tablets (1 tablet is sufficient for 50 ml lysis buffer). One tablet can be dissolved in 1 ml water to obtain a 50× stock solution that can be stored at −20 °C.

Phenylmethanesulfonyl fluoride (PMSF, Sigma): prepare 500 mM stock in DMSO. Add to the buffer just before use as PMSF has a short half-life in aqueous solutions.

Phosphatase inhibitors:

β-Glycerophosphate (inhibits Ser/Thr phosphatases). Prepare 1 M stock solution in water. The working concentration is 100 mM.

Sodium fluoride (inhibits Ser/Thr and acidic phosphatases). Prepare 1 M stock solution in water. The working concentration is 50 mM.

Sodium orthovanadate (inhibits protein-phosphotyrosine phosphatases and alkaline phosphatases) [10, 11]. Prepare a 0.2 M stock solution in water (*see* **Note 5**). The working concentration is 0.2 mM.

Alkaline phosphatase (Sigma) or Lambda phosphatase (New England Biolabs).

2.4 In Vitro Phosphorylation Assays

1. Adenosine 5′-triphosphate disodium salt hydrate (ATP, Sigma).
2. ATP, [γ-^{32}P]-3000 Ci/mmol 10 m Ci/ml EasyTide, 250 μCi (PerkinElmer).
3. Colloidal Coomassie staining [12] (*see* **Notes 2–4** for preparation details).
4. Bovine Serum Albumin powder (Sigma).
5. Histone H1 (from calf thymus; Sigma).
6. Myelin Basic Protein (Sigma).
7. Casein (from bovine brain; Sigma).

3 Methods

3.1 Sample Preparation for Detection of In Vivo Phosphorylation During Cytokinesis by Phosphorylation Shift

3.1.1 Yeast Cell Synchronization and Sample Collection

To detect the phosphorylation profile of proteins during cytokinesis, we employ a two-step synchronization protocol, in which yeast cells are first arrested in the G1-phase of the cell cycle with mating pheromone (α-factor) and subsequently arrested in metaphase using microtubule depolymerizing drug nocodazole (*see* **Note 6**). This procedure generates a highly synchronous population after nocodazole wash-out. To arrest cells with α-factor, we employ the protocol originally described by Breeden et al. [13] with slight modifications.

1. Inoculate yeast cells in YPDA medium 2 days before analysis and keep cells in the logarithmic growth phase (OD_{600} < 1.0) (*see* **Note 7**).
2. One day before the experiment, inoculate an appropriate volume of cell culture, which is sufficient for the experiment (*see* **Note 7**).
3. On the day of the experiment, measure the OD_{600} of the culture and re-adjust it to 0.2 if necessary (*see* **Note 8**). Add α-factor to the cell culture to a final concentration of 1 μg/ml (for *bar1Δ* cells) or 10 μg/ml (for *BAR1* cells) and place cultures in a shaker with controlled temperature. Keep cells in *α-factor-containing* medium for approximately 2 h 15 min at 30 °C or 3 h 45 min at 23 °C (*see* **Note 9**).
4. Wash yeast cells twice with YPDA by centrifugation. For this, centrifuge yeast cell culture (e.g. 30 ml) in a 50 ml tube at

1000 × *g* for 2 min, decant the supernatant, resuspend the cell pellet in 30 ml YPDA and repeat the washing procedure once. After the second washing step, resuspend the cell pellet in 30 ml YPDA.

5. After the final wash in YPDA, transfer the cell suspension into a clean conical glass flask and add nocodazole the cell culture to a final concentration of 15 μg/ml. Place cultures in the incubation shaker until more than 90 % of cells are arrested in metaphase (i.e. large budded cells with one DAPI-stained DNA region). To check the arrest, stain ethanol-fixed cells with DAPI as described below. Inspect cells under the fluorescence microscope (excitation 350/50 nm; emission 470/30 nm) (*see* **Note 10**).
6. Release cells from nocodazole arrest by washing cells twice with YPDA as described in **step 4**.
7. After the final wash in YPDA, transfer the cell suspension into a clean conical glass flask and initiate sample collection (this will be the first time point, $t = 0$). After collecting the first sample, return the flask to the incubator (*see* **Note 11**). Samples should be taken accordingly to time points to be studied. Different samples are required for each time point:
   - Samples for protein extract preparation (TCA-precipitation, see below): pipette 1.5 ml of cell culture in a 2 ml tube. Centrifuge at 16,000 × *g* for 2 min at room temperature. Aspirate off the supernatant and place the tube on dry ice or liquid nitrogen to rapidly freeze the cell pellet. Frozen pellets can be stored at −20 °C or −80 °C freezer.
   - Samples for determination of the optical density at 600 nm (*see* **Note 12**).
   - Samples for DAPI staining and budding index analysis: mix 300 μl of cell culture with 700 μl of absolute ethanol (final ethanol concentration should be around 70 %, *see* **Note 13**). Keep cells on ice or proceed immediately with the DAPI staining. For DAPI staining, centrifuge cells (5000 × *g*, 3 min) and carefully remove supernatant. Centrifuge again to remove the excess of ethanol. Resuspend the cell pellet in 10–20 μl of 0.1 μg/ml DAPI solution in PBS.
8. Confirm synchronization by analysis and quantification of the budding index and DNA separation (*see* **Note 14**). Samples should be taken every 15 min until the bulk of cells enter a new cell cycle, which normally takes 1.5–2 h for wild-type cells growing in YPDA medium at 30 °C.

3.1.2 Protein Precipitation Using TCA

1. Resuspend the frozen cell pellets in 800 μl ice-cold distilled water. Keep tubes on ice.
2. Add 150 μl 1.85 M NaOH and vortex for 3 s to lyse the cells.
3. Add 150 μl of 55 % TCA solution and vortex for 3 s for protein precipitation. Keep on ice for 15 min.
4. Centrifuge cells for 20 min at 16,000 × *g* at 4 °C to pellet the precipitated protein. Remove supernatant carefully to avoid disturbing the pellet. Centrifuge briefly (30 s, 16,000 × *g*) and remove remaining supernatant (*see* **Note 15**).
5. Resuspend pellet in HU-DTT using a Vortex mixer or similar (*see* **Notes 16–18**).
6. Incubate mixture for 15 min at 65 °C (*see* **Note 19**).
7. Centrifuge at room temperature for 2–3 min at 16,000 × *g* (*see* **Note 20**).
8. Load 15 μl on a SDS-PAGE Gel (*see* **Notes 21** and **22**).

3.2 Western Blots to Monitor Protein Phosphorylation

3.2.1 SDS-PAGE Gel Preparation and Running Conditions

SDS-PAGE gels should be prepared fresh or 1 day before usage (*see* **Note 23**). The optimal acrylamide percentage and ratio of acrylamide and bisacrylamide of the SDS-PAGE gel depends on the size and characteristics of the protein of interest (Table 3 and *see* **Note 24**). Different gel conditions can significantly change the ability to separate phosphorylation forms of one protein (Fig. 1). Gel conditions should be optimized for each phospho-protein. Here, we describe three different conditions, which are commonly used in our laboratory (Tables 1, 2, and 3 and Fig. 1). Please see in addition following protocols for basic principles of SDS-PAGE and Western blotting [14–17].

1. Assemble the glass plates according to the manufacturer.
2. Mix and pour the separating gel (*see* **Note 25**).
3. Overlay the gel with isopropanol to prevent the formation of an uneven edge and drying of the gel.
4. After polymerization, remove isopropanol by tilting the plate and drying the liquid with a piece of Whatman® paper. Rinse carefully with water and remove the excess of water by carefully inserting a small piece of Whatman® paper between the plates (do not touch the gel surface). Mix and poor the stacking gel and attach the comb (*see* **Note 26**). Wait until the gel is completely polymerized.
5. Remove the comb and carefully wash the wells with water. Assemble the gel in the running chamber (follow the manufacturer's instructions), fill with running buffer and allow the gel to equilibrate for 15 min.
6. Load gel with a protein ladder marker (dilute in HU-DTT if necessary) and 15 μl of each sample of interest (*see* **Note 27**).

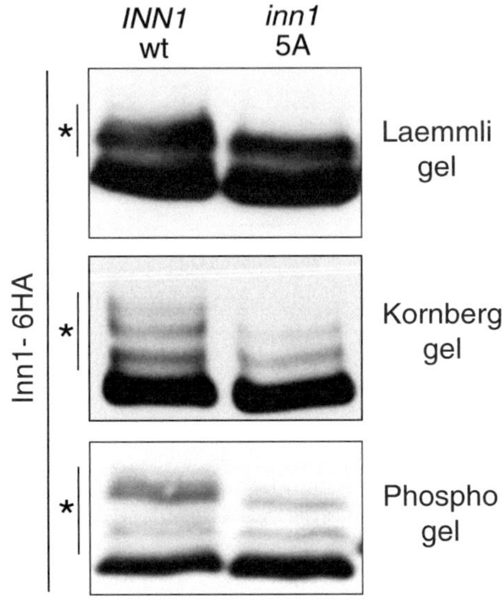

Fig. 1 Mobility shift of phosphorylated Inn1 with different SDS-PAGE conditions. The migration of phosphorylated wild-type Inn1-6HA and phospho-mutant Inn1-5A-6HA was analyzed by SDS-PAGE and Western blotting. The same cell lysate of wild-type and mutant was loaded on a Laemmli, Kornberg and Phospho gel and probed with anti-HA antibodies. Each gel shows different separating characteristics of the phosphorylated forms of Inn1 (*asterisks*). Please note that the Inn1-5A mutant, which cannot be phosphorylated by Cdk1 [18], misses distinct phosphorylation bands

Table 2
Composition of the stacking gel[a]

	Volume in μl[b]
Acrylamide stock (30 %, 37.5:1)	1300
0.5 M Tris–HCl pH 6.8	2500
10 % SDS	100
Water	6000
10 % APS	50
TEMED	10

[a]Final acrylamide/bisacrylamide concentration is 4 %
[b]The final volume is 10 ml, which is sufficient to cast two mini-protean gels using the BioRad system (plate dimensions: 100 × 80 mm with 1 mm spacer)

7. Connect the gel chamber to a power supply and run gels at 20 mA. Limit the voltage to 200 V per gel (*see* **Notes 28** and **29**).
8. Transfer proteins onto nitrocellulose membrane using an apparatus for semi-dry or wet Western blotting. Follow the manufacturer's instructions for assembly and transfer.

Table 3
Composition of separating gels

	Separating gel				
Protein size in kDa	>100	70–120	45–100	30–70	10–40
Final gel concentration	6 %	8 %	10 %	12 %	15 %
	Volume in μl[a]				
30 % Acrylamide stock[b]	2000	2666	3333	4000	5000
Water	5345	4679	4012	3345	2345
1.5 M or 3 M Tris–HCl pH 8.8[c]	2500				
10 % SDS	100				
10 % APS	50				
TEMED	5				

[a]The final volume is 10 ml, which is sufficient to cast two mini-protean gels using the BioRad system (plate dimensions: 100 × 80 mm with 1 mm spacer)
[b]For Laemmli gels, use acrylamide/bisacrylamide at 37.5:1 ratio, for phospho-gels 200:1 and for Kornberg gels 111.7:1 ratios
[c]Use 1.5 M for Laemmli and Phospho gels and 3 M for Kornberg gels

9. Incubate the membrane with Ponceau S solution for 5 min. Wash with water until the excess of Ponceau S is removed and protein bands are clearly visible. This staining allows you to confirm the successful transfer of the proteins to the membrane. At this stage, you can also use a pen to mark irregularities that occurred during the blotting (e.g. no transfer due to air bubbles) and the protein bands of the ladder for reference (if no pre-stained marker is used).
10. Incubate the membrane with 30–50 ml of blocking solution for 1 h at room temperature or overnight at 4 °C on a rocking shaker.
11. Probe the protein of interest with a suitable antibody (*see* **Note 30**). For this, dilute the antibody in blocking solution and incubate with the membrane for 1–2 h at room temperature or overnight at 4 °C. In general, 1 μg/ml antibody diluted in blocking buffer worked well for most of the antibodies used in the lab. Useful cell cycle markers, which should be probed in parallel to the protein of interest, include Clb2 (degradation during mitotic exit) and Sic1 (expression during mitotic exit). Cytokinesis occurs during Clb2 degradation and Sic1 accumulation. Phosphorylation or de-phosphorylation events that are involved in cytokinesis should occur during this transition.
12. Wash the membrane three times for 5 min with PBS-T on a rocking shaker.

13. Incubate the membrane for 1 h at room temperature with a suitable antibody against the primary antibody and fused to a reporter like horseradish peroxidase (diluted in blocking solution). These antibodies are commercially available. Follow the supplier's guidelines for dilutions.
14. Wash the membrane three times for 5 min with PBS-T on a rocking shaker.
15. Incubate the membrane with a substrate for horseradish peroxidase (e.g. enhanced chemiluminescence substrate, ECL) and use a suitable gel documentation station to visualize the chemiluminescence.

Figure 1 shows the separation of phosphorylation bands of Inn1 with three different gel conditions (*see* Tables 1, 2, and 3). Phosphorylation-deficient mutants (Inn1-A) are shown to confirm that the mobility shift is at least partially dependent on phosphorylation [18]. Alternatively, if the phosphorylation sites are not known, dephosphorylation assays can be used to confirm that the mobility shift depends on phosphorylation (*see* Subheading 3.3). *See* also **Note 31**.

3.3 Dephosphorylation Assays

Dephosphorylation assays should be performed to test whether the observed mobility shift is caused by phosphorylation and not other post-translational modifications.

1. Synchronize yeast cells as described in Subheading 3.1.1 and harvest the cells by centrifugation ($1000 \times g$, 2 min, room temperature) when most of the cells are in the cell cycle stage of interest (*see* **Notes 32** and **33**). Before harvesting the cells, measure the OD_{600} of the culture (1 OD_{600} corresponds to approximately 2×10^7 cells/ml). Harvest 2×10^9 cells in 50 ml tubes.
2. Resuspend cell pellet in 10 ml of ice-cold PBS and centrifuge as before. Discard the supernatant, resuspend cell pellet in 1 ml of ice-cold PBS and transfer cell suspension into a 1.5 ml conical screw cap tube (*see* **Note 34**). Centrifuge at $16{,}000 \times g$ for 2 min at 4 °C, discard supernatant and freeze the cell pellet at −20 °C.
3. Add 300 μl ice-cold lysis buffer containing phosphatase and protease inhibitors to the frozen cell pellet and add acid-washed glass beads until the liquid surface is covered.
4. Place the tubes in the ribolyser and break the cells using six cycles of homogenization (speed 5 m/s and run time 20 s). Allow cells to cool down between each cycle for 1 min (*see* **Note 35**). Check the efficiency of cell lysis under the microscope (lysed cells have a dark appearance in phase-contrast microscopy).

5. Transfer the cell lysate to a new tube, add 1 % Triton X100 (final concentration, *see* **Note 36**) and incubate tubes for 20 min at 4 °C in a rotating shaker.
6. Centrifuge cell lysate at 16,000 × *g* for 10 min at 4 °C to separate cell debris from the cell lysate.
7. Carefully transfer the cell lysate into a new tube (do not disturb the cell pellet). Mix the cell lysate with sepharose beads coupled to an antibody against the protein of interest (*see* **Note 37**) [19]. Place tubes on a rotating shaker (use a gentle rotation).
8. Incubate for 2 h at 4 °C (*see* **Note 37**).
9. After incubation, centrifuge the tube at 500 × *g* for 1 min at 4 °C. Place the tube on ice for a few minutes to make sure that beads are completely pelleted. Discard the supernatant and wash the beads twice with lysis buffer and once with phosphatase buffer (supplied with the phosphatase).
10. Split the beads into three tubes, each containing approx. 4 μg of the immunoprecipitated protein-bound to beads (*see* **Note 38**). To one tube, add 1 unit of alkaline phosphatase [20] or 1 unit of lambda phosphatase [21, 22]. Adjust the volume with the supplied phosphatase buffer. To the second tube, add buffer only. To the third tube, add the phosphatase together with phosphatase inhibitor(s) (e.g. 10 mM sodium orthovanadate or 50 mM EDTA) [10]. The final reaction volume is 50 μl. Incubate reactions for 30 min at 30 °C.
11. Stop reactions by adding 5× sample buffer and boiling for 2 min at 95 °C.
12. Run samples on a SDS-PAGE gel of choice and perform a Western blot as described. The amount of protein to be loaded on the SDS-PAGE gel depends on the amount of protein of interest in the lysate and on the sensitivity of the antibody used to detect the protein on Western blots. Phosphorylated bands will disappear after treatment with phosphatase (but not in the buffer-only control or in the reaction containing phosphatase inhibitors), indicating that the mobility shift was due to phosphorylation.

3.4 Detection of Phosphorylation of Cytokinesis Proteins In Vitro

In vitro phosphorylation can be used to validate the activity of a specific kinase on your protein of interest. This method, in combination with mass spectrometry, also permits one to determine the kinase dependent phosphorylation sites contained within your protein of interest.

1. Purify the kinase of interest as described [23–25]. If possible, also purify a kinase dead or kinase analog-sensitive mutant [26] to be used as a negative control for all reactions. Keep the purified kinases in small aliquots (enough for 2–4 kinase reactions) at −80 °C (*see* **Note 39**).

Table 4
Buffer conditions used for different mitotic kinases [27–30]

Kinase(-regulatory subunit)	Cdk1(-Clb2)	Cdc5	Dbf2(-Mob1)
Buffer	25 mM HEPES pH 7.5	50 mM Tris–HCl pH 7.5	50 mM Tris–HCl pH 7.5
Salt	100–150 mM NaCl[a]	0–100 mM NaCl[a]	0–100 mM NaCl Or 75 mM $C_2H_3KO_2$[a]
Add 0.25 μl ATP γ-^{32}P or 4 μl 100 mM ATP, 50 mM glycerophosphate, 10 mM NaF, 10 mM $MgCl_2$ to all reactions			

[a]The corresponding kinase was reported to work in this range of molarity and/or salt

2. Purify protein of interest to be used as a substrate (*see* **Note 40**). Alternatively, Histone H1, Casein or Myelin Basic Protein can be used as substrates for many kinases (*see* **Note 41**).
3. Incubate protein of interest (around 0.5–1 μg) with the purified kinase (*see* **Note 42**) in the presence of ATP γ-^{32}P for autoradiography (*see* **Note 43**). ATP is the last ingredient to be added as it starts the reaction. Prepare control reactions that lack the protein of interest or the kinase. Control reactions should also be performed with kinase dead or kinase analog-sensitive mutants, as previously stated (*see* **step 1** in Subheading 3.4). Buffer conditions depend on the kinase. *See* Table 4 for three examples of buffers used for MEN kinases involved in cytokinesis. Prepare 20 μl reactions.
4. The incubation time depends on the kinase activity and can range between 5 min and 60 min (*see* **Note 44**).
5. Stop reactions by adding 5 μl 5× sample buffer. Incubate samples for 2 min at 95 °C.
6. Load samples on an appropriate SDS-PAGE gel and run gel as described.
7. Apply gel onto Whatman® filter paper and cover with transparent foil.
8. Dry gel on a vacuum gel dryer for 60 min at 95 °C.
9. Expose dried gel to a phospho-imaging screen in a cassette for 1 h–3 days (incubation times vary depending on the signal intensity).
10. Read out the screen with a phospho-imager.
11. Re-hydrate gel with water (*see* **Note 45**).
12. Gel can be subject to Western blot analysis or Coomassie staining to control for the amount of the substrate and kinase used in the different reactions.

4 Notes

1. Several acrylamide stock solutions are commercially available (e.g. 37.5:1, 29:1 or 19:1 from Roth or Bio-Rad). To our knowledge, stock solutions for the ratios 111.7:1 and 200:1 are not commercially available. The ratio between acrylamide and bisacrylamide can be further optimized for each specific phospho-protein if the herein described conditions do not work.
2. Ingredients of colloidal Coomassie: water, methanol, 117 ml H_3PO_4; 100 g ammonium sulfate; 1.2 g Coomassie Brilliant Blue G-250. Mix 100 ml dH_2O with 117 ml H_3PO_4. Add ammonium sulfate to the solution and stir until completely dissolved. Add Coomassie and water to a total volume of 800 ml. Add 200 ml methanol and mix (*see* also **Notes 3** and **4**).
3. Since methanol tends to evaporate, we normally store the CBB solution without methanol and add it just before use.
4. Fifty milliliters of this solution (i.e. 40 ml of CBB plus 10 ml methanol added before use) are sufficient to stain 1 SDS-PAGE gel.
5. Sodium orthovanadate has to be activated before use. The activation procedure depolymerizes the vanadate and converts it into a more potent inhibitor. Adjust the pH of the sodium orthovanadate stock solution to ten using either NaOH or HCl. The pH of the solution varies according to the quality of the chemical and can be higher or lower than 10. Boil the sodium orthovanadate stock solution for 10 min or until it turns colorless. Re-adjust the pH to 10 with HCl. The sodium orthovanadate stock solution turns yellow after adding HCl. Repeat boiling, cooling down and re-adjusting the pH until the pH stabilizes at 10.
6. Alternatively, yeast cells can be robustly arrested in metaphase by repressing the expression of the anaphase-promoting complex activator, *CDC20* [31]. This requires the exchange of the endogenous *CDC20* promoter by a repressible one (e.g. Gal1 or Met promoter) through genetic modification of the yeast genome.
7. Start with a 5 ml culture. In the morning, inoculate around 2×10^7 cells in 5 ml YPDA medium. This corresponds to an optical density of 0.2 at the wavelength 600 nm (OD_{600}, 1 $OD_{600} = 2 \times 10^7$ cells/ml). Measure the OD_{600} of the culture in the evening again (dilute the culture for the OD measurement, if necessary. Keep in mind that the measured value has to be between 0.1 and 0.8 to be accurate. In this range, the relationship between optical density and cell concentration is linear). Taking into consideration the doubling time of the cells, dilute the cells to obtain a culture of $OD_{600} = 0.5$ in the next morning.

Example for a yeast culture grown at 30 °C in YPDA medium with a doubling time of 1.5 h: in 15 h, the cell culture doubles 10 times. Thus, you have to divide 0.5 (desired OD_{600} at the next day) by 1024 ($=2^{15h/1.5h}=2^{10}$) to obtain the concentration of the cell culture you have to inoculate. In this case, the cell culture should have an OD_{600} of 0.0005. The day before the experiment, inoculate cells in the volume required for your experiment, which can be estimated considering the amount of culture required for each sample (for Western blotting, DAPI staining and OD_{600} measurement) and the number of time points. For the initial experiment, we recommend taking samples every 15 min for 2 h after release from the metaphase block (i.e. 9 time points: 0 min, 15 min, 30 min, …, 120 min). For this experiment, we suggest to start with 30 ml of cell culture.

8. In our hands, the cell cycle arrest procedure works best when the starting OD_{600} of the culture is between 0.2 and 0.3 (i.e. around 4×10^6 cells per milliliter).
9. We incubate the cells for 1.5 generation times with α-factor. This allows all cells to complete the previous cell cycle and to arrest in the G1 phase as unbudded cells with a mating projection (shmoo). Check for this morphology under the microscope. Incubate with α-factor until more than 90 % of the cells form a shmoo. The time of incubation with α-factor may vary depending on the genetic background of the strain.
10. After release from α-factor arrest, cells need about 1 h 30 min at 30 °C or 2 h 30 min at 23 °C to arrest in metaphase upon nocodazole treatment. Release cells from metaphase arrest when more than 90 % of the cells are large budded and contain one DAPI-stained region. Prolonged incubation with nocodazole leads to adaptation of the cells, which allows them to eventually escape from the metaphase arrest. Escape is indicated by the appearance of cells with two DAPI-stained regions. Do not allow more than 5 % of the cells to escape.
11. It is important to work fast at this point because the arrested cells may proceed with the cell cycle during a prolonged washing procedure. It is also advisable to collect samples within a short timeframe to ensure that all samples represent the same time point. Therefore, prepare all tubes for sample collection in advance and take samples close to the incubator. Proceed immediately with sample processing to stop cell cycle progression (e.g. by freezing or fixation).
12. The OD_{600} should be between 0.3 and 0.8. If necessary, dilute the cell culture accordingly. Dilute all samples in the same way.
13. The cell suspension can be kept in ethanol for several days at 4 °C. Once cells are re-suspended in PBS/DAPI, they should be inspected on the same day.

14. The budding index is the percentage of cells with no bud (G1 phase), small bud (G1-S phase), and large bud (M phase).
15. The second brief centrifugation is required to completely remove the TCA solution. Remove the supernatant carefully as soon as the centrifugation stops because the pellet tends to detach from the tube over time. The remaining supernatant can be removed by aspiration but we normally use a micropipette to prevent unwanted pellet loss.
16. The amount of HU-DTT should be adjusted experimentally for your protein of interest. We usually normalize the amount of HU-DTT per cell pellet accordingly to the OD_{600} and volume of the cell culture. For most proteins we have analyzed, 150 μl of HU-DTT per 1 OD_{600} pellet (corresponding to approx. 2×10^7 cells) worked well. For this, for example, we resuspend the TCA-precipitate of 1 ml cell culture of $OD_{600} = 1$ in 150 μl of HU-DTT. If 2 ml of the same culture is used (total of 2 OD_{600}), the TCA precipitate is resuspended in 300 μl of HU-DTT. For proteins that are less abundant, a ratio of 1 OD_{600} to 75 μl of HU-DTT works better.
17. The HU-DTT buffer turns yellow if the TCA is not completely removed. In this case, re-adjust the pH by adding stepwise 1 μl of a 2 M Tris solution (without pH adjustment) until the buffer turns blue again.
18. Process the samples at room temperature after addition of HU-DTT because urea solidifies in the cold.
19. Alternatively, samples could be heated up for 5 min at 95 °C but 15 min at 65 °C is preferable. Urea breaks down to isocyanate at higher temperatures over time. This causes carbamylation of proteins that can lead to artifact on SDS-PAGE gels.
20. Usually a white pellet of insoluble cell debris is visible at the bottom of the tube. Avoid resuspending this pellet prior to SDS-PAGE gel loading. Cell debris may disturb the migration and separation of proteins during electrophoresis.
21. We recommend loading 15 μl of the lysate per well. However, the optimal amount depends on the concentration of the protein of interest in the cell lysate and thus needs optimization for each protein.
22. Samples can be stored for several months at −20 °C. However, some phospho-proteins are sensitive to freezing and a phosphorylation-dependent shift might be lost after freezing the samples.
23. SDS-PAGE gels can be stored at 4 °C wrapped in wet tissue paper and foil to prevent drying. The separation of phospho-bands can be compromised when using older SDS-PAGE gels.

24. The ratio between acrylamide and bisacrylamide can be varied. Bisacrylamide, which is responsible to crosslink the acrylamide network, may be increased or decreased to favor the separation of the phospho-bands. However, a general rule is not applicable.
25. The volume of the mix will depend on the size of the plates and thickness of spacers being used; check the manufacturer's instructions or determine it empirically. Mix all ingredients in a 50 ml tube. The mix will stay liquid until APS and TEMED are added. Mix gently by inverting the tube. Do not vortex to prevent bubble formation. The gel will solidify within 30–60 min at room temperature depending on the percentage of acrylamide/bisacrylamide. Leave some gel mix in the 50 ml tube to check the progress of solidification.
26. For the stacking gel, we normally leave 2–4 mm of distance between the bottom of the well and the separating gel. In our experience, larger stacking gels lead to inefficient separation of phosphorylation bands in several cases.
27. Pre-stained protein markers are used to monitor the migration of proteins according to molecular weight during the SDS-PAGE gel run. We normally stop running the gel when the molecular marker, which is closest to the weight of the protein of interest, migrates half to two-thirds of the separating gel. At this point, the phosphorylation bands should be well separated. Running the gel for longer periods can compromise the resolution of the protein bands.
28. Separation of phosphorylation bands of high molecular weight or in high percentage gels may be improved by running gels with less current and/or in the cold (i.e. in a cold room or using an ice bath).
29. Run Kornberg gel and Phopho Gel with constant 100 V in a cold room or using an ice bath.
30. Some primary antibodies can be reused. In this case store the antibody-milk solution at −20 °C. If no antibody is available for your protein of interest, you may consider using C- or N-terminal fusions of your protein with epitopes, e.g. hemaglutinin (HA) or Myc, against which established commercial antibodies are available.
31. Phos-tag gels can be used if no phospho-shift is observed with the described method [32]. Alternatively, a phospho-specific peptide antibody can be raised against a predicted phosphorylation site.
32. It is useful to set up a small-scale experiment to determine the exact time point at which you want to determine the phosphorylation sites of your protein of interest (e.g. time point

with the highest phosphorylation shift or specific time point of the cell cycle).

33. The necessary amount of cells depends on the abundance of the protein. Cell culture volumes can vary from 100 ml to 2 l.
34. Please note that the tubes should be compatible with the ribolyser.
35. Alternatively, cells can be lysed using a Vortex mixer. To do so, harvest 2×10^9 cells in a 50 ml tube. Add 3 ml lysis buffer and enough glass beads to touch the meniscus of the lysate without covering the liquid surface. Vortex for 1 min and place the tube on ice for 1 min. Repeat this cycle until more than 80 % of the cells are lysed (dark/ghost appearance under the light microscope).
36. To transfer the cell lysate to a new tube, cut a slit with a scalpel into the lid. Place the tube upside down in a 15 ml tube and centrifuge for 2 min at $1000 \times g$ at 4 °C. Remove the screw cap tube and transfer the cell lysate into a 2 ml tube.
37. If no specific antibody against the protein of interest is available, the protein of interest can be fused to standard epitopes, like HA or Myc, using standard gene fusion protocols [33, 34]. Commercially available antibodies against the epitope tags are then used for the coupling reaction with beads. We normally use Protein-A or Protein-G Sepharose beads and standard protocols for the coupling reactions [19]. You should also plan controls for unspecific binding. Therefore, if using an antibody against your protein, a strain in which the corresponding gene was deleted could be used as negative control. If using tagged strains, also perform the immunoprecipitations with strains carrying untagged proteins.
38. We normally estimate the amount of protein bound to the beads by comparing it with different amounts of bovine serum albumin (BSA) of known concentration. For this, we run a SDS-PAGE gel with a defined amount of beads (10 μl) and BSA standard (e.g. 0.25, 0.5, 1.0, 2.0, 4.0 and 8 μg BSA). Both, BSA and beads are resuspended in HU-DTT and incubated for 15 min at 65 °C before loading to the gel. Centrifuge for 2 min at full speed (room temperature) in a benchtop centrifuge. Make sure that the beads are not loaded onto the gel, as they disturb the migration behavior of the loaded protein.
39. Most kinases retain their activity after freezing. However, multiple thawing and freezing steps may reduce the kinase activity.
40. We recommend recombinant protein purified from bacteria as a source of non-phosphorylated substrate. Full-length or truncated forms of the protein of interest can be used as a substrate.

41. Histone H1, Casein, and Myelin Basic Protein are unspecifically phosphorylated by most kinases and can be used as substrates to confirm that the purified kinase is active. Use 10–20 μg of these substrates per reaction. Each substrate may work better for different kinases. Thus, we recommend initially testing the kinase with all three substrates.
42. The amount of kinase varies and depends on the activity of the purified kinase. We recommend to initially test different amounts of the purified kinase to determine the lowest amount of the kinase that gives the maximum phosphorylation in a defined reaction time (e.g. 30 min). An excess of kinase can lead to unspecific phosphorylation. We recommend using 100 ng or less of the kinase per reaction.
43. This protocol can also be used to determine phosphorylation sites by mass spectrometry. In this case, use ATP instead of ATP γ-^{32}P in the kinase reaction. To stop the kinase reaction, add 5 μl of 5× sample buffer and incubate samples for 2 min at 95 °C. Load samples on an appropriate SDS-PAGE gel and stain with colloidal Coomassie. Send for mass spectrometry analysis. Note, the herein described colloidal Coomassie staining is compatible with mass spectrometry.
44. We recommend a preliminary incubation of 1 h. To determine phosphorylation kinetics, several incubation times need to be analyzed before reaching the kinase-substrate saturation level.
45. Remove foil and pour water quickly and equally on the dried gel. Unequal re-hydration can cause tearing of the gel.

References

1. Barford D, Hu SH, Johnson LN (1991) Structural mechanism for glycogen phosphorylase control by phosphorylation and AMP. J Mol Biol 218(1):233–260
2. Seet BT, Dikic I, Zhou MM, Pawson T (2006) Reading protein modifications with interaction domains. Nat Rev Mol Cell Biol 7(7):473–483, doi: 10.1038/nrm1960
3. Strickfaden SC, Winters MJ, Ben-Ari G, Lamson RE, Tyers M, Pryciak PM (2007) A mechanism for cell-cycle regulation of MAP kinase signaling in a yeast differentiation pathway. Cell 128(3):519–531, doi: 10.1016/j.cell.2006.12.032
4. Meitinger F, Palani S, Pereira G (2012) The power of MEN in cytokinesis. Cell Cycle 11(2):219–228
5. Barr FA, Gruneberg U (2007) Cytokinesis: placing and making the final cut. Cell 131(5):847–860
6. Bouchoux C, Uhlmann F (2011) A quantitative model for ordered Cdk substrate dephosphorylation during mitotic exit. Cell 147(4):803–814, doi: 10.1016/j.cell.2011.09.047
7. Sherman F (1991) Getting started with yeast. Methods Enzymol 194:3–21
8. Laemmli UK (1970) Cleavage of structural proteins during the assembly of the head of bacteriophage T4. Nature 227(5259):680–685
9. Thomas JO, Kornberg RD (1975) An octamer of histones in chromatin and free in solution. Proc Natl Acad Sci U S A 72(7):2626–2630
10. Gordon JA (1991) Use of vanadate as protein-phosphotyrosine phosphatase inhibitor. Methods Enzymol 201:477–482
11. Huyer G, Liu S, Kelly J, Moffat J, Payette P, Kennedy B, Tsaprailis G, Gresser MJ, Ramachandran C (1997) Mechanism of inhibition of protein-tyrosine phosphatases by

vanadate and pervanadate. J Biol Chem 272(2):843–851

12. Candiano G, Bruschi M, Musante L, Santucci L, Ghiggeri GM, Carnemolla B, Orecchia P, Zardi L, Righetti PG (2004) Blue silver: a very sensitive colloidal Coomassie G-250 staining for proteome analysis. Electrophoresis 25(9):1327–1333, doi: 10.1002/elps.200305844
13. Breeden LL (1997) Alpha-factor synchronization of budding yeast. Methods Enzymol 283:332–341
14. Harlow E, Lane D (2006) Immunoblotting: antigen detection using chemiluminescence. CSH Protoc 2006(1), doi: 10.1101/pdb.prot4271
15. Harlow E, Lane D (2006) Immunoblotting: semi-dry electrophoretic transfer of proteins from gels to membranes. CSH Protoc 2006(1), doi: 10.1101/pdb.prot4301
16. Sambrook J, Russell DW (2006) SDS-polyacrylamide gel electrophoresis of proteins. CSH Protoc 2006(4), doi: 10.1101/pdb.prot4540
17. Towbin H, Staehelin T, Gordon J (1979) Electrophoretic transfer of proteins from polyacrylamide gels to nitrocellulose sheets: procedure and some applications. Proc Natl Acad Sci U S A 76(9):4350–4354
18. Palani S, Meitinger F, Boehm ME, Lehmann WD, Pereira G (2012) Cdc14-dependent dephosphorylation of Inn1 contributes to Inn1-Cyk3 complex formation. J Cell Sci 125(Pt 13):3091–3096
19. Harlow E, Lane D (2006) Immunoaffinity purification: coupling antibodies to protein a or g bead columns. CSH Protoc 2006(1), doi: 10.1101/pdb.prot4303
20. Labugger R, Organ L, Collier C, Atar D, Van Eyk JE (2000) Extensive troponin I and T modification detected in serum from patients with acute myocardial infarction. Circulation 102(11):1221–1226
21. Cohen PT, Cohen P (1989) Discovery of a protein phosphatase activity encoded in the genome of bacteriophage lambda. Probable identity with open reading frame 221. Biochem J 260(3):931–934
22. Zhuo S, Clemens JC, Hakes DJ, Barford D, Dixon JE (1993) Expression, purification, crystallization, and biochemical characterization of a recombinant protein phosphatase. J Biol Chem 268(24):17754–17761
23. Holt LJ, Tuch BB, Villen J, Johnson AD, Gygi SP, Morgan DO (2009) Global analysis of Cdk1 substrate phosphorylation sites provides insights into evolution. Science 325(5948):1682–1686
24. Konig C, Maekawa H, Schiebel E (2010) Mutual regulation of cyclin-dependent kinase and the mitotic exit network. J Cell Biol 188(3):351–368
25. Geymonat M, Spanos A, Sedgwick SG (2007) A Saccharomyces cerevisiae autoselection system for optimised recombinant protein expression. Gene 399(2):120–128
26. Bishop AC, Ubersax JA, Petsch DT, Matheos DP, Gray NS, Blethrow J, Shimizu E, Tsien JZ, Schultz PG, Rose MD, Wood JL, Morgan DO, Shokat KM (2000) A chemical switch for inhibitor-sensitive alleles of any protein kinase. Nature 407(6802):395–401
27. Ubersax JA, Woodbury EL, Quang PN, Paraz M, Blethrow JD, Shah K, Shokat KM, Morgan DO (2003) Targets of the cyclin-dependent kinase Cdk1. Nature 425(6960):859–864
28. Meitinger F, Boehm ME, Hofmann A, Hub B, Zentgraf H, Lehmann WD, Pereira G (2011) Phosphorylation-dependent regulation of the F-BAR protein Hof1 during cytokinesis. Genes Dev 25(8):875–888
29. Mah AS, Elia AE, Devgan G, Ptacek J, Schutkowski M, Snyder M, Yaffe MB, Deshaies RJ (2005) Substrate specificity analysis of protein kinase complex Dbf2-Mob1 by peptide library and proteome array screening. BMC Biochem 6:22
30. Geymonat M, Spanos A, Walker PA, Johnston LH, Sedgwick SG (2003) In vitro regulation of budding yeast Bfa1/Bub2 GAP activity by Cdc5. J Biol Chem 278(17):14591–14594, doi: 10.1074/jbc.C300059200
31. Shirayama M, Toth A, Galova M, Nasmyth K (1999) APC(Cdc20) promotes exit from mitosis by destroying the anaphase inhibitor Pds1 and cyclin Clb5. Nature 402(6758):203–207, doi: 10.1038/46080
32. Kinoshita E, Kinoshita-Kikuta E, Koike T (2014) Advances in Phos-tag-based methodologies for separation and detection of the phosphoproteome. Biochim Biophys Acta, doi: 10.1016/j.bbapap.2014.10.004
33. Janke C, Magiera MM, Rathfelder N, Taxis C, Reber S, Maekawa H, Moreno-Borchart A, Doenges G, Schwob E, Schiebel E, Knop M (2004) A versatile toolbox for PCR-based tagging of yeast genes: new fluorescent proteins, more markers and promoter substitution cassettes. Yeast 21(11):947–962
34. Knop M, Siegers K, Pereira G, Zachariae W, Winsor B, Nasmyth K, Schiebel E (1999) Epitope tagging of yeast genes using a PCR-based strategy: more tags and improved practical routines. Yeast 15(10B):963–972

Chapter 17

Studying Protein–Protein Interactions in Budding Yeast Using Co-immunoprecipitation

Magdalena Foltman and Alberto Sanchez-Diaz

Abstract

Understanding protein–protein interactions and the architecture of protein complexes in which they work is essential to identify their biological role. Protein co-immunoprecipitation (co-IP) is an invaluable technique used in biochemistry allowing the identification of protein interactors. Here, we describe in detail an immunoaffinity purification protocol as a one-step or two-step immunoprecipitation from budding yeast *Saccharomyces cerevisiae* cells to subsequently detect interactions between proteins involved in the same biological process.

Key words Co-immunoprecipitation, Immunoaffinity purification, Magnetic beads, Antibody, Protein interactions, Protein complexes, Cell extract, Freezer/Mill grinder

1 Introduction

Proteins are important building blocks in cellular architecture and the vast majority of them interact with other proteins to form multi-protein complexes and perform particular cellular tasks. Biochemical analysis of protein complexes and identification of their components has been fundamental for the biological understanding of their function [1, 2].

Protein–protein interactions are mediated by a variety of bonds, namely the combination of hydrophobic bonding, van der Waal's forces, and salt bridges and might involve specific binding domains on each protein or the whole protein itself. Interaction between proteins might be stable or transient, which is reflected by whether such proteins bind in a strong or weak fashion. Stable interactions are those associated with proteins that are purified as subunits of multiprotein complexes, such as the core RNA polymerase I [3]. On the other hand, dynamic interactions are expected to control the majority of cellular processes and it is the dynamic nature of these protein–protein interactions that makes them more challenging to be detected.

Alberto Sanchez-Diaz and Pilar Perez (eds.), *Yeast Cytokinesis: Methods and Protocols*, Methods in Molecular Biology, vol. 1369, DOI 10.1007/978-1-4939-3145-3_17, © Springer Science+Business Media New York 2016

Here, we describe a protocol based on immunoaffinity purification, as an efficient way to isolate protein complexes under physiological conditions and subsequently analyze their components. For this, budding yeast cells need to express the protein of interest with a purification tag, which is then used as a bait to capture the interacting proteins. One of the key feature of this protocol is that cell lysis is performed while cells are immersed under liquid nitrogen to preserve protein–protein interactions. Commercially available antibodies are coupled to magnetic beads and then antibody-coupled beads are used to immunoprecipitate the protein of interest, together with its binding partners. Finally, the purified material can be studied in different ways to characterize and identify all components of the protein complexes. The protocol can be divided into several steps:

- Growth, collection, and subsequent freezing of cells in liquid nitrogen.
- Cell lysis and preparation of cell extract.
- Pre-clearing the protein extract using high-speed centrifugation steps.
- Protein immunoprecipitation from cell extract using antibodies coupled to magnetic beads.
- Extensive wash of the protein complexes bound to the magnetic beads.
- Recovery of protein complexes and downstream analysis, such as Western blot or mass spectrometry.

The protocol we report has been extensively used to study protein complexes and their biological role. We have applied this co-immunoprecipitation to study different aspects of cellular biology, such as chromosome replication [4–10] or cell division [11]. This protocol can be efficiently adapted to study specific protein–protein interactions, for example, to study the weaker and transient interactions by cross-linking the proteins with formaldehyde [8] in order to preserve the transiently occurring interactions and enable their analysis.

2 Materials

Prepare all solutions using ultrapure water and analytical grade reagents. Store all reagents at room temperature unless specified otherwise.

2.1 Preparation of Antibody-Coupled--> Magnetic Beads

1. Magnetic rack.
2. Rotating wheel.
3. M-270 Epoxy magnetic beads (Life Technologies), store at 4 °C.
4. Dimethylformamide.

5. 0.1 M Sodium phosphate pH 7.4.
6. 3 M Ammonium sulfate.
7. Antibody of your choice to couple to the magnetic beads (*see* **Note 1**).
8. PBS.
9. 10 % IGEPAL CA-630 (NP-40).
10. 20 % Sodium azide.

2.2 One-Step Co-immunoprecipitation

Store all reagents at room temperature unless specified otherwise.

1. Ice, dry ice, liquid nitrogen.
2. SPEX Sample Prep Freezer/Mill, grinding vials.
3. Ultracentrifuge and appropriate tubes.
4. High-speed centrifuge and appropriate tubes.
5. Benchtop centrifuge.
6. 100 ml plastic beakers (one per each sample).
7. Thin long spatulas (one per each sample).
8. 50 and 250 ml tubes.
9. 1.5 ml tubes.
10. Magnetic rack for 1.5 ml tubes.
11. Rotating wheel.
12. Heating block.
13. 5 ml syringes with 20.0-gauge needles.
14. Antibody-coupled beads (*see* Subheading 3.1).
15. 1 M HEPES-KOH pH 7.9, pH needs to be adjusted at 4 °C and solution stored at 4 °C.
16. 0.5 M EDTA pH 7.9, store at 4 °C.
17. 50 % Glycerol.
18. 10 % IGEPAL CA-630 (NP-40).
19. 5 M Potassium acetate.
20. 1 M Magnesium acetate.
21. 1 M Dithiothreitol (DTT), store at –20 °C.
22. 0.2 M Sodium β-glycerophosphate pentahydrate, store at –20 °C.
23. 0.2 M Sodium fluoride, store at –20 °C.
24. Protease inhibitor cocktail (Sigma), store at –20 °C.
25. Complete protease inhibitor cocktail, EDTA-free (Roche) – stock solution 25× made by dissolving 1 tablet in 1 ml of water, store at –20 °C.
26. PBS.

27. 20 % Sodium dodecylsulfate.
28. 1 M Tris–HCl pH 6.7.
29. β-Mercaptoethanol.
30. Bromophenol blue.
31. Lysis buffer (store at 4 °C) (*see* **Note 2**): 100 mM HEPES-KOH pH 7.9, 100 mM potassium acetate, 10 mM magnesium acetate, 2 mM EDTA (*see* **Note 3**). Prior to use buffer needs to be supplemented with 1 mM DTT and inhibitors: 2 mM sodium β-glycerophosphate pentahydrate, 2 mM sodium fluoride, 1 % protease inhibitor cocktail, 1× complete protease inhibitor cocktail.
32. Glycerol mix buffer (store at 4 °C) (*see* **Note 2**): 100 mM HEPES-KOH pH 7.9, 50 % glycerol, 100 mM potassium acetate, 10 mM magnesium acetate, 2 mM EDTA, 0.5 % IGEPAL CA-630 (*see* **Note 3**). Prior to use buffer needs to be supplemented with 1 mM DTT and inhibitors: 2 mM sodium β-glycerophosphate pentahydrate, 2 mM sodium fluoride, 1 % protease inhibitor cocktail, 1× complete protease inhibitor cocktail.
33. Wash buffer (store at 4 °C) (*see* **Note 2**): 100 mM HEPES-KOH pH 7.9, 100 mM potassium acetate, 10 mM magnesium acetate, 2 mM EDTA, 0.1 % IGEPAL CA-630 (*see* **Note 3**). Prior to use buffer needs to be supplemented with inhibitors: 2 mM sodium β-glycerophosphate pentahydrate, 2 mM sodium fluoride, 1 % protease inhibitor cocktail, 1× complete protease inhibitor cocktail.
34. 3× Laemmli buffer (dilute buffer with water to make corresponding 1.5× and 1× buffers that are required for the immunoprecipitation protocol): 5 % (w/v) sodium dodecylsulfate, 666 mM Tris–HCl pH 6.7, 30 % glycerol, 715 mM β-mercaptoethanol, 0.0125 % bromophenol blue (*see* **Note 4**).

2.3 Large-Scale Two-Step TAP-MYC Co-immunoprecipitation

Store all reagents at room temperature unless specified otherwise.

1. Ice, dry ice, liquid nitrogen.
2. SPEX Sample Prep Freezer/Mill, grinding vials.
3. Ultracentrifuge and appropriate tubes.
4. High-speed centrifuge and appropriate tubes.
5. Benchtop centrifuge.
6. 100 ml plastic beakers (one per each sample).
7. Thin long spatulas (one per each sample).
8. 50 and 250 ml tubes.
9. 1.5 ml tubes.
10. Magnetic rack for 1.5 ml tubes.

11. Rotating wheel.
12. Heating block.
13. Shaking thermomixer.
14. 5 ml syringes with 20.0-gauge needles.
15. Antibody-coupled beads (*see* Subheading 3.1).
16. 1 M HEPES-KOH pH 7.9, pH needs to be adjusted at 4 °C and solution stored at 4 °C.
17. 0.5 M EDTA pH 7.9, store at 4 °C.
18. Glycerol.
19. 10 % IGEPAL CA-630 (NP-40).
20. 5 M Potassium acetate.
21. 1 M Magnesium acetate.
22. 1 M Dithiothreitol (DTT), store at –20 °C.
23. 0.2 M Sodium β-glycerophosphate pentahydrate, store at –20 °C.
24. 0.2 M Sodium fluoride, store at –20 °C.
25. Protease inhibitor cocktail (Sigma), store at –20 °C.
26. Complete protease inhibitor cocktail (Roche) – stock solution 25× made by dissolving 1 tablet in 1 ml of water, store at –20 °C.
27. TEV protease (Life Technologies), store at –20 °C.
28. PBS.
29. 20 % Sodium dodecylsulfate.
30. 1 M Tris–HCl pH 6.7.
31. β-mercaptoethanol.
32. Bromophenol blue.
33. Lysis buffer 1 (store at 4 °C) (*see* **Note 2**): 100 mM HEPES-KOH pH 7.9, 100 mM potassium acetate, 10 mM magnesium acetate, 2 mM EDTA (*see* **Note 3**). Prior to use buffer needs to be supplemented with 4 mM DTT and inhibitors: 8 mM sodium β-glycerophosphate pentahydrate, 8 mM sodium fluoride, 4 % protease inhibitor cocktail, 4× complete protease inhibitor cocktail (*see* **Note 5**).
34. Lysis buffer 2 (store at 4 °C) (*see* **Note 2**): 100 mM HEPES-KOH pH 7.9, 100 mM potassium acetate, 10 mM magnesium acetate, 2 mM EDTA (*see* **Note 3**). Prior to use buffer needs to be supplemented with 1 mM DTT and inhibitors: 2 mM sodium β-glycerophosphate pentahydrate, 2 mM sodium fluoride, 1 % protease inhibitor cocktail, 1× complete protease inhibitor cocktail.
35. Glycerol mix buffer (store at 4 °C) (*see* **Note 2**): 100 mM HEPES-KOH pH 7.9, 50 % glycerol, 100 mM potassium acetate, 10 mM magnesium acetate, 2 mM EDTA, 0.5 % IGEPAL CA-630 (*see* **Note 3**). Prior to use buffer needs to be supplemented with

1 mM DTT and inhibitors: 2 mM sodium β-glycerophosphate pentahydrate, 2 mM sodium fluoride, 1 % protease inhibitor cocktail, 1× complete protease inhibitor cocktail.

36. Wash buffer 1 (store at 4 °C) (*see* **Note 2**): 100 mM HEPES-KOH pH 7.9, 100 mM Potassium acetate, 10 mM Magnesium acetate, 2 mM EDTA, 0.1 % IGEPAL CA-630 (*see* **Note 3**). Prior to use buffer needs to be supplemented with 1 mM DTT and phosphatase inhibitors: 2 mM sodium β-glycerophosphate pentahydrate, 2 mM sodium fluoride (we do not add protease inhibitors as that would block TEV protease cleavage).
37. Wash buffer 2 (store at 4 °C) (*see* **Note 2**): 100 mM HEPES-KOH pH 7.9, 100 mM potassium acetate, 10 mM magnesium acetate, 2 mM EDTA, 0.1 % IGEPAL CA-630 (*see* **Note 3**). Prior to use buffer needs to be supplemented with inhibitors: 2 mM sodium β-glycerophosphate pentahydrate, 2 mM sodium fluoride, 1 % protease inhibitor cocktail, 1× complete protease inhibitor cocktail.
38. 3× Laemmli buffer (dilute buffer with water to make corresponding 1.5× and 1× buffers that are required for the immunoprecipitation protocol): 5 % (w/v) sodium dodecylsulfate, 666 mM Tris–HCl pH 6.7, 30 % glycerol, 715 mM β-mercaptoethanol, 0.0125 % bromophenol blue (*see* **Note 4**).

3 Methods

3.1 Preparation of the Antibody-Coupled--> Magnetic Beads

1. Add appropriate amount of dimethylformamide to the vial containing M-270 Epoxy beads as per the manufacturer's instruction (*see* **Note 6**).
2. Vortex the vial with beads vigorously and store at 4 °C (*see* **Note 7**).
3. Remove 425 μl of beads to a 1.5 ml tube (*see* **Note 8**). Place the tube on a magnetic rack and wait 30 s for the magnet in the rack to gently capture the beads on the side of the tube. Remove the supernatant.
4. Add 1 ml of 0.1 M sodium phosphate pH 7.4 and agitate the tube for 10 min on a rotating wheel.
5. Place on magnetic rack and remove the supernatant and repeat wash as in **step 4**.
6. Place on magnetic rack and remove the supernatant and add reagents and antibody in the following order (*see* Table 1).
7. Leave the beads on rotating wheel for 2 days at 4 °C (*see* **Note 9**).
8. After the incubation, spin tubes briefly to collect any liquid in the cap and then place tubes containing beads in magnetic rack

Table 1
Preparation of antibody-coupled magnetic beads

Reagent	Volume	Comment
3 M Ammonium sulfate	300 μl	
0.1M Sodium phosphate pH 7.4	X μl	Adjust depending on volume of antibody
Antibody	Y μl	Calculate volume that contains 300 μg antibody
Total volume	900 μl	

for 4 min, discard supernatant carefully and wash four times with 1 ml of ice-cold PBS at room temperature (*see* **Note 10**).

9. Add 1 ml of PBS containing 0.5 % IGEPAL, agitate the tube on rotating wheel for 10 min place in magnetic rack and remove supernatant (*see* **Note 10**).
10. Add 1 ml of PBS and leave on rotating wheel for 5 min, place in magnetic rack and then remove supernatant.
11. Repeat the wash with a further 1 ml of PBS for 5 min. Place in magnetic rack and discard the supernatant (*see* **Note 10**).
12. Resuspend the beads in 900 μl of PBS and add sodium azide to a final concentration of 0.02 % (*see* **Note 11**). Store beads at 4 °C.

3.2 One-Step Co-immunoprecipitation

Important: Please follow all health and safety rules relating to work using liquid nitrogen and always wear appropriate protective equipment (mask, gloves, etc.).

3.2.1 Preparation of Frozen Cells Pellet: "Popcorn"

Keep all the buffers and reagents on ice.

1. Grow 250 ml yeast culture of your choice (*see* **Note 12**) and centrifuge culture at 200 × *g* for 3 min using 250 ml tubes.
2. Resuspend cells with 50 ml of ice-cold 20 mM HEPES-KOH pH 7.9 buffer, transfer to 50 ml tube and pellet cells at 200 × *g* for 3 min. Take off the supernatant.
3. Wash cells with 10 ml of ice-cold Lysis buffer WITHOUT the inhibitors and centrifuge cells at 200 × *g* for 3 min, discard supernatant.
4. Resuspend the cell pellet in three volumes of cold Lysis buffer (supplemented with inhibitors) (1 g of cell pellet equals 1 ml of buffer therefore, to resuspend 1 g of cell pellet add 3 ml of cold lysis buffer with inhibitors).
5. Place an empty 50 ml tube on dry ice, fill with liquid nitrogen and freeze the resuspended pellet by dispensing dropwise into the liquid nitrogen (*see* **Note 13**). Leave the tubes with cells

on dry ice until all remaining liquid nitrogen has evaporated (*see* **Note 14**).

6. Store frozen yeast cells (hereafter referred to as "popcorn") at −80 °C.

3.2.2 Co-immunoprecipitation

Important: Please follow all health and safety rules relating to work using liquid nitrogen and always wear appropriate protective equipment (mask, gloves, etc.).

Keep all the buffers and reagents on ice. Pre-cool all centrifuges and centrifuge tubes.

1. Pre-cool the SPEX Sample Prep Freezer/Mill by filling it with liquid nitrogen (follow the manufacturer's instruction). Pre-cool the grinding vials in the filled freezer Mill (*see* **Note 15**).
2. Weigh 1–3.5 g of frozen popcorn for every sample and place into fresh 50 ml tube previously kept on dry ice (*see* **Note 16**).
3. Place popcorn into appropriate grinding vials, seal tightly and place securely inside the grinder (follow the manufacturer's instruction). To break efficiently budding yeast cells use two cycles of grinding, each one with the following settings:

   2 min pre-cool.

   2 min run.

   2 min cool.

   Rate 14.
4. After the run, collect the ground material carefully using a different spatula for each sample (*see* **Note 17**). Place the powder (ground yeast) into a plastic beaker and let it thaw at room temperature. This will take approximately 15–20 min.
5. Transfer thawed extract from the plastic beaker into the appropriate centrifuge tubes (we use 50 ml tubes). Take note of the volume of the extract transferred, as the volume of cell extract is important to calculate the amount of buffer needed in the next step.
6. Add one-quarter volume of Glycerol mix buffer supplemented with inhibitors to the same plastic beaker from which you have collected the extract (therefore for each 1 ml of recovered cell extract add 0.25 ml of Glycerol mix buffer with inhibitors). Vortex well, collect the remaining liquid and transfer to the centrifuge tube with the rest of the extract. This allows recovery of all the remaining extract from the wall of plastic beaker (*see* **Note 18**).
7. Pellet the insoluble cell debris by centrifugation at 25,000 × *g* for 30 min at 4 °C.
8. Collect the supernatant and transfer into an ultracentrifuge tube, top up with a thin layer of mineral oil ensuring the tubes are balanced and centrifuge at 100,000 × *g* for 1 h at 4 °C.

9. During the centrifugation wash the antibody-coupled beads (prepared previously, *see* Subheading 3.1). Mix antibody-coupled beads thoroughly and aliquot 100 μl into 1.5 ml tubes (prepare two aliquots of beads for each extract). Wash beads twice with 1 ml of PBS at room temperature (*see* **Note 10**).
10. After the ultracentrifuge spin remove the tube carefully to avoid disrupting the separated phases. Using a 5 ml syringe with a needle, carefully pierce the ultracentrifuge tube just above the pellet. Make sure that the bevel of the needle is facing up to avoid disturbing the pellet. Collect the supernatant into a fresh 5 ml tube. At this stage prepare a cell extract sample for later analysis. Remove 50 μl of cell extract and add 100 μl of 1.5× Laemmli buffer. Mix and heat cell extract at 95 °C for 5 min and store at −80 °C.
11. Split remaining cell extract into two by adding half of the extract to one of the two tubes of antibody-coupled beads (prepared in **step 9**) after having removed any leftovers of remaining PBS from the beads (*see* **Note 9**).
12. Incubate at 4 °C for 2 h on a rotating wheel.
13. After the incubation, spin tubes briefly to collect any liquid in the cap and then place tubes on a magnetic rack and remove the supernatant. Once beads are captured on the side of the tube, wash protein complexes bound to the antibody-coupled magnetic beads four times with 1 ml of Wash buffer (*see* **Note 10**).
14. After the last wash, discard the supernatant. *Important:* make sure all the washing buffer is completely removed. Add 50 μl of 1× Laemmli buffer and heat samples at 95 °C for 5 min.
15. Place tubes in a magnetic rack and collect supernatant that contains the purified material into a fresh 1.5 ml tube. Mix the eluates from the corresponding samples (split in **step 11**), together, aliquot the material accordingly (*see* **Note 19**) and snap-freeze on dry ice and store at −80 °C.
16. Run purified material on an SDS-PAGE gel to subsequently perform downstream analysis (*see* **Note 20**).

3.3 Large-Scale Two-Step TAP-MYC Co-immunoprecipitation

A protein can interact with different partners and be involved in different cellular processes. In order to try and isolate specific protein complexes we present one possible variation of the standard co-immunoprecipitation protocol, which consist of a two-step purification procedure. First, a protein of interest is immunoprecipitated together with its binding partners. One of these interactor proteins needs to be known, in order to be used as bait for second round of immunoprecipitation using the material purified in the first step. We have efficiently used TAP and MYC tags to perform this type of procedure. We have immunoprecipitated TAP-tagged protein of interest and then we have used purified

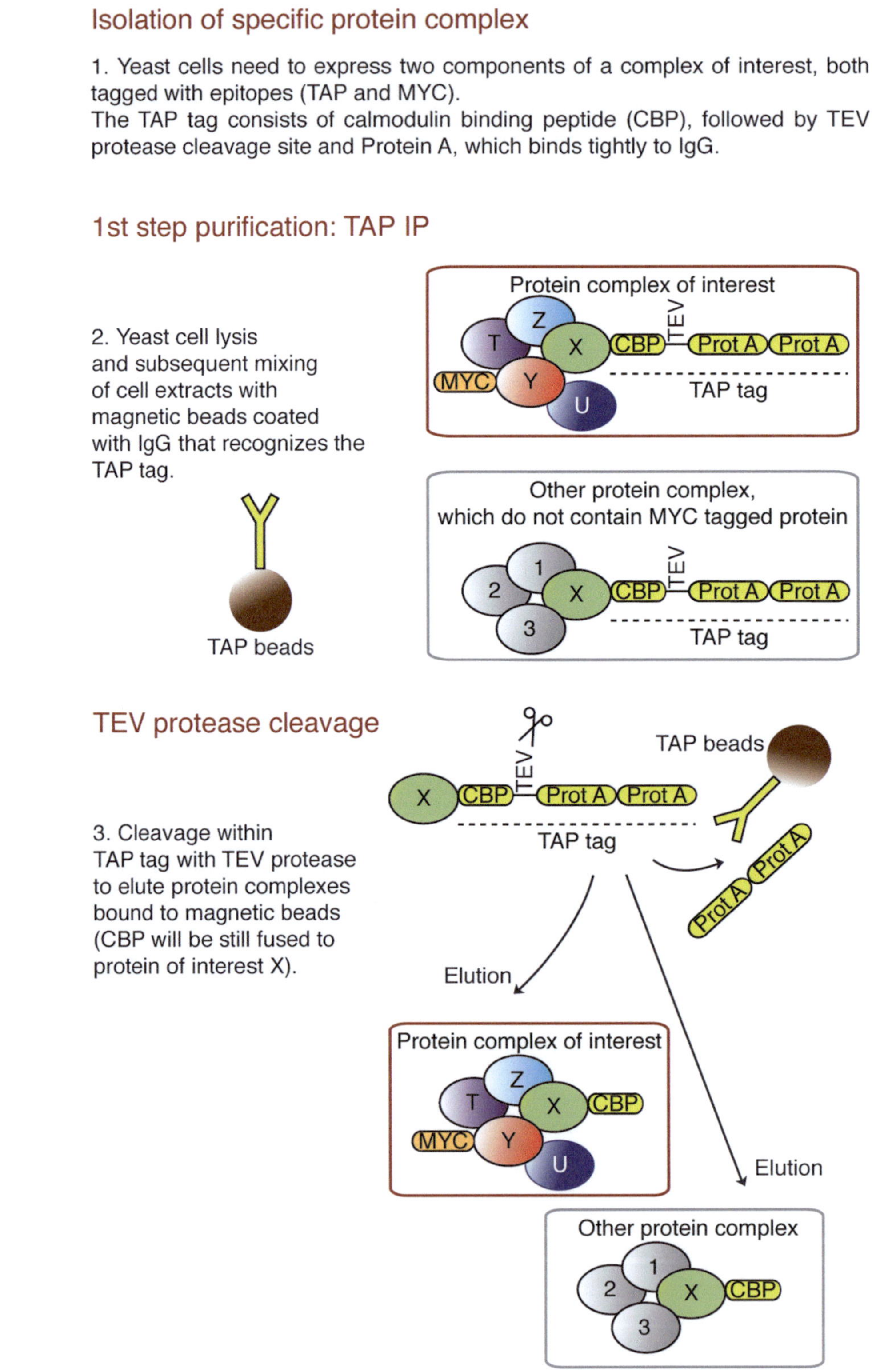

Fig. 1 Overview of a two-step TAP-MYC co-immunoprecipitation

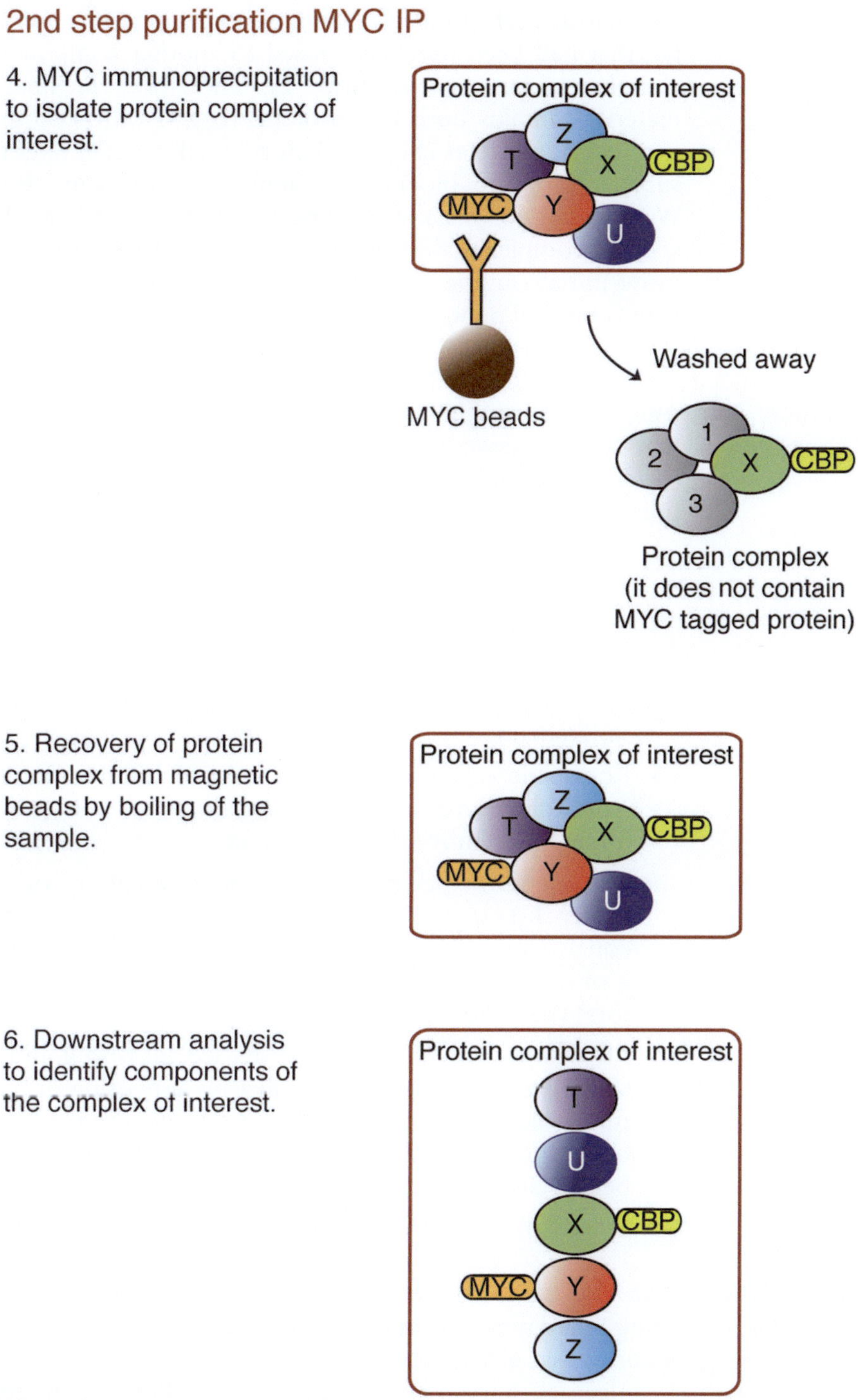

Fig. 1 (continued)

material to perform a second immunoprecipitation using a MYC tag that had been previously fused to another component of the complex of interest (*see* **Note 1**). Therefore, the final isolated material contains both bait proteins tagged with TAP and MYC and their interactors (Fig. 1). This purified material might be then subjected to the biochemical application of your choice, e.g. Western blot or mass spectrometry. As we aim to enrich for the specific complexes that we are trying to isolate we recommend using initially higher amounts of cells and prepare what we term "concentrated popcorn."

3.3.1 Preparation of Concentrated Frozen Cells Pellet: "Concentrated Popcorn"

Important: Please follow all health and safety rules relating to work using liquid nitrogen and always wear appropriate protective equipment (mask, gloves, etc.).

Keep all the buffers and reagents on ice. Pre-cool all centrifuges and centrifuge tubes.

1. Grow 1000 ml of yeast culture expressing proteins of interest tagged with TAP and MYC (*see* **Note 12**) and centrifuge culture at 200 × *g* for 3 min using a 250 ml tube (*see* **Note 21**).
2. Wash cells with 100 ml of ice-cold 20 mM HEPES-KOH pH 7.9 buffer and pellet cells at 200 × *g* for 3 min. Remove the supernatant.
3. Resuspend cells with 20 ml of ice-cold Lysis buffer 1 without inhibitors, transfer to 50 ml tube and pellet cells at 200 × *g* for 3 min and remove the supernatant.
4. Resuspend the pellet in one-quarter of a volume of cold Lysis buffer 1 (supplemented with inhibitors) (1 g of cell pellet equals 1 ml of buffer with inhibitors, therefore, to resuspend 1 g of cell pellet add 0.25 ml of cold Lysis buffer 1 with inhibitors) (*see* **Note 22**).
5. Place an empty 50 ml tube on dry ice, fill with liquid nitrogen and freeze the resuspended pellet by dispensing dropwise into the liquid nitrogen (*see* **Note 13**). Leave the tubes with cells on dry ice until all remaining liquid nitrogen has evaporated before closing the tubes (*see* **Note 14**).
6. Store frozen yeast cells ("concentrated popcorn") at –80 °C.

3.3.2 Two-Step TAP-MYC Co-immunoprecipitation

Important: Please follow all health and safety rules relating to work using liquid nitrogen and always wear appropriate protective equipment (mask, gloves, etc.).

Keep all buffers and reagents on ice. Pre-cool all centrifuges and centrifuge tubes.

1. Pre-cool the SPEX Sample Prep Freezer/Mill by filling it with liquid nitrogen (follow the manufacturer's instruction) (*see* **Note 15**).

2. Weigh out equal amounts of frozen popcorn for every sample and place into fresh 50 ml tube previously kept on dry ice (*see* **Note 16**).
3. Place popcorn into appropriate pre-cooled grinding vials (*see* **Note 23**), seal tightly and place securely inside the grinder (follow the manufacturer's instruction). To efficiently break budding yeast cells use two cycles of grinding, each one with the following settings:

   2 min pre-cool.

   2 min run.

   2 min cool.

   Rate 14.
4. After the run collect the ground material carefully using a different spatula for each sample (*see* **Note 17**). Place powder (ground yeast) into a plastic beaker and let it thaw at room temperature. This will take approximately 15–20 min.
5. After thawing add 1 ml of Lysis buffer 2 supplemented with inhibitors and transfer melted extract from plastic beaker into the appropriate centrifuge tubes (we use 50 ml tubes). Vortex well, collect the remaining liquid and transfer to the centrifuge tube with the rest of the extract. Take note of the volume of the extract in the process, as the volume of cell extract is important to calculate the amount of buffer needed in the next step.
6. Add one-quarter volume of Glycerol mix buffer supplemented with inhibitors to the cell extract (*see* **Note 18**).
7. Pellet the insoluble cell debris by centrifugation at 25,000 × *g* for 30 min at 4 °C.
8. Collect the supernatant and transfer into an ultracentrifuge tube, top up with a thin layer of mineral oil ensuring the tubes are balanced and centrifuge at 100,000 × *g* for 1 h at 4 °C.
9. Wash the IgG antibody-coupled beads (to pull down the TAP-tagged protein) during the ultracentrifugation spin (prepared previously, *see* Subheading 3.1). Mix the antibody-coupled beads thoroughly and aliquot 100 μl of antibody-coupled beads into 1.5 ml tubes (prepare two aliquots of beads for each 1000 ml of original yeast culture). Wash beads twice with 1 ml of PBS at room temperature (*see* **Note 10**).
10. After the ultracentrifuge spin remove the tube carefully to avoid disrupting the separated phases. Using a 5 ml syringe with a needle, carefully pierce the ultracentrifuge tube just above the pellet. Make sure that the bevel of the needle is facing up to avoid disturbing the pellet. Collect the supernatant into a fresh 5 ml tube. At this stage prepare a cell extract sample for later analysis. Remove 50 μl of cell extract and add 100

µl of 1.5× Laemmli buffer. Boil cell extracts at 95 °C for 5 min and store at −80 °C.

11. Split remaining cell extract into two adding each half of the extract to one of the two tubes of antibody-coupled beads (prepared in **step 9**) after having removed any leftovers of remaining PBS from magnetic beads (*see* **Note 9**).
12. Incubate at 4 °C for 2 h on a rotating wheel.
13. After the incubation, spin tubes briefly to collect any liquid in the cap and then place tubes on a magnetic rack and remove the supernatant. Once the beads are captured on the side of the tube, wash the protein complexes bound to the antibody-coupled magnetic beads four times with 1 ml of Wash buffer 1 (*not* containing any protease inhibitors) (*see* **Note 10**).
14. After the last wash discard supernatant. *Important:* make sure that all the washing buffer is removed and there is no liquid remaining on the side of the tube (*see* **Note 24**).
15. Add 50 µl of Wash buffer 1 and 10 U of TEV protease for each 1000 ml of original yeast culture (*see* **Note 25**).
16. Incubate for 2 h with agitation at room temperature, e.g. in shaking thermomixer.
17. Wash the MYC antibody-coupled beads (to pull down the MYC-tagged protein) during the TEV incubation (prepared previously, *see* Subheading 3.1). Aliquot 100 µl of antibody-coupled beads into 1.5 ml tubes (prepare two aliquots of beads). Wash beads twice with 1 ml of PBS at room temperature (*see* **Note 10**).
18. After incubation with TEV protease, combine the extracts together, adjust with Lysis buffer 2 to a final volume of 2 ml and split the extracts again into the tubes containing the MYC antibody-coupled magnetic beads.
19. Incubate at 4 °C for 2 h on a rotating wheel.
20. After the incubation, spin tubes briefly to collect any liquid in the cap and then place the tubes on a magnetic rack and remove the supernatant. Once the beads are captured on the side of the tube, wash protein complexes bound to the antibody-coupled magnetic beads four times with 1 ml of Wash buffer 2 (*see* **Note 10**).
21. After the last wash, discard the supernatant. *Important:* make sure all the washing buffer is removed and there is no liquid remaining on the side of the tube. Add 50 µl of 1× Laemmli buffer and heat samples at 95 °C for 5 min.
22. Place tubes in a magnetic rack and collect supernatant that contains the purified material into a fresh 1.5 ml tube. Mix the eluates from the corresponding samples (split in **step 18**)

together, aliquot the material accordingly (*see* **Note 19**) and snap-freeze on dry ice and store at −80 °C.

23. Run purified material on a SDS-PAGE gel to subsequently carry out Western blot analysis (*see* **Note 20**). Alternatively, to identify components of protein complexes, mass spectrometry can be performed on purified material (*see* **Note 26**).

4 Notes

1. We do not recommend using the polyclonal antibodies raised against particular proteins. In our experience the most efficient way to pull down the protein of your choice is to use widely available epitopes, we have extensively used: TAP, MYC, HA, and FLAG. When performing two-step purification including TAP tag on one of the protein it is important to use TAP purification always as a first step since TAP-tagged proteins can bind some antibody-coupled beads and would ruin second-step purification by binding unspecifically.
2. We find that it is best to prepare this fresh each time.
3. When calculating the amount of water that needs to be added to the buffer, the volume of inhibitors that you need to add prior to use needs to be taken into account.
4. SDS precipitates at 4 °C therefore the buffer needs to be warmed up to room temperature prior to use.
5. Note that the amounts of inhibitors used when making popcorn and concentrated popcorn (from concentrated extracts) are different.
6. Add dimethylformamide in a fume hood and follow all waste disposal regulations accordingly.
7. You can store beads for up to 6 months at 4 °C.
8. We use 3.4×10^9 of antibody-coupled magnetic beads for each cell extract.
9. Placing parafilm around the tube lid will reduce the risk of the tube suddenly opening during the incubation.
10. To perform an extensive wash each time place tube on magnetic rack, remove magnet from rack and shake the rack for a minimum of twenty times. Place back the magnet and allow it to capture the beads on the side of the tube.
11. Antibody-coupled beads can be stored for up to 3–4 months at 4 °C. It is possible to omit sodium azide and store the coupled beads up to 1 week at 4 °C.
12. We have experience using yeast cultures with a density ranging from 0.7×10^7 cells/ml up to 2×10^7 cells/ml. We grow strains of interest together with corresponding controls.

13. The most efficient way to create frozen cell extract is to use a 1 ml pipette and keep adding drops of yeast extract into a 50 ml tube placed on dry ice that has been filled with liquid nitrogen, they will freeze instantly. It is important to do this slowly in order to obtain separate drops of frozen yeast cells. Add drops from above and in the middle of the 50 ml tube so drops do not stick to the sides of the tube. Avoid placing the pipette tip too close to the liquid nitrogen otherwise the extract will freeze inside the pipette tip.
14. Make sure all liquid nitrogen is evaporated before closing the tube. Secure caps loosely on the tubes overnight at −80 °C whilst all traces of liquid nitrogen evaporate to prevent the tube from "popping." Tighten the lids fully the next day.
15. Depending on the type of the cryogenic grinder and the number of samples to be lysed, more liquid nitrogen might be needed to be added to the grinder, always to be filled up in between the cycles of grinding. Always follow the health and safety rules when handling liquid nitrogen.
16. When analyzing few extracts that belong to the same experiment, it is important to lyse the same amount of popcorn for all of them. Start weighing the samples from the one that weighed the least (you will know which one as you have to calculate the weight of cell extracts before making the popcorn) and adjust the weight of remaining samples accordingly.
17. The ground material should form a fine powder. If not fine, white powder is observed after grinding, consider extending the number or the length of grinding cycles. You can check efficiency of the lysis by placing 2 μl of material under the microscope and checking the breakage of cells. Conditions that we use give us 90–95 % breakage efficiency. When collecting the ground material, it is essential to work quickly as once the powder starts to thaw it becomes more difficult to recover it from the vials.
18. The study of protein interactions within complexes built around DNA requires digestion of DNA in order to release proteins. Add 400 U/ml Universal Nuclease (Fisher) (or 1600 U/ml when preparing "concentrated popcorn") at this point of the protocol and incubate for 30 min at 4 °C while mixing on rotating wheel prior to the first centrifugation step.
19. We routinely make aliquots of 1× 50 μl and 2× 25 μl but depending on the downstream application you might need to reconsider these amounts. We do not recommend the repeated freeze–thaw of immunoprecipitated extracts.
20. To identify proteins present in the purified material perform Western blot analysis. We routinely load 5 μl of the cell extract sample and 12 μl of the purified immunoprecipitated complex on a SDS-PAGE gel in order to resolve the proteins and we transfer them routinely onto nitrocellulose membrane.

21. In our experience, to improve visualization and increase the yield of purified complexes we have grown anything between 1 and 5 l of a single yeast culture.
22. For large-scale experiments we have adapted the protocol so we produce more concentrated cell extracts by adding a quarter of a volume of a lysis buffer compared to the three times volume of lysis buffer for the standard protocol.
23. Depending on the manufacturer's instruction for the cryogenic grinder of your choice there might be a weight limit for the amount of sample that can be ground at once. More than one vial per sample might be used, depending on the scale of the original culture (*see* **Note 21**).
24. Make sure that the entire amount of washing buffer is collected and there is no remaining liquid on the side of the tube.
25. TEV protease cuts within TEV recognition site that is included in the TAP tag (Fig. 1). Therefore, the tagged protein and its interacting partners will be released from the magnetic beads. It is important to remember that TAP tag will be lost after this step and thereafter the protein of interest will be fused only to the remaining Calmodulin Binding Protein (CBP) (details in Fig. 1), thus anti-CBP or native antibodies against the protein must be used in order to detect the protein on Western blots. Efficiency of TEV protease digestion will depend on the protein thus more or less TEV protease might be required.
26. When more concentrated protein samples are required you can elute your protein complexes bound to magnetic beads in 30 μl of 1× Laemmli buffer. We have routinely eluted the proteins in 30 μl and used the whole eluate to run in SDS-PAGE gels. To identify components of the purified protein complexes in an unbiased fashion, mass spectrometry can be used.

Acknowledgements

We are grateful for teaching and scientific advice to Professor Karim Labib. Methods described in this chapter were developed in Labib's laboratory and we would like to thank members of his group past and present who contributed to our current understanding of methods presented here. We especially thank Dr. Frederick van Deursen, Dr. Sugopa Sengupta and Dr. Giacomo De Piccoli for comments on the manuscript. ASD is a recipient of a Ramon y Cajal contract and received funding from the Cantabria International Campus and via grant BFU2011-23193 from the Spanish "Ministerio de Economia y Competitividad" (co-funded by the European Regional Development Fund).

References

1. Gavin AC, Bosche M, Krause R, Grandi P, Marzioch M, Bauer A, Schultz J, Rick JM, Michon AM, Cruciat CM, Remor M, Hofert C, Schelder M, Brajenovic M, Ruffner H, Merino A, Klein K, Hudak M, Dickson D, Rudi T, Gnau V, Bauch A, Bastuck S, Huhse B, Leutwein C, Heurtier MA, Copley RR, Edelmann A, Querfurth E, Rybin V, Drewes G, Raida M, Bouwmeester T, Bork P, Seraphin B, Kuster B, Neubauer G, Superti-Furga G (2002) Functional organization of the yeast proteome by systematic analysis of protein complexes. Nature 415(6868):141–147
2. Ho Y, Gruhler A, Heilbut A, Bader GD, Moore L, Adams SL, Millar A, Taylor P, Bennett K, Boutilier K, Yang L, Wolting C, Donaldson I, Schandorff S, Shewnarane J, Vo M, Taggart J, Goudreault M, Muskat B, Alfarano C, Dewar D, Lin Z, Michalickova K, Willems AR, Sassi H, Nielsen PA, Rasmussen KJ, Andersen JR, Johansen LE, Hansen LH, Jespersen H, Podtelejnikov A, Nielsen E, Crawford J, Poulsen V, Sorensen BD, Matthiesen J, Hendrickson RC, Gleeson F, Pawson T, Moran MF, Durocher D, Mann M, Hogue CW, Figeys D, Tyers M (2002) Systematic identification of protein complexes in Saccharomyces cerevisiae by mass spectrometry. Nature 415(6868):180–183. doi:10.1038/415180a
3. Knutson BA, Luo J, Ranish J, Hahn S (2014) Architecture of the Saccharomyces cerevisiae RNA polymerase I Core Factor complex. Nat Struct Mol Biol 21(9):810–816. doi:10.1038/nsmb.2873
4. Foltman M, Evrin C, De Piccoli G, Jones RC, Edmondson RD, Katou Y, Nakato R, Shirahige K, Labib K (2013) Eukaryotic replisome components cooperate to process histones during chromosome replication. Cell Rep 3(3):892–904. doi:10.1016/j.celrep.2013.02.028
5. Gambus A, Jones RC, Sanchez-Diaz A, Kanemaki M, van Deursen F, Edmondson RD, Labib K (2006) GINS maintains association of Cdc45 with MCM in replisome progression complexes at eukaryotic DNA replication forks. Nat Cell Biol 8(4):358–366
6. Gambus A, van Deursen F, Polychronopoulos D, Foltman M, Jones RC, Edmondson RD, Calzada A, Labib K (2009) A key role for Ctf4 in coupling the MCM2-7 helicase to DNA polymerase alpha within the eukaryotic replisome. EMBO J 28(19):2992–3004, doi: emboj2009226 [pii] 10.1038/emboj.2009.226
7. Sengupta S, van Deursen F, de Piccoli G, Labib K (2013) Dpb2 integrates the leading-strand DNA polymerase into the eukaryotic replisome. Curr Biol 23(7):543–552. doi:10.1016/j.cub.2013.02.011
8. van Deursen F, Sengupta S, De Piccoli G, Sanchez-Diaz A, Labib K (2012) Mcm10 associates with the loaded DNA helicase at replication origins and defines a novel step in its activation. EMBO J 31(9):2195–2206. doi:10.1038/emboj.2012.69
9. De Piccoli G, Katou Y, Itoh T, Nakato R, Shirahige K, Labib K (2012) Replisome stability at defective DNA replication forks is independent of S phase checkpoint kinases. Mol Cell 45(5):696–704. doi:10.1016/j.molcel.2012.01.007
10. Maric M, Maculins T, De Piccoli G, Labib K (2014) Cdc48 and a ubiquitin ligase drive disassembly of the CMG helicase at the end of DNA replication. Science 346(6208):1253596. doi:10.1126/science.1253596
11. Sanchez-Diaz A, Marchesi V, Murray S, Jones R, Pereira G, Edmondson R, Allen T, Labib K (2008) Inn1 couples contraction of the actomyosin ring to membrane ingression during cytokinesis in budding yeast. Nat Cell Biol 10(4):395–406. doi:10.1038/ncb1701

Chapter 18

Conditional Budding Yeast Mutants with Temperature-Sensitive and Auxin-Inducible Degrons for Screening of Suppressor Genes

Asli Devrekanli and Masato T. Kanemaki

Abstract

The conditional control of protein expression is useful to characterize the function of proteins, especially of those that are essential for cell viability. Two degron-based systems, temperature-sensitive and auxin-inducible degrons, can be used to generate conditional mutants of budding yeast, simply by transforming appropriate cells with PCR-amplified DNA. We describe a protocol for the generation of temperature-sensitive and auxin-inducible degron mutants. We also show that a conditional mutant with few spontaneous revertants was generated by combining two degron systems for the Inn1 protein. Finally, we describe a suppressor screening method that uses the dual degron-Inn1 mutant to identify mutant proteins that suppress Inn1-K31A, which has a defect in cytokinesis.

Key words Temperature-sensitive degron, Auxin-inducible degron, Protein degradation, Conditional mutant, Suppressor screening, Inn1

1 Introduction

Conditional depletion or inactivation of proteins of interest (POIs) is useful to understand their role in vivo, especially to characterize the role of essential proteins in cell viability. Two conditional degron-based technologies, the temperature-sensitive (ts) degron and the auxin-inducible degron (AID) systems, allow us to deplete POIs in a short period of time under restrictive conditions (usually within 30 min) [1]. We showed previously that the fast depletion of POIs is important to assess the direct consequences of their function, because loss of POIs might impair the functional network within cells, causing a secondary effect [2].

The ts-degron technology was originally developed by Varshavsky and colleagues [3]. The fusion of a cassette composed of ubiquitin (UBI) and temperature-sensitive dihydrofolate reductase (called ts-degron: td) at the N terminus of a POI leads to the

Alberto Sanchez-Diaz and Pilar Perez (eds.), *Yeast Cytokinesis: Methods and Protocols*, Methods in Molecular Biology, vol. 1369, DOI 10.1007/978-1-4939-3145-3_18, © Springer Science+Business Media New York 2016

expression of a fusion protein (Fig. 1a, 24 °C). The ubiquitin moiety is cleaved by endogenous processing, thus exposing an arginine residue at the N terminus. The exposed arginine residue is recognized by the Ubr1–Ubc2 ubiquitin ligase complex; however, the fusion proteins are usually stable at the permissive temperature of 24 °C. At the restrictive temperature of 37 °C, a conformational change in td allows the ligase to place poly-ubiquitin. The fusion protein is, therefore, rapidly degraded by proteasomes (Fig. 1a, 37 °C). The original ts-degron technology was improved by overproducing Ubr1 from the conditional *GAL1-10* (*PGAL*) promoter before the temperature shift [4]. Importantly, td can be fused only at the N terminus of the POI because the degradation is dependent on the N-end rule [5, 6].

As an alternative degron system, we developed the AID technology by transplanting a plant-specific degradation pathway controlled by the phytohormone auxin to budding yeast [7].

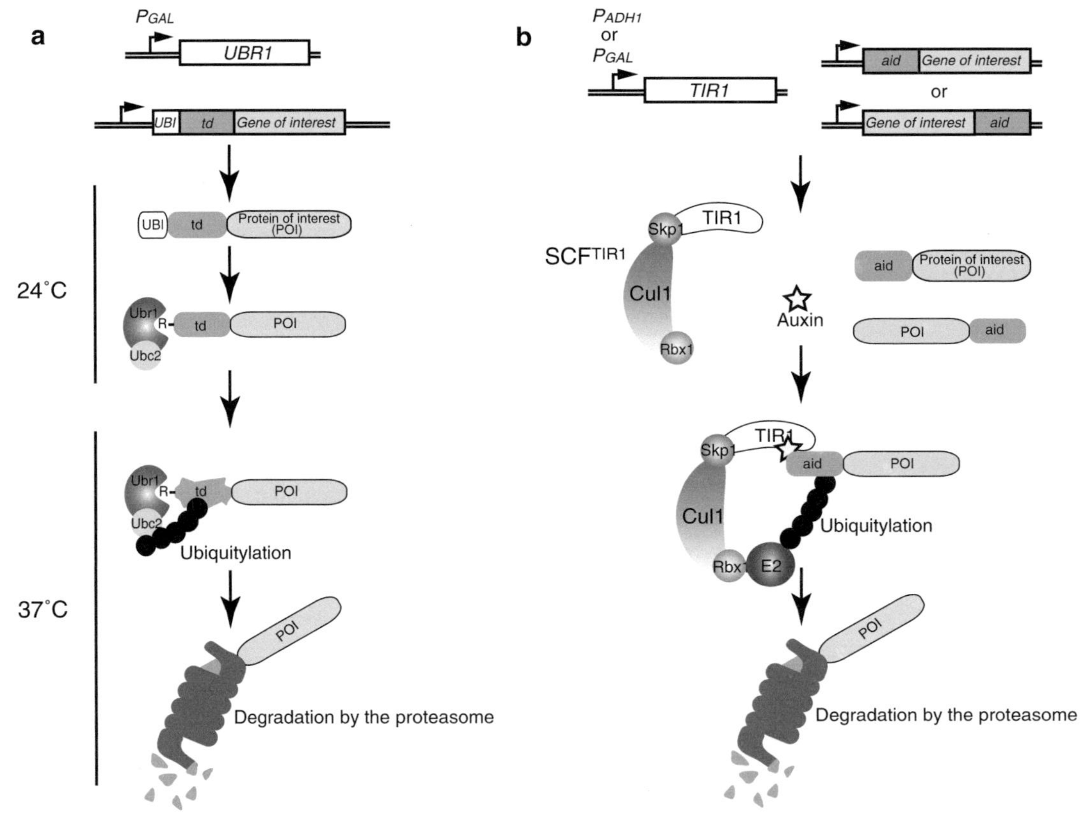

Fig. 1 Principle of the ts-degron and AID systems. (**a**) Schematic illustration showing the mechanism of the ts-degron (td) system. The ubiquitin (UBI) moiety is cleaved by endogenous processing, exposing an arginine (R) residue at the N terminus. The arginine is recognized by Ubr1–Ubc2, which promotes the poly-ubiquitylation of td only at 37 °C. (**b**) Schematic illustration showing the mechanism of the AID system. The degron (aid) fused at either the N or C terminus of the POI is recognized and poly-ubiquitylated by the SCF^{TIR1} E3 ubiquitin ligase in the presence of auxin

For this purpose, the expression of the auxin-perceptive F-box protein, TIR1, of *Oryza sativa* is required. Together with the endogenous SCF subunits, TIR1 forms the SCF^{TIR1}–E3 ligase complex in cells (Fig. 1b). By fusing a plant-derived degron (aid) either at the N or C terminus of the POI, the degradation of the fusion protein via the ubiquitin–proteasome system can be rapidly induced. The original degron derived from the *Arabidopsis* IAA17 protein is 25 kD [7]. We recently prepared a smaller (8 kD) degron as well as highly active degrons by fusing multiple smaller degrons [8, 9].

Degron mutants can be used not only for the functional characterization of POIs, in theory, but also for the identification of suppressor mutations in genes that encode proteins that are functionally related to POIs [10, 11]. However, the degron system has never been used in suppressor screenings. This is because revertants can be generated simply by losing the integrated *UBR1* or *TIR1* gene. To overcome this problem, i.e., to obtain a very tight conditional allele, we used both ts-degron and AID on the same protein (Fig. 3a). This enables the depletion of proteins via two independent pathways, which reduces the possibility of reversion.

In this article, we describe protocols for the construction of ts-degron and AID mutants using PCR-based tagging. Subsequently, we describe a method to screen for mutations that suppress the lethal effects of mutating the essential cytokinesis protein Inn1 [12]. We found that a positively charged amino acid at position 31 of Inn1 was essential for cytokinesis, which may be a key site of interaction with an unidentified factor that is important for cytokinesis. Thus, we aimed to identify such a factor that was able to suppress the cytokinesis defect of the non-functional *inn1-K31A* allele when functional Inn1 was depleted by the dual-degron system. In this example, we generated a very tight conditional allele of *INN1* (*td-inn1-aid*) with a very low reversion frequency, combined this allele with *inn1-K31A* and screened for suppressor mutations that were generated by random mutagenesis that restored growth under the restrictive condition.

2 Materials

2.1 Yeast Strains

All yeast strains described in this protocol are listed in Table 1 and are based on the W303-1a or -1b genetic background.

1. The yeast strain YKL200 (*see* Table 1) expressing Ubr1 under the control of the conditional *GAL1-10* promoter (*PGAL*) is available from EUROSCARF (http://web.uni-frankfurt.de/fb15/mikro/euroscarf/data/degron.htmL).
2. The yeast strains YMK726 and YMK728 (*see* Table 1) expressing codon-optimized TIR1 under the control of the conditional

Table 1
Yeast strains used in this protocol

Strain name	Genotype	Source
W303--1a/b	*MATa/α ade2-1 ura3-1 his3-11,15 trp1-1 leu2-3,112 can1-100*	R. Rothstein
YKL200	*MATa UBR1::PGAL-UBR1 (HIS3)*	[13]
YMK726	*MATa ura3-1::PGAL-TIR1 (URA3)*	[9]
YMK728	*MATa ura3-1::PADH1-TIR1 (URA3)*	[9]
YAD240	*MATa ubr1Δ::PGAL-HA-UBR1 (HIS3) ura3-1:: PADH1-TIR1-9MYC (URA3)*	[12]
YAD257	*MATa ubr1Δ::PGAL-HA-UBR1 (HIS3) ura3-1::PADH1-TIR1-9MYC (URA3) leu2-3,112::pRS305-inn1-K31A-GFP(LEU2) inn1::PCUP1-td-inn1 -aid (K.l.TRP1-kanMX)*	[12]
YASD522	*MATa ubr1Δ::PGAL-HA-UBR1 (HIS3) inn1::PCUP1-td-inn1 (kanMX)*	[19]
YAD236	*MATa ubr1Δ::PGAL-HA-UBR1 (HIS3) ura3-1::pRS306-PGAL-PKC1-R398P (URA3) leu2-3,112::pRS305-PGAL-GPA2-G132V (LEU2)*	[12]
YAD245	*MATa ubr1Δ::PGAL-HA-UBR1 (HIS3) ura3-1::PADH1-Skp1-TIR1-9MYC (URA3) inn1::PCUP1-td-inn1-aid (K.l.TRP1&kanMX)*	[12]
YAD258	*MATa ubr1Δ::PGAL-HA-UBR1 (HIS3) ura3-1:: PADH1-TIR1-9MYC (URA3) leu2-3,112::pRS305-inn1-K31A-GFP(LEU2) inn1::PCUP1-td-inn1 -aid (K.l.TRP1&kanMX)*	[12]
YAD276	*MATa ubr1Δ::PGAL-HA-UBR1 (HIS3) ura3-1::PADH1-TIR1-9MYC (URA3) leu2-3,112::pRS305-inn1-K31A-GFP(LEU2) inn1::PCUP1-td-inn1 -aid (K.l.TRP1&kanMX) suppressor-d21*	[12]

PGAL and the constitutive *ADH1* promoter (*PADH1*), respectively, are available from National BioResource Project-Yeast (http://yeast.lab.nig.ac.jp/nig/index_en.htmL).

2.2 Plasmids

1. pKL187 for tagging with ts-degron at the N terminus of POIs is available from EUROSCARF [13].
2. pMK38 and pMK43 for tagging with aid at the N and C termini of POIs, respectively, are available from National BioResource Project-Yeast. Other related plasmids encoding mini- or tandem mini-aid with various selection markers are also available from National BioResource Project-Yeast [9].

2.3 Reagents

1. 20 % Glucose: Dissolve 100 g of glucose in water to a final volume of 500 mL. Sterilize the solution by autoclaving.
2. 20 % Raffinose and galactose: Dissolve 100 g of raffinose or galactose in water to a final volume of 500 mL. Note that heating

is needed to dissolve raffinose or galactose powder. Do not boil when heating. Sterilize the solutions by filtration.

3. 10× TE, pH 7.5: Mix 20 mL of 1 M Tris–HCl, pH 7.5 and 4 mL of 0.5 M EDTA, pH 8.0 by adding water to a final volume of 200 mL; autoclave and store at room temperature.
4. 1 M LiAc, pH 7.5: Dissolve 10.2 g of lithium acetate in water to a final volume of 100 mL by adjusting the pH 7.5 with acetic acid. Autoclave and store for up to 6 months at room temperature.
5. 50 % PEG 4000: Dissolve 50 g of PEG 4000 in water to a final volume of 100 mL. Autoclave and store for up to 1 month at room temperature.
6. TE/LiAc: Mix 1 M LiAc, pH 7.5, 10× TE, pH 7.5 and sterilized water at 1:1:8. Prepare just before use.
7. TE/LiAc/PEG: Mix 1 M LiAc, pH 7.5, 10× TE, pH 7.5, 50 % PEG 4000 at 1:1:8. Prepare just before use.
8. 10 mg/mL salmon sperm DNA.
9. 0.1 M $CuSO_4$: Dissolve 2.5 g of $CuSO_4 \cdot 5H_2O$ in water to a final volume of 100 mL. Autoclave and store at room temperature.
10. 50 mg/mL G418: Dissolve in water at a concentration of 50 mg/mL. Filter-sterilize and store at −20 °C.
11. DMSO.
12. β-Glucoronidase.
13. Phusion DNA polymerase.
14. 0.5 M indole-3-acetic acid (IAA) (Sigma-Aldrich, 45533): Dissolve in 100 % ethanol or DMSO. Aliquots can be kept for 6 months in the dark at −20 °C.
15. 0.5 M 1-naphthaleneacetic acid (NAA) (Sigma-Aldrich, 35745): Dissolve in 100 % DMSO. Aliquots can be kept for 6 months in the dark at −20 °C.
16. Ethyl methanesulfonate (EMS).
17. 0.1 M sodium phosphate buffer.
18. 5 % sodium thiosulfate.
19. 1× TE, pH 8.0: Mix 2 mL of 1 M Tris–HCl, pH 8.0 and 0.4 mL of 0.5 M EDTA by adding water to a final volume of 200 mL; autoclave.
20. STET: 8 % sucrose, 5 % TritonX-100, 50 mM Tris–HCl, pH 8.0 and 50 mM EDTA.
21. 7.5 M ammonium acetate.
22. Sau3AI restriction enzyme.
23. BamHI restriction enzyme.

24. Qiagen Genomic DNA kit with 500/G tips.
25. 0.5 mm glass beads: 0.5 mm glass beads should be acid bathed in HCl for 2 h. Subsequently, wash three times each in water and 1 M Tris–HCl, pH 8.0. Wash again extensively in water before drying.
26. Antibody to detect aid-tagged proteins via western blotting, a specific monoclonal antibody against aid is commercially available from MBL (M214-3).

2.4 Media

1. YPD, YPG, or YPR medium: 2 % peptone and 1 % yeast extract in water. Sterilize by autoclaving before the addition of sterile 2 % glucose (YPD), 2 % galactose (YPG), or 2 % raffinose (YPR).
2. YPD or YPG plates: Weigh 10 g peptone, 5 g yeast extract and 10 g agar. Add water to a final volume of 450 mL. Sterilize by autoclaving before the addition of 50 mL of 20 % glucose (YPD) or 20 % galactose (YPG). To prepare YPD plates containing G418 (YPD + G418 plate), add 3 mL of 50 mg/mL G418 (final concentration of 300 μg/mL). To prepare YPD or YPG plates with auxin, add 500 μL of 0.5 M IAA or NAA (final concentration of 0.5 mM). All supplements should be added when the temperature is at around 50 °C.
3. Synthetic complete (SC) drop-out plates with or without auxin. Prepare the SC plates as described previously [14] by dropping out the amino acid as required in the experiment (in our case, we used SC-His plates supplemented with 0.125 mM NAA to plate cells after transforming the genomic library with a *HIS3* yeast selection marker).

3 Methods

3.1 Design of Oligonucleotides for Making Ts-degron Mutants

Ts-degron mutants expressing td-POI from its endogenous locus can be generated by transforming YKL200 cells with PCR-amplified DNA from pKL187 (Fig. 2, upper panel) [13]. Note that td-POI is expressed from a *CUP1* promoter (*PCUP1*), instead of its own promoter. To maintain expression of td-POI, all media after transformation should contain 100 μM $CuSO_4$.

1. To design two oligonucleotides for PCR using pKL187, retrieve the genomic sequence of the target gene and the surrounding 1 kb from the Saccharomyces Genome Database (http://www.yeastgenome.org).
2. Design HPLC-purified "5′ and 3′ 70-mers" as follows. The "5′ 70-mer" should begin with 50 bases chosen from the promoter region of the target gene, followed by 20 bases (5′-ATTAAGGCGCGCCAGATCTG-3′) that are identical to

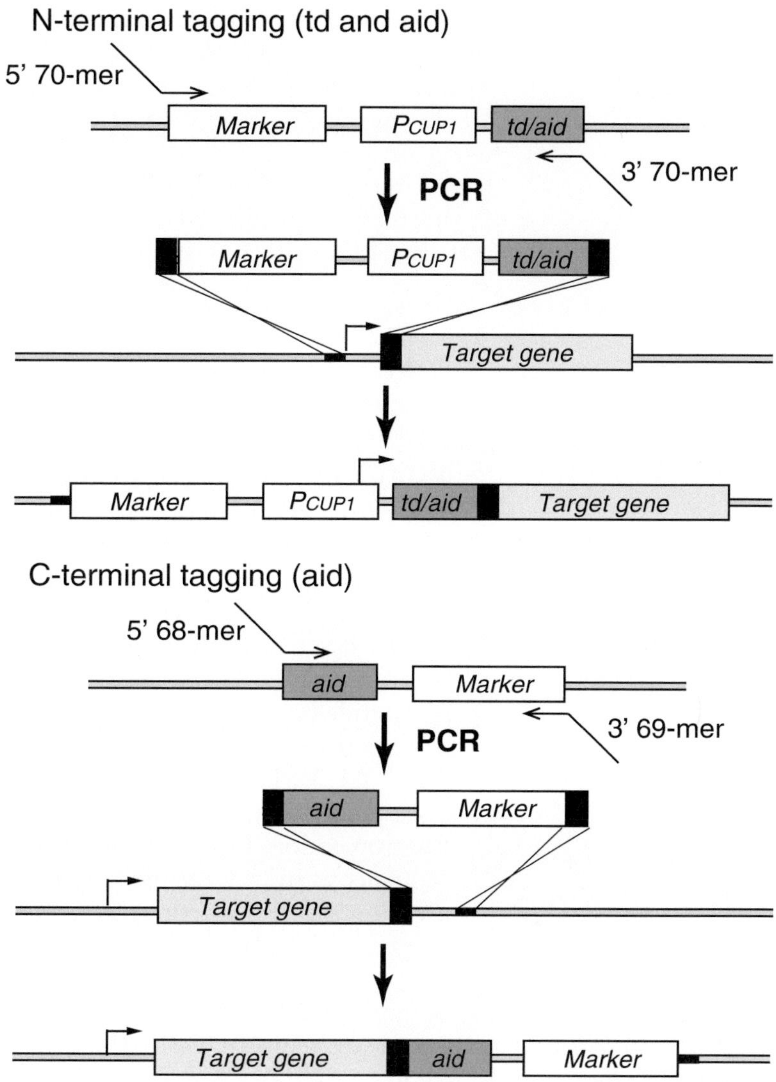

Fig. 2 PCR-based degron tagging to generate ts-degron and AID mutants. The ts-degron (td) can be placed only at the N terminus, whereas aid can be fused at either the N or C terminus of the POI. Note that the expression of N-terminally tagged proteins is driven by *PCUP1*

the 5′ region of pKL187. We usually choose the first 50 bases from the promoter region, located 50–100 bp upstream from the first ATG sequence, which do not contain repeats or many G/Cs. To design the "3′ 70-mer," prepare an oligonucleotide beginning with the first 20 bases (5′-CAGGCGCTGGAGCGGTGCC-3′) followed by the first 50 bases chosen from the open reading frame of the target gene, including the first ATG. Order the anti-parallel oligonucleotide as the "3′ 70-mer."

3.2 Design of Oligonucleotides to Prepare AID Mutants

AID mutants can be made by transforming YMK726 (conditionally expressing TIR1 from *PGAL*) or YMK728 (constitutively expressing TIR1 from *PADH1*) with PCR-amplified DNA (Fig. 2) [9]. In the case of fusing aid at the N terminus of the POI, the fusion protein is expressed from *PCUP1*. As in the case of td-degron mutants, all media should contain 100 μM $CuSO_4$ after transformation. In the case of fusing aid at the C terminus of the POI, the expression of the protein is driven by its endogenous promoter. Therefore, $CuSO_4$ is not required for expression.

1. To place aid at the N terminus of the POI using PCR-amplified DNA from pMK38, the same oligonucleotides as those designed for the construction of ts-degron mutants can be used.
2. To place aid at the C terminus of the POI using PCR-amplified DNA from pMK43, prepare two HPLC-purified oligonucleotides, "5′ 68-mer" and "3′ 69-mer." Design "5′ 68 mer" using the first 50 bases from the 3′ end of the open reading frame, without the stop codon. This sequence is followed by the 18 bases 5′-CGTACGCTGCAGGTCGAC-3′, which are identical to the 5′ end of the degrons. Design a 69-mer oligonucleotide beginning with the 19 bases that are identical to the 3′ region of the template plasmids, 5′-CGAGCTCGAATTC ATCGAT-3′. The 3′ end is followed by 50 bases retrieved from a sequence located between 50 and 200 bp downstream from the stop codon. Order the anti-parallel oligonucleotide of this sequence as the "3′ 69-mer." Note that these oligonucleotides are compatible with the plasmids used to generate epitope-tagged strains [15, 16].

3.3 Preparation of PCR-Amplified DNA for Transformation

1. Set up a 100 μL PCR reaction.
   10 ng/μL template DNA 1 μL.
   20 μM 5′ 70- or 68-mer 2.5 μL.
   20 μM 3′ 70- or 69-mer 2.5 μL.
   5× HF buffer 20 μL.
   2.5 mM dNTPs 8 μL.
   Phusion DNA polymerase 1 μL.
   Water up to 100 μL.
2. Run PCR using the following thermal cycling conditions:
   98 °C, 30 s → (98 °C, 5 s → 55 °C, 10 s → 72 °C, 1.5 min), 30 times → 72 °C, 5 min → 4 °C, ∞.
3. Apply 1–2 μL of the PCR reaction to 0.8 % agarose gel electrophoresis to check DNA amplification.
4. Purify the amplified DNA using a PCR purification kit and elute DNA with 30 μL of TE. The eluted DNA can be stored at −20 °C.

3.4 Yeast Transformation and Selection of Degron Mutants

Budding yeast cells are transformed with PCR-amplified DNA following the standard LiAc/SS-DNA/PEG protocol before growing on selection plates [17].

1. Choose appropriate yeast cells. To prepare ts-degron mutants, use YKL200, which express Ubr1 from *PGAL*. To prepare AID mutants, use either YMK726 or YMK728, which expresses TIR1 from the conditional *PGAL* or constitutive *PADH1* promoter, respectively.
2. Count the cells and dilute the culture to 0.4×10^7 cells/mL in YPD medium. One transformation requires 10 mL of culture.
3. Grow the cells at 24 °C until they reach a density of 0.9–1.1×10^7 cells/mL.
4. Pellet the cells by centrifugation at $1600 \times g$ for 3 min in a 50 mL tube.
5. Resuspend the cells in the same volume of sterilized water and pellet them by centrifugation at $1600 \times g$ for 3 min.
6. Resuspend the cells in 1 mL of sterilized water before transferring into a 1.5 mL microtube.
7. Pellet the cells by centrifugation at $15,000 \times g$ for a few seconds and remove all liquid by aspiration.
8. Resuspend the cells in 1 mL of TE/LiAc before centrifugation at $15,000 \times g$ for a few seconds. Remove all liquid by aspiration.
9. Resuspend the cells in 50 μL of TE/LiAc per 10 mL of culture. The cells are now ready for transformation.
10. Transfer 50 μL of the cell suspension into a new tube containing 5 μL of 10 mg/mL of salmon sperm DNA and 5 μL of purified PCR-amplified DNA. Mix well by vortexing.
11. Add 0.3 mL of TE/LiAc/PEG and mix by vortexing.
12. Incubate the cells at 24 °C for 30–60 min with agitation.
13. Add 40 μL of 100 % DMSO to the cell suspension and mix by vortexing.
14. Incubate the cell suspension at 42 °C for 15 min and cool it on ice for 2 min.
15. The cells have to be recovered in 1 mL of YPD medium at 24 °C with agitation before selection with G418. We usually allow 3 h for recovery. Note that all media should contain 100 μM $CuSO_4$ from this point on, to drive the expression of POI from *PCUP1* if the degrons are placed at the N terminus of the POI.
16. Collect the cells by centrifugation at $900 \times g$ for a few seconds and remove all liquid by aspiration. Resuspend the cells in 300 μL of TE, and plate 100 μL of the cell suspension on YPD + G418 plate (total of three plates). Incubate the plates at 24 °C.

17. Cells without integration also grow for a few days. To reduce the background, make replica plates 1 or 2 days after transformation.
18. Four to five days after transformation, colonies should be visible on the plates. Pick eight resistant clones and re-streak them on a fresh selection plate to confirm their growth. Incubate the re-streaked plate for 2–3 days, until colonies are formed.
19. Check integration of the PCR-amplified cassette by genomic PCR [9, 13].

3.5 Testing the Growth of Strains Containing the Degron Fusion Proteins (Serial-Dilution and Liquid Culture Assays)

In the case of preparing degron mutants for proteins that are essential for cell viability, test the growth of the strains under both restrictive and permissive conditions (*see* **Note 1**).

First, a serial-dilution assay can be performed to compare the growth of cells on different media (YPD, YPG, YPD + auxin, or YPG + auxin) and at different temperatures (24 °C or 37 °C) (Fig. 3b). At this step, it is also important to test the auxin type to assess whether NAA or IAA works best for the strain. In this experiment, it is possible to determine under which conditions the revertants do not appear after prolonged incubation times (Fig. 3b, also *see* **Note 2**).

1. Prepare 10 cm square YPD, YPG, YPD + auxin, and YPG + auxin plates.
2. Pick 2–3 growing colonies of the yeast strains and resuspend each in 1 mL of phosphate-buffered saline (PBS).
3. Determine the cell density using a hemocytometer and dilute the cells to a concentration of 3.33×10^6 cells/mL. Prepare three further tenfold dilutions.
4. Drop 15 μL of a cell suspension from each dilution onto appropriate plates.
5. Incubate the plates for 2–4 days at 24 °C or 37 °C, as required.

   A liquid culture assay can be performed to monitor defects under a restrictive condition. Cell-cycle defects can be monitored by flow cytometry as explained below (Fig. 4). The expression level of degron-fused proteins can be checked by Western blotting [9].
6. Inoculate cells in YPR medium (supplemented with 100 μM $CuSO_4$ in case the degron–POI is expressed from *PCUP1*) and grow overnight at 24 °C.
7. On the following morning, determine the cell density using a hemocytometer and dilute the cultures to 0.4×10^7 cells/mL.
8. Grow the cells for several hours until the cell density reaches $0.7–0.8 \times 10^7$ cells/mL.

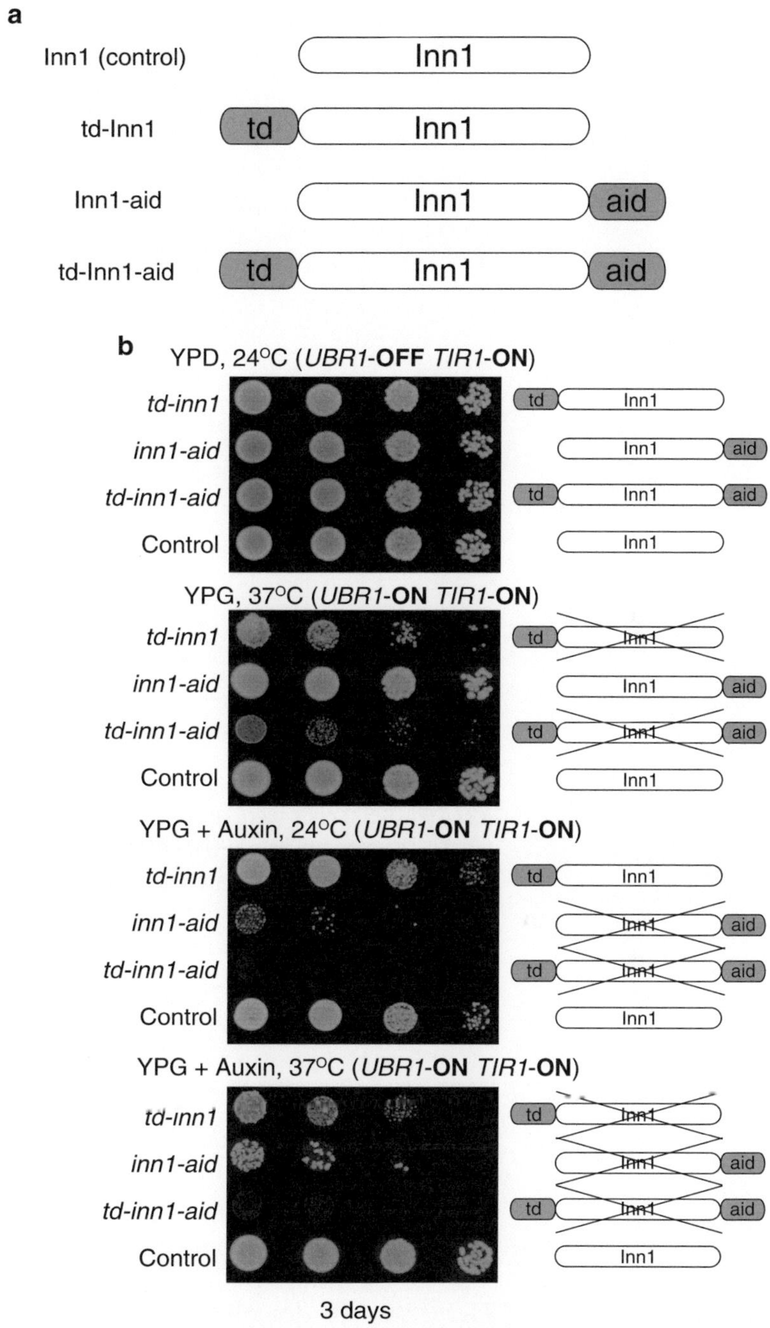

Fig. 3 Degron systems used to modify the target locus. (**a**) Schematic representations in which the *INN1* locus was modified by adding the indicated degron cassettes. (**b**) Strains that carry both degron cassettes on the target protein do not yield revertant colonies after prolonged incubation under restrictive growth conditions. The *td-inn1* (YASD522), *inn1-aid* (YAD236), *td-inn1-aid* (YAD245), and control cells (YAD240) were grown at 24 °C on YPD medium before serial dilutions of 50,000, 5000, 500, and 50 cells were plated on the indicated media and incubated at the indicated temperatures for 3 days

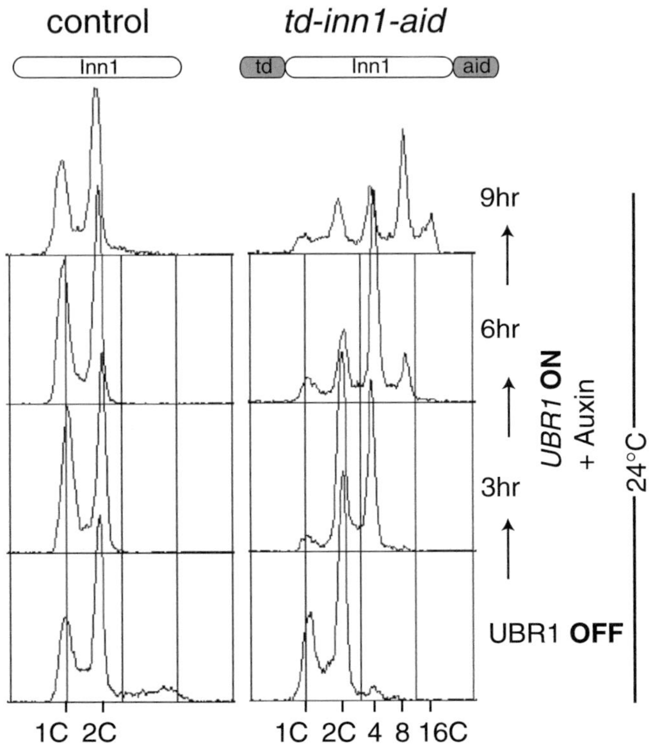

Fig. 4 Asynchronous cultures of control (W303-1a) and *td-inn1-aid leu2::inn1-K31A* (YAD257) cells were grown at 24 °C in YPR medium, followed by replacement of the medium with YPG medium supplemented with 0.5 mM NAA. Samples were collected at the indicated time points and DNA content was measured by flow cytometry

9. Pellet the cells at 200×*g*, resuspend in YPG+auxin medium and continue to grow cells at 24 °C (either 0.5 mM NAA or 0.5 mM IAA can be used, depending on the response of the strain).
10. Collect 1 or 10 mL of cell culture for flow cytometry or for protein extraction for Western blotting [7].
11. It is also possible to capture phase-contrast images of the cells at appropriate time points.

3.6 EMS Mutagenesis

Hereafter, we describe a screening that was performed for the identification of genomic mutations that suppress the lethality of YAD258 cells expressing Inn1-K31A in the *td-inn1-aid* background. As shown in Fig. 3b, the *td-inn1-aid* strain did not show spontaneous revertants in the presence of auxin at both 24 and 37 °C. We introduced the *inn1-K31A* gene into the *td-inn1-aid* strain. Subsequently, the resultant YAD258 cells were mutagenized using EMS to identify suppressor mutations under the restrictive condition. Dual-degron mutants showing few spontaneous revertants can also be used for other genetic screenings, such as dosage-suppressor screening, by introducing multicopy plasmids containing a genomic library [11].

3.6.1 Optimizing the Conditions

1. Inoculate an appropriate dual-degron strain in 100 mL of YPD medium, and grow the cells overnight at 24 °C.
2. On the following morning, determine the cell density using a hemocytometer and dilute the cells to 0.4×10^7 cells/mL in fresh YPD medium.
3. Grow the cells until the cell density reaches a concentration of $0.8–1 \times 10^7$ cells/mL.
4. Transfer a culture corresponding to 10^8 cells into each of several 15 mL conical tubes (the number of tubes depends on the number of incubation times to be tested, e.g., seven tubes in total required: one for the control and six for incubation with EMS for an amount of time of 10, 15, 20, 25, 30, or 45 min).
5. Pellet the cells at $200 \times g$ and resuspend each in 5 mL water.
6. Pellet the cells at $200 \times g$ and resuspend each in 5 mL of 0.1 M sodium phosphate buffer.
7. Pellet the cells at $200 \times g$ and resuspend each in 1 mL of 0.1 M sodium phosphate buffer.
8. Add 30 μL of EMS to each tube, with the exception of the one that will be used as a control. Note that EMS is a strong mutagen; therefore, handle it in a fume hood with appropriate protection.
9. Incubate the cells with agitation for varying amounts of time (e.g., 10, 15, 20, 25, 30, or 45 min).
10. For each tube, pellet the cells when the corresponding incubation time is completed. Resuspend the cells in 500 μL of 5 % sodium thiosulfate, to inactivate EMS.
11. Wash the cells twice with 500 μL of 5 % sodium thiosulfate and, after the second wash, resuspend the cells in water.
12. For each time point, plate the cells on four YPDCu plates, to give about 500 cells/plate.
13. Incubate the plates at 24 °C for 2–3 days and count the number of growing colonies on each plate for each time point.
14. Determine the time point corresponding to 50 % survival at 24 °C, and use this in the subsequent mutagenesis experiment.

3.6.2 Mutagenesis with EMS

The number of cells that will be screened can be changed depending on the experiment. The protocol included below indicates the conditions to screen 2×10^8 cells. Refer to Fig. 5a for a summary of the protocol.

a

Day 1

Grow an overnight culture in YPD

Day 2

Dilute the culture to 0.4×10^7 cells/ml

Regrow the culture to 10^7 cells/ml

Mutagenise the cells with 3% ethyl methanesulfonate(EMS) corresponding to 50% survival

Inactivate EMS by adding 5% sodium thiosulfate

Resuspend cells in YPD and incubate 3 hours for recovery

Plate out cells on selective plates (YPG + auxin)

Day 3

Next day replica plate on selective plates to avoid background

Day 5

Look for growing colonies

Check each marker by replica plating to selective medium

Day 6-7

Pick up and make stocks of growing colonies for further analysis

b

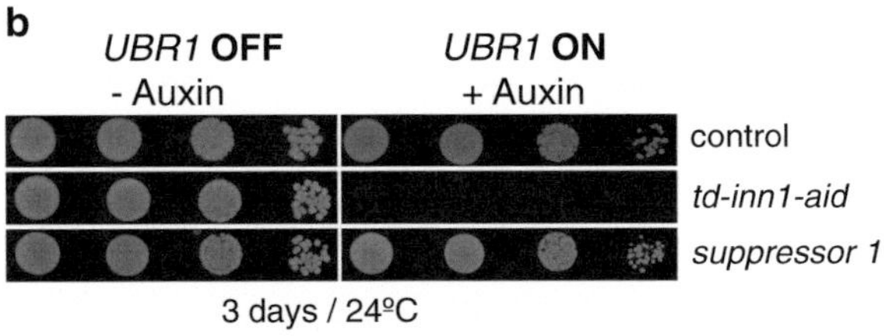

Fig. 5 Isolation of suppressor mutations. (**a**) Overview and timeline of the steps used for the isolation of suppressor mutations. (**b**) Serial dilutions of the *td-inn1-aid* (YAD258) and suppressor clone (YAD276) cells were plated on YPD or YPG medium supplemented with 0.5 mM NAA and incubated for 3 days at 24 °C

1. Inoculate an appropriate dual-degron strain in 100 mL of YPD medium and grow an overnight culture at 24 °C. To express the dual-degron protein from *PCUP1*, all media under permissive conditions should contain 100 μM $CuSO_4$, to drive its expression.
2. On the following morning, determine the cell density using a hemocytometer, and dilute the cells to 0.4×10^7 cells/mL in fresh YPD medium.
3. Grow the cells until the cell density reaches a concentration of $0.8\text{–}1 \times 10^7$ cells/mL.

4. With the aim of spreading 10^7 cells/plate, pellet the required amount of cells at 200 × *g* (e.g., for 20 plates, 2×10^8 cells are required; therefore, pellet 20 mL of the culture at a density of 1×10^7 cells/mL).
5. Resuspend the cells in 10 mL of water and pellet by centrifugation at 200 × *g*.
6. Wash the cells once with 10 mL of 0.1 M sodium phosphate buffer and resuspend in 2 mL of 0.1 M sodium phosphate buffer.
7. Add 30 μL of EMS per 10^8 cells to the culture and incubate with agitation for the amount of time determined above at 24 °C.
8. Wash the cells twice with 1 mL of 5 % sodium thiosulfate and, after the second wash, resuspend the cells in 2 mL of YPD medium.
9. Incubate the cultures for 3 h at 24 °C with agitation, for recovery.
10. Plate the cells on selective plates (100 μL per plate for a total of 20 plates) and incubate at 24 °C.
11. On the following day, prepare replica plates of selective plates, to avoid a high background.
12. When the colonies become visible, pick all growing colonies and re-streak them on both selective and YPD plates. It is important to confirm that the revertants obtained carry *TIR1*, *UBR1*, and dual-degron genes.
13. Prepare replica plates of all necessary plates, to check the markers in the selected colonies.
14. Prepare glycerol stocks of the mutants that carry all the relevant markers, for further analysis. To prepare a glycerol stock, mix a 1 mL aliquot of overnight-grown YPD culture with 1 mL of sterile 50 % glycerol in a cryotube, and store at −80 °C.
15. Perform a serial-dilution assay to compare the growth of wild-type, parental and suppressor strains under both permissive and restrictive growth conditions, to determine the effectiveness of suppression and facilitate greatly the genetic analysis described below (Fig. 5b).

3.7 Genetic Crosses and Dominance Test

This process will eliminate revertants, in which td-POI-aid was not degraded because of a defect in the degron systems (Fig. 6-1).

1. Cross each of the mutant clones with WT cells of the opposite mating type (W303-1a or -1b).
2. Perform a tetrad analysis of the resultant diploids.
3. Assess whether any of the markers associated with *PGAL-UBR1*, *PADH1-TIR1*, and *td-POI-aid* always segregate with

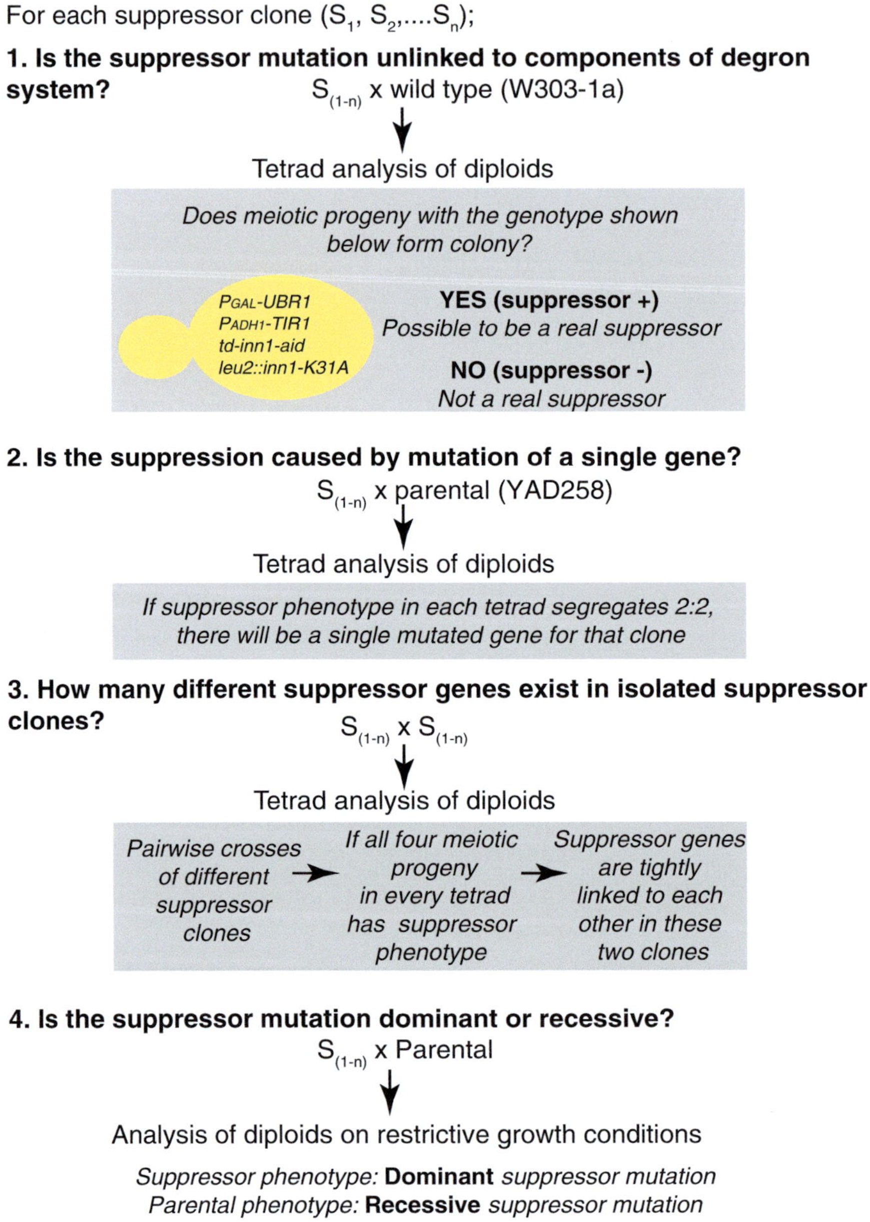

Fig. 6 Overview of the steps used for linkage analysis and the dominance test of suppressor clones

the suppressor phenotype. If this is the case, the suppressor may represent a mutation at one of these loci.

4. Conversely, free segregation of the suppressor phenotype will indicate that the suppressor mutation is located elsewhere in the genome. Pursue the analysis using those clones.

   This process will determine whether the suppressor mutation in each clone is derived from a single gene (Fig. 6-2).

5. Cross each of the mutant clones with the parental strain of the opposite mating type.
6. Observe the segregation of the suppressor phenotype in each tetrad.
7. The presence of two colonies with the suppressor phenotype in each tetrad (2:2 segregation) will indicate the mutation of a single gene for that suppressor clone.
   This process will determine the number of different suppressor genes represented in the clones obtained (Fig. 6-3).
8. Divide the suppressors into groups and cross each member of the group to the other members in the same group.
9. Determine whether members of each group are closely linked to each other or not (the observation that all four meiotic progeny in a tetrad have a suppressor phenotype will indicate mutation in a single gene for the two clones crossed).
10. If there is such a linkage, cross one suppressor from each group to another suppressor in each of the other groups and analyze the meiotic progeny.
11. Determine the total number of different suppressor genes for all the suppressed clones. *See* **Note 3** before moving the next step.
    This process will determine whether the mutant allele is dominant or recessive (Fig. 6-4).
12. Cross each of the suppressor clones with the parental strain of the opposite mating type.
13. Examine the diploid cells obtained on selective plates (*UBR1* ON, *TIR1* ON, + auxin).
14. The observation of growth of diploid cells on selective medium will indicate that the mutation is dominant. The opposite observation will indicate that the suppressor mutation is recessive. Depending on the type of mutation, the strategy used for cloning the mutated gene will change (Fig. 7, also *see* **Note 4**).

3.8 Cloning of Suppressors

As shown in Fig. 7, if the suppressor is dominant and there is no additional phenotype, a genomic DNA library from the suppressor clone is generated and transformed to the parental strain, which is subsequently examined on selective medium (+auxin). Because cells containing the suppressor plasmid will grow on the selective medium, the identification and sequencing of that plasmid will reveal the DNA sequence that causes suppression.

3.8.1 Preparing a Library of Yeast Genomic DNA Fragments

1. Grow the suppressor clone in 400 mL of YPD medium overnight.
2. Isolate genomic DNA using Qiagen Genomic DNA kit with 500/G tips.

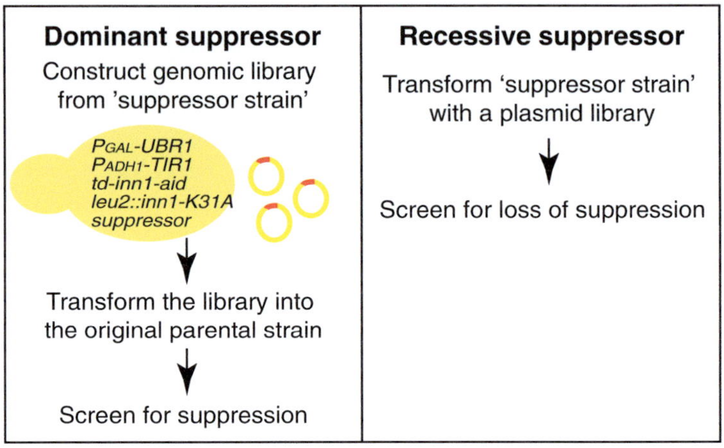

Fig. 7 Overview of the steps used for cloning suppressors

3. Partially digest the 100 μg of genomic DNA with the Sau3AI restriction enzyme (0.03125 U), to give fragments between 3 and 10 kb after determining the conditions in a small-scale experiment. Firstly, digest 1 μg of DNA with serial dilutions of Sau3AI: 1, 0.5, 0.25, 0.125, 0.0625, 0.03125, 0.0156, 0.0078, 0.0039, and 0.002 U. Secondly, run each reaction on a 0.8 % agarose gel together with a 1 kb DNA Ladder, to assess the size distribution of the digested DNA fragments. Finally, select the reaction that yields a smear of digested DNA between 3 and 10 kb.
4. Purify the digested DNA using Qiagen Genomic Tip 500/G.
5. Precipitate the DNA first with isopropanol, and wash with 70 % ethanol.
6. Resuspend the DNA in 100 μL of TE, pH 8.0.
7. Apply 100 μL of the purified DNA to a wide well on a 0.8 % agarose gel (prepare a wide comb by connecting six teeth using Scotch tape) and run the gel.
8. Cut the fragments in the range of 3–10 kb from the agarose gel. Purify the DNA using a gel extraction kit and elute in 40 μL of TE, pH 8.0.
9. Digest one of the RS centromere vectors (we use pRS313, which has the *HIS3* selection marker) with BamHI and treat with calf intestinal phosphatase (3 μg of pRS313 should be digested with 30 U of BamHI for 3 h at 37 °C, followed by treatment with 1 U of phosphatase for 30 min at 37 °C).
10. Perform several ligation reactions, each using 20 ng of digested vector and 20 ng of digested genomic DNA.
11. Transform the ligation reactions into competent DH10B *E. coli* cells by electroporation (2 μL aliquots of ligation reactions

can be electroporated into 50 μL of competent cells) and spread on large square plates (245 × 245 mm) containing LB-ampicillin medium (we usually use seven plates).

12. Grow transformants overnight at 37 °C.
13. On the following day, scrape off the *E. coli* cells from plates, pellet the cells and purify the plasmid DNA, which represents the genomic library (*see* **Note 5**).

3.8.2 Determination of the Suppressor Plasmids from the Genomic Library

1. Transform approximately 500 ng of the genomic library into the parental strain according to the standard LiAc/SS-DNA/PEG protocol [17].
2. A total of 10^8 cells should be transformed; plate the cells on SC plates lacking the appropriate amino acid (for pRS313, use SC-His plates), to give about 5×10^7 cells/plate. Incubate the plates at 24 °C overnight.
3. On the following day, prepare replica plates of fresh SC plates lacking the appropriate amino acid but supplemented with 0.125 mM NAA (*see* **Note 6**).
4. When the colonies become visible, pick them and check the presence of all relevant markers by replica plating onto selective medium.
5. Pick single colonies with all the relevant markers and streak them on YPD plates supplemented with 0.5 mM NAA, to check whether the loss of plasmids on YPD medium causes loss of the suppression phenotype.
6. Loss of the suppression phenotype on YPD + 0.5 mM NAA plates indicates that those plasmids carry the relevant sequences that cause suppression. The plasmid should be recovered and used to transform *E. coli*.

3.8.3 Recovery of Plasmids from Yeast

Here, we describe a standard protocol for plasmid purification from yeast [18].

1. Grow cells overnight at 24 °C in 10 mL of SC medium lacking the appropriate amino acid (SC-His if the library is made with pRS313).
2. Pellet the cells (density must be very high) at 200 × *g* and resuspend in 100 μL of STET with 0.2 g of 0.5 mm glass beads in a microfuge tube.
3. Vortex the mixture for 5 min.
4. Add 100 μL of STET to the tube and incubate at 100 °C for 3 min.
5. Cool on ice briefly and spin in a microfuge at 13,000 × *g* for 10 min at 4 °C.
6. Transfer the supernatant to a fresh tube containing 50 μL of 7.5 M ammonium acetate.

7. Incubate the tube for 60 min at −20 °C.
8. Spin in a microcentrifuge at 13,000 × *g* for 10 min at 4 °C.
9. Add 100 μL of the supernatant to 200 μL of ice-cold ethanol and spin again at 13,000 × *g* for 10 min at 4 °C.
10. Wash the pellet twice with 70 % ethanol.
11. Resuspend the DNA pellet in 20 μL of water and use 10 μL for transformation of competent *E. coli* cells (e.g. DH5α, XL10-Gold).
12. After transformation into competent *E. coli* cells, isolate the plasmids and sequence the insert, to identify the suppressor gene.

If the suppressor is recessive, a plasmid library is transformed into the suppressed strain and the loss of the suppression phenotype is screened among the transformants (Fig. 7).

4 Notes

1. To prepare the degron strain for a suppressor screen, it is important to check the growth of cells under permissive conditions. Cells should grow well, similar to wild-type cells. As the two ends of the protein are tagged, it may not function as the wild-type counterpart, which would interfere with its functioning. Moreover, the expression and localization of the degron protein must be compared with those of the wild-type protein. We observed that the fact that the expression level of the td-inn1-aid protein was much lower (probably related to the larger size of the protein after degron fusion) compared with the wild-type protein did not interfere with the growth of cells.
2. Similar to that described in Fig. 3, the wild-type and degron strains must be compared in all conditions via a serial-dilution assay. In our case, even when using the heat-inducible degron system, we chose 24 °C as a cell-growth temperature for both restrictive and permissive conditions, as no revertants appeared after a prolonged incubation time at that temperature. The conditions can change depending on the protein of interest.
3. Because the genetic screen described here is based on the degradation of the endogenous degron protein in cells, before trying to clone the suppressor gene, it is advised to test whether the protein is still degraded in the suppressor clones under restrictive growth conditions, as is expected for a true suppressor.
4. After obtaining suppressor clones, whether the suppressor strain has an additional phenotype should also be assessed [10]. Such an additional phenotype will facilitate the identification of

the suppressor. If there is no additional phenotype, then the method described in this protocol should be used.

5. The size of the genomic library used to clone the suppressor is important. The genomic library that we prepared from the suppressor strain consisted of approximately 35,000 clones with an average insert size of 3900 bp. This means that the library covers the genome about 11 times.

6. Always use NAA when using SC plates, as IAA causes a growth defect in wild-type cells. In our case, we also reduced the amount of NAA and supplemented the SC plates with 0.125 mM NAA.

References

1. Kanemaki MT (2013) Frontiers of protein expression control with conditional degrons. Pflugers Arch 465(3):419–425. doi:10.1007/s00424-012-1203-y
2. Kanemaki M, Sanchez-Diaz A, Gambus A, Labib K (2003) Functional proteomic identification of DNA replication proteins by induced proteolysis in vivo. Nature 423(6941):720–724. doi:10.1038/nature01692
3. Dohmen RJ, Wu P, Varshavsky A (1994) Heat-inducible degron: a method for constructing temperature-sensitive mutants. Science 263(5151):1273–1276
4. Labib K, Tercero JA, Diffley JF (2000) Uninterrupted MCM2-7 function required for DNA replication fork progression. Science 288(5471):1643–1647
5. Sriram SM, Kim BY, Kwon YT (2011) The N-end rule pathway: emerging functions and molecular principles of substrate recognition. Nat Rev Mol Cell Biol 12(11):735–747. doi:10.1038/nrm3217
6. Varshavsky A (1996) The N-end rule: functions, mysteries, uses. Proc Natl Acad Sci U S A 93(22):12142–12149
7. Nishimura K, Fukagawa T, Takisawa H, Kakimoto T, Kanemaki M (2009) An auxin-based degron system for the rapid depletion of proteins in nonplant cells. Nat Methods 6(12):917–922. doi:10.1038/nmeth.1401
8. Kubota T, Nishimura K, Kanemaki MT, Donaldson AD (2013) The Elg1 replication factor C-like complex functions in PCNA unloading during DNA replication. Mol Cell 50(2):273–280. doi:10.1016/j.molcel.2013.02.012
9. Nishimura K, Kanemaki MT (2014) Rapid depletion of budding yeast proteins via the fusion of an auxin-inducible degron (AID). Curr Protoc Cell Biol 64:20.9.1–20.9.16. doi:10.1002/0471143030.cb2009s64
10. Forsburg SL (2001) The art and design of genetic screens: yeast. Nat Rev Genet 2(9):659–668. doi:10.1038/35088500
11. Appling DR (1999) Genetic approaches to the study of protein-protein interactions. Methods 19(2):338–349. doi:10.1006/meth.1999.0861
12. Devrekanli A, Foltman M, Roncero C, Sanchez-Diaz A, Labib K (2012) Inn1 and Cyk3 regulate chitin synthase during cytokinesis in budding yeasts. J Cell Sci 125(Pt 22):5453–5466. doi:10.1242/jcs.109157
13. Sanchez-Diaz A, Kanemaki M, Marchesi V, Labib K (2004) Rapid depletion of budding yeast proteins by fusion to a heat-inducible degron. Sci STKE 2004(223):PL8. doi:10.1126/stke.2232004pl8
14. Sherman F (2002) Getting started with yeast. Methods Enzymol 350:3–41
15. Knop M, Siegers K, Pereira G, Zachariae W, Winsor B, Nasmyth K, Schiebel E (1999) Epitope tagging of yeast genes using a PCR-based strategy: more tags and improved practical routines. Yeast 15(10B):963–972. doi:10.1002/(SICI)1097-0061(199907)15:10B<963::AID-YEA399>3.0.CO;2-W
16. Janke C, Magiera MM, Rathfelder N, Taxis C, Reber S, Maekawa H, Moreno-Borchart A, Doenges G, Schwob E, Schiebel E, Knop M (2004) A versatile toolbox for PCR-based tagging of yeast genes: new fluorescent proteins, more markers and promoter substitution cassettes. Yeast 21(11):947–962. doi:10.1002/yea.1142
17. Gietz RD, Schiestl RH (2007) Quick and easy yeast transformation using the LiAc/SS carrier

DNA/PEG method. Nat Protoc 2(1):35–37. doi:10.1038/nprot.2007.14
18. Robzyk K, Kassir Y (1992) A simple and highly efficient procedure for rescuing autonomous plasmids from yeast. Nucleic Acids Res 20(14):3790
19. Sanchez-Diaz A, Marchesi V, Murray S, Jones R, Pereira G, Edmondson R, Allen T, Labib K (2008) Inn1 couples contraction of the actomyosin ring to membrane ingression during cytokinesis in budding yeast. Nat Cell Biol 10(4):395–406. doi:10.1038/ncb1701

Chapter 19

Synchronization of the Budding Yeast *Saccharomyces cerevisiae*

Magdalena Foltman, Iago Molist, and Alberto Sanchez-Diaz

Abstract

A number of model organisms have provided the basis for our understanding of the eukaryotic cell cycle. These model organisms are generally much easier to manipulate than mammalian cells and as such provide amenable tools for extensive genetic and biochemical analysis. One of the most common model organisms used to study the cell cycle is the budding yeast *Saccharomyces cerevisiae*. This model provides the ability to synchronise cells efficiently at different stages of the cell cycle, which in turn opens up the possibility for extensive and detailed study of mechanisms regulating the eukaryotic cell cycle. Here, we describe methods in which budding yeast cells are arrested at a particular phase of the cell cycle and then released from the block, permitting the study of molecular mechanisms that drive the progression through the cell cycle.

Key words Alpha factor, Hydroxyurea, Nocodazole, *cdc15-2*, G1 phase, G2 and M phases, Early S phase, Mitosis, Budding yeast, *Saccharomyces cerevisiae*, Synchronization

1 Introduction

A significant part of our current understanding of the eukaryotic cell cycle and its regulation has come from studies involving the budding yeast, *Saccharomyces cerevisiae* (extensively reviewed, e.g., [1–3]). The use of budding yeast as a model organism in research has tremendous advantages, as it allows powerful genetics to be combined with a multitude of biochemical analyses and advanced microscopic studies. One of the most valuable advantages of using this model organism comes from the ability to efficiently synchronise budding yeast cells in different stages of the cell cycle [4, 5]. Upon arrest, cells can be synchronously released from the block and allowed to progress through the cell cycle synchronously (this type of experiment is called: block and release), which enables the analysis of the mechanisms regulating the eukaryotic cell cycle.

As the name might suggest, daughter cells of budding yeast cells grow via the formation of a 'bud' as they advance through the cell cycle and therefore cell cycle position can be simply determined

Alberto Sanchez-Diaz and Pilar Perez (eds.), *Yeast Cytokinesis: Methods and Protocols*, Methods in Molecular Biology, vol. 1369, DOI 10.1007/978-1-4939-3145-3_19,

microscopically by comparing the size of the bud to the mother cell. Cells in G1 phase are unbudded. At the end of G1, a small bud emerges when cells have past the restriction point (named START in yeast) at which point they are committed for a new round of cell division. Buds grow in parallel with cell progression through the cell cycle; therefore, one can easily follow cell synchrony by observing the size of the bud in the population of cells [6, 7]. The synchrony of the experimental culture in block and release experiments can be assayed by flow cytometry and microscopy (by calculating the budding index, as well as the percentage of divided nuclei).

Here, we describe two ways to synchronise cells at different stages of the cell cycle, followed by synchronous release. One method involves the use of chemical agents that block cell cycle progression. By removing the chemical from the culture medium, cells can be released and progress synchronously through the cell cycle. We report on how to synchronise cells in the G1 phase using mating pheromone (alpha factor), in early S phase by the addition of ribonucleotide reductase inhibitor, hydroxyurea (HU), or induce G2/M arrest by inhibiting microtubule polymerization with addition of nocodazole to the growth medium [5, 8].

Another efficient way in which synchrony of yeast cells can be achieved is through the use of temperature-sensitive mutants where a cell-division cycle gene (cdc) is modified in such a way that it contains a mutation which makes the protein dysfunctional when the temperature of the cell culture is changed to restrictive conditions. The advantage of using this type of arrest is that experiments can be scaled up in a simple and cheap manner without using large amounts of chemical reagents. We describe in detail the use of a temperature sensitive mutant, *cdc15-2*, which arrests cells in mitosis under restrictive conditions [9].

2 Materials

Prepare all solutions using ultrapure water and analytical grade reagents. Store all reagents and media at room temperature unless specified otherwise.

2.1 Cell Cycle Block and Release

1. Shaking water bath or shaking incubator.
2. Sonicator.
3. Microscope.
4. Counting chamber.
5. Bench centrifuge and microcentrifuge.
6. 50 and 1.5 ml tubes.
7. 100 ml culture flasks.

8. Haploid yeast cells (IMPORTANT arrest with alpha factor requires MAT**a** strain).
9. YPD growth medium (1 % yeast extract, 2 % peptone, 2 % glucose). Glucose should be prepared separately as 20 % stock solution and autoclaved.
10. 2x YP concentrated growth medium (2 % yeast extract, 4 % peptone).
11. Mating pheromone (alpha factor)—prepare stock solution of 5 mg/ml dissolved in water and store at −20 °C (*see* **Note 1**).
12. 0.6 M Hydroxyurea (HU), store at 4 °C.
13. YPDHU growth medium, for 500 ml use 250 ml 2× YP, 50 ml glucose, 167 ml 0.6 M hydroxyurea, 33 ml sterile water. Glucose should be prepared separately as 20 % stock solution and autoclaved (*see* **Note 2**).
14. Nocodazole—prepare stock solution of 2 mg/ml in DMSO and store at −20 °C.

2.2 Monitoring Synchrony of the Cell Culture

1. Flow cytometer.
2. Microscope.
3. 5 ml FACS tubes.
4. 1.5 ml tubes.
5. Poly-L-lysine-coated slides and cover slides.
6. 70 % ethanol.
7. RNase A—prepare 10 mg/ml stock solution in 10 mM Tris–HCl pH 7.5, 15 mM NaCl and boil for 15 min, aliquot and store at −20 °C.
8. 50 mM Sodium citrate—prepare 0.5 M stock solution.
9. 50 mM HCl—prepare 0.5 M stock solution.
10. Porcine pepsin (AMS Biotechnology).
11. Propidium iodide (PI) (Sigma)—prepare stock solution of 0.5 mg/ml in water, protect from light and store at 4 °C.

3 Methods

3.1 Cell Cycle Block and Release

In block and release experiments budding yeast cells are arrested at the appropriate stage of the cell cycle and subsequently released from that particular block. Experimental samples can be collected at different times after the release to study the molecular mechanism associated with a particular phase of the cell cycle. In order to achieve successful synchrony yeast cultures are used in early logarithmic phase (log phase) of growth [10]. The easiest way to monitor yeast growth is by counting the cell number.

For early logarithmic phase a density of cells between 5 and 10×10^6 cells/ml is required. Therefore any synchronization experiment requires determination of cell density as the first step in the protocol.

3.1.1 Mating Pheromone (Alpha Factor) Arrest and Release

Mating pheromone, called alpha factor, is a peptide produced by mating type alpha (MATα) cells (*see* **Note 1**) that binds to its corresponding receptors on mating type a (MAT**a**) cells. Binding of alpha factor to the receptor on a recipient cell leads to the inhibition of Cln-Cdc28 kinase activity; thus cells are arrested in G1 phase of the cell cycle with 1C DNA content and present a characteristic projection called "shmoo" (Fig. 1a) [4, 11].

1. Inoculate 1–2 yeast colonies of MAT**a** strain of your choice (yeast colonies have been grown on a plate for 3 days at 24 °C) in 50 ml of YPD medium (in 100 ml flask) and culture overnight at 24 °C in shaking water bath or incubator.
2. Take 1 ml of cell culture the following morning, sonicate for 4 s, count cells, and dilute to concentration of 4×10^6 cells/ml in 50 ml of YPD medium (*see* **Note 3**).

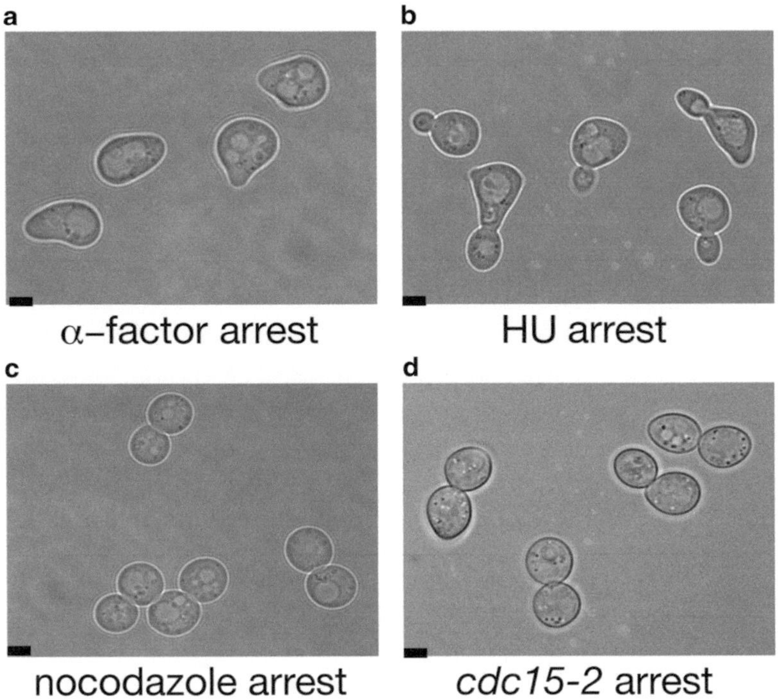

Fig. 1 Examples of yeast cells arrested at different stages of the cell cycle. (**a**) Budding yeast cells arrested in G1 phase with mating pheromone in YPD medium at 24 °C. (**b**) Budding yeast cells arrested in early S phase with hydroxyurea in YPD medium at 24 °C. (**c**) Budding yeast cells arrested in G2 and M phases with nocodazole in YPD medium at 24 °C. (**d**) Temperature-sensitive *cdc15-2* cells arrested in mitosis in YPD medium at 37 °C. The bar represents 2 μm

3. Grow cell culture at 24 °C until cells reach density of 7×10^6 cells/ml (*see* **Note 4**).
4. When cells reach 7×10^6 cells/ml, add 7.5 μg/ml of alpha-factor and grow culture for further 2 h. Then, start adding 2.5 μg/ml of alpha factor every 20 min (*see* **Note 5**).
5. After two and a half hours of first alpha factor addition start monitoring cell cycle arrest by microscopy. Take 1 ml of the culture, sonicate for 4 s, and examine cells under the microscope; alpha factor arrest is successful when the number of cells presenting a characteristic projection, called "shmoo," is higher than 90 % (Fig. 1a) (*see* **Note 6**).
6. To release cells from alpha factor arrest, spin down the culture at $200 \times g$ for 3 min in 50 ml tube, take off supernatant, and wash the pellet with 10 ml of YPD medium.
7. Spin down the culture at $200 \times g$ for 3 min, take off supernatant, and repeat the wash as in **step 6**.
8. Resuspend the pellet in 50 ml of YPD medium and release cells from alpha factor arrest by transferring the culture back to 24 °C. Start collecting experimental samples for downstream applications at the beginning of the release and then every 15 min to monitor the synchrony and cell progression.

3.1.2 Hydroxyurea (HU) Arrest and Release

Hydroxyurea is a chemical reagent that blocks DNA metabolism of yeast cells in early S phase of the cell cycle by inhibiting the enzyme ribonucleotide reductase [4], subsequently cells accumulate with small buds (Fig. 1b) and DNA content between 1C and 2C. In our experience the best way to synchronise the culture efficiently is to synchronise cells first in G1 phase of the cell cycle with alpha factor and then synchronously shift cells to medium containing hydroxyurea for efficient early S phase arrest (*see* **Note 2**).

1. Inoculate 1–2 yeast colonies of MAT**a** strain of your choice (yeast colonies have been grown on a plate for 3 days at 24 °C) in 50 ml of YPD medium (in 100 ml flask) and culture overnight at 24 °C in shaking water bath or incubator.
2. Take 1 ml of cell culture the following morning, sonicate for 4 s, count cells, and dilute to a concentration of 4×10^6 cells/ml in 50 ml of YPD medium (*see* **Note 3**).
3. Grow cell culture at 24 °C until cells will reach concentration of 7×10^6 cells/ml (*see* **Note 4**).
4. When cells reach 7×10^6 cells/ml, add 7.5 μg/ml of alpha-factor and grow culture for further 2 h. Then, start adding 2.5 μg/ml of alpha factor every 20 min (*see* **Note 5**).
5. After two and a half hours of alpha factor start monitoring the cell cycle arrest by microscopy. Take 1 ml of the culture, sonicate for 4 s, and examine cells under microscope; alpha factor

arrest is successful when the number of cells presenting a characteristic projection called "shmoo" is higher than 90 % (Fig. 1a) (*see* **Notes 6** and 7).

6. Once arrested in G1 phase the culture needs to be synchronously released into YPDHU medium containing 0.2 M hydroxyurea (*see* **Note 2**). Spin down cells at 200 × *g* for 3 min in 50 ml tube, take off supernatant, and wash the pellet with 10 ml of YPDHU medium.
7. Spin down the culture at 200 × *g* for 3 min, take off supernatant and repeat the wash as in **step 6**.
8. Resuspend cells in 50 ml of fresh YPDHU medium and incubate the culture for additional 90 min at 24 °C. Monitor cell cycle arrest by microscopy, take 1 ml of the culture, sonicate for 4 s, and examine cells under microscope; hydroxyurea arrest is successful when the number of cells presenting small buds is higher than 90 % (Fig. 1b).
9. To release cells from hydroxyurea arrest in early S phase, spin down the culture at 200 × *g* for 3 min in 50 ml tube, take off supernatant, and wash the pellet with 10 ml of YPD medium.
10. Spin down the culture at 200 × *g* for 3 min, take off supernatant, and repeat the wash as in **step 9**.
11. Resuspend the pellet in 50 ml of YPD medium and release cells from hydroxyurea arrest by transferring the culture back at 24 °C. Start collecting experimental samples for downstream applications at the beginning of the release and then every 15 min to monitor the synchrony and cell progression.

3.1.3 Nocodazole Arrest and Release

Nocodazole is a chemical agent that inhibits microtubule polymerization, and thus blocks cells in G2 and M phases. After addition of nocodazole to the growth media, cells accumulate as large budded cells (Fig. 1c) with 2C DNA content and undivided nuclei.

1. Inoculate 1–2 yeast colonies of MAT**a** or MATα strain of your choice (yeast colonies have been grown on a plate for 3 days at 24 °C) in 50 ml of YPD medium (in 100 ml flask) and culture overnight at 24 °C in shaking water bath or incubator.
2. Take 1 ml of cell culture the following morning, sonicate for 4 s, count cells, and dilute to concentration of 4×10^6 cells/ml in 50 ml of YPD medium (*see* **Note 3**).
3. Grow cell culture at 24 °C until cells will reach concentration of 7×10^6 cells/ml (*see* **Note 4**).
4. When cells reach 7×10^6 cells/ml, add 5 μg/ml of nocodazole and grow culture for further 3 h at 24 °C.
5. After two and a half hours of incubation with nocodazole start monitoring cell cycle arrest by microscopy. Take 1 ml of the culture, sonicate for 4 s, and examine cells under microscope;

nocodazole arrest is successful when number of cells presenting large buds is higher than 90 % (Fig. 1c).

6. To release cells from nocodazole arrest in G2/M, spin down the culture at 200 × *g* for 3 min in 50 ml tube, take off supernatant, and wash the pellet with 10 ml of YPD medium.
7. Spin down the culture at 200 × *g* for 3 min, take off supernatant, and repeat the wash as in **step 6**.
8. Resuspend the pellet in 50 ml of YPD medium and release cells from nocodazole arrest by transferring the culture back to 24 °C. Start collecting experimental samples for downstream applications at the beginning of the release and then every 15 min to monitor the synchrony and cell progression.

3.1.4 *cdc15-2* Arrest and Release

A widely used temperature-sensitive mutant is the allele *cdc15-2* which contains the temperature-sensitive mutation in a protein kinase. Cdc15 is required for budding yeast cells to exit from mitosis, so when grown at the restrictive temperature (37 °C) the mutated allele *cdc15-2* blocks cells in mitosis, precisely in late anaphase with a characteristic morphology of cells with large buds (Fig. 1d) [9].

1. Inoculate 1–2 yeast colonies of MAT**a** or MATα strain of your choice (yeast colonies have been grown on a plate for 3 days at 24 °C) in 50 ml of YPD medium (in 100 ml flask) and culture overnight at 24 °C in shaking water bath or incubator.
2. Take 1 ml of cell culture the following morning, sonicate for 4 s, count cells, and dilute to a concentration of 4×10^6 cells/ml in 50 ml of YPD medium (*see* **Note 3**).
3. Grow cell culture at 24 °C until cells will reach concentration of 7×10^6 cells/ml (*see* **Note 4**).
4. Spin down the culture at 200 × *g* for 3 min in 50 ml tube then resuspend the pellet in 50 ml of YPD medium, which was prewarmed at 37 °C for at least 1 h.
5. Incubate the culture at 37 °C for 3 h to arrest.
6. Monitor the cell cycle arrest by microscopy, each time take 1 ml of the culture, sonicate for 4 s, and examine cells under microscope; *cdc15-2* arrest is successful when number of cells presenting large buds is higher than 90 % (Fig. 1d).
7. To release cells from *cdc15-2* arrest, spin down the culture at 200 × *g* for 3 min in 50 ml tube and then resuspend the pellet in 50 ml of YPD medium kept previously at 24 °C.
8. Release cells from *cdc15-2* arrest by transferring the culture back at 24 °C. Start collecting experimental samples for downstream applications at the beginning of the release and then every 15 min to monitor the synchrony and cell progression.

3.2 Monitoring the Synchrony of Cell Culture

Achieving synchrony is critical for the success of any experiment, thus it needs to be precisely monitored. The percentage of budded cells (budding index) can be easily determined. In addition, DNA content can be studied by flow cytometry; analysis of the histogram plot can determine percentage of cells at the G1 phase (1C DNA content), G2 and M phases (2C DNA content) or cells undergoing chromosome replication, as the valley between 1C and 2C DNA peaks (Fig. 2a). Finally, progression through mitosis can be determined by counting divided nuclei of stained cells under a fluorescence microscope.

3.2.1 Percentage of Budded Cells (Budding Index)

Cell cycle stage of budding yeast cells can be easily assigned using phase contrast microscopy. Budding yeast divide by budding so the progression through cell cycle is assessed by the size of the growing bud. Budding index represents the percentage of budded cells in the population and gives the indication of the synchrony of the culture, as well as helping to determine whether subsequent progression through cell cycle following release was synchronous.

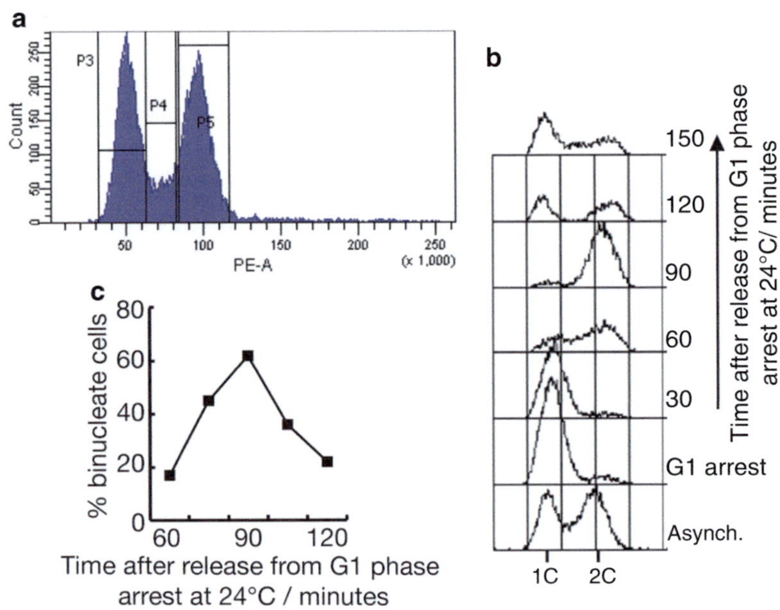

Fig. 2 Monitoring the synchrony of the cell culture. (**a**) Standard histogram representing an asynchronous culture. Here, it shows three populations of cells: P3 containing 1C DNA content, P5 for 2C DNA content and the intermediate P4, which corresponds to cells undergoing chromosome replication. (**b**) Standard histogram overlay for G1 arrest and release experiment. Wild-type cells were grown in YPD medium at 24 °C. Samples were collected from the asynchronous culture, then from G1 arrest and every 30 min upon the release from G1 phase. (**c**) Typical analysis of binucleate cells in G1 arrest and release experiment. Wild-type cells were grown in YPD medium at 24 °C. Samples were collected and binucleate cells counted every 15 min upon the release from G1 phase

1. Take 1 ml of the yeast culture, sonicate for 4 s to ensure separation of all mother and daughter cells, and keep cells on ice (*see* **Note 8**).
2. Spin down cells at full speed in microcentrifuge for 30 s.
3. Take off the supernatant leaving 20 μl at the bottom of the tube.
4. Resuspend cells by pipetting up and down for a minimum of 40 times.
5. Examine 3 μl of cells under phase-contrast microscope and determine the budding index, counted from a minimum of 100 cells.

3.2.2 Flow Cytometry

Flow cytometry allows the study of cell cycle progression. DNA of fixed cells is stained with a fluorescent dye, here propidium iodide, and subsequently cells are passed through a laser flow cytometer [12–14]. A histogram plot, in which DNA content is shown, provides information about the synchrony of the culture, as well as helping to determine whether subsequent progression through the cell cycle after release was synchronous. The protocol described here is based on the use of Becton-Dickinson FACScan or FACSort cytometer.

1. During block and release experiments described above, take 1 ml of the yeast culture, transfer to 1.5 ml tube, and spin down at 17,000 × *g* (full speed in microcentrifuge) for 30 s. Take off supernatant.
2. Resuspend cell pellet in 1 ml of cold 70 % ethanol and vortex vigorously to fix cells (*see* **Note 9**).
3. After completion of the experiment, add 100 μl of fixed cells from every time point to a 1.5 ml tube containing 1 ml of 50 mM sodium citrate to rehydrate cells.
4. Spin down cells at 17,000 × *g* for 2 min and take off supernatant carefully (*see* **Note 10**).
5. Resuspend the pellet in 500 μl of 50 mM sodium citrate containing 0.1 mg/ml RNase A and incubate at 37 °C for 2 h (*see* **Note 11**).
6. During this incubation prepare 5 mg/ml pepsin porcine resuspended in 50 mM HCl (*see* **Note 12**).
7. After incubation spin down cells at 17,000 × *g* and resuspend cells in 500 μl of 5 mg/ml pepsin in 50 mM HCl solution. Incubate samples at 37 °C for 30 min (*see* **Note 13**).
8. Spin down cells at 17,000 × *g* for 2 min and take off supernatant.
9. Resuspend cells in 1 ml of 50 mM sodium citrate buffer supplemented with 2 μg/ml propidium iodide and protect samples from light (*see* **Note 14**).
10. Sonicate samples for 5 s and transfer them to 5 ml FACS tube.

11. Settings of the flow cytometer need to be adjusted before every experiment using an asynchronous sample taken at the beginning of each experiment and need to be determined for each experiment separately (*see* **Note 15**). A representative example of the histogram for an asynchronous culture stained with propidium iodide is presented in Fig. 2a. The diagram represents the number of cells in the culture against their fluorescent signal of propidium iodide (PE-A). The parameters that need to be changed manually in order to achieve this type of graph are detector voltage and amplifier gain. It is recommended to analyze 100,000 cells for each time point.
12. Vortex samples briefly just before mounting the tube at the cytometer and pass each sample on the low speed setting (500–1000 cells/s).
13. After collecting data from all experimental time-points we choose to present them as an overlay of histogram plots as shown in Fig. 2b (*see* **Note 16**).

3.2.3 Percentage of Divided Nuclei (Binucleate Cells Counting)

A helpful way of monitoring the synchrony of the culture and progression through cell cycle is based on following cell nuclei. Flow cytometry determines DNA content, but is unable to address whether cells with 2C DNA content have undergone mitosis: cells will have their nuclei divided but they will still share their cytoplasms. Therefore counting binucleate cells is a useful way to determine progression through mitosis. Once sample has been analysed by flow cytometry, the same sample can be used to determine divided nuclei. Alternatively, samples can be prepared using this method to count binucleate cells, as it guarantees separation between cells, which undoubtedly facilitates their identification. An example of the binucleate cell analysis is presented in Fig. 2c. If cells have been used previously for flow cytometry analysis, then proceed directly to **step 10**.

1. Take 1 ml of the yeast culture and spin down at 17,000 × *g* (full speed in microcentrifuge) for 30 s. Take off supernatant.
2. Resuspend cell pellet in 1 ml of cold 70 % ethanol and vortex vigorously to fix cells (*see* **Note 9**).
3. Add 100 μl of fixed cells to 1.5 ml tube containing 1 ml of 50 mM sodium citrate.
4. Spin down cells at 17,000 × *g* for 2 min and take off supernatant (*see* **Note 10**).
5. Resuspend the pellet in 1 ml of 50 mM sodium citrate containing 0.1 mg/ml RNase A and incubate at 37 °C for 2 h (*see* **Note 11**).
6. During this incubation prepare 5 mg/ml porcine pepsin resuspended in 50 mM HCl (*see* **Note 12**).

7. After incubation spin down cells at 17,000 × *g* and resuspend cells in 500 μl of 5 mg/ml pepsin in 50 mM HCl solution. Incubate samples at 37 °C for 30 min (*see* **Note 13**).
8. Spin down cells at 17,000 × *g* for 2 min and take off supernatant.
9. Resuspend cells in 1 ml of 50 mM sodium citrate buffer supplemented with 2 μg/ml propidium iodide and protect samples from light (*see* **Note 14**).
10. Spin down cells at full speed in microcentrifuge for 30 s.
11. Take off the supernatant leaving 20 μl at the bottom of the tube.
12. Resuspend cells by pipetting up and down for a minimum of 40 times.
13. Examine 1,8 μl of cells using rhodamine-specific filter set on fluorescence microscope. For each time point we determine the number of binucleates from a minimum of 100 cells.

4 Notes

1. The sequence of the alpha factor peptide is as follows Trp-His-Trp-Leu-Gln-Leu-Lys-Pro-Gly-Gln-Pro-Met-Tyr (WHWLQLKPGQPMY).
2. Hydroxyurea needs to be used at high concentration (final concentration of hydroxyurea required for experiment is 0.2 M) and hydroxyurea powder is difficult to dissolve, so the use of concentrated 2x YP medium is recommended. For small volumes it is possible to weigh the required amount of hydroxyurea powder and adding it directly to the growing medium however one needs to be sure that hydroxyurea powder is completely dissolved before using the medium.
3. We routinely count cells microscopically using widely available counting chambers. In this way, apart from getting to know number of cells per millilitre, we can observe cells directly and detect any growth or contamination problems.
4. The growth of the culture might take up to 1.5–2 h at 24 °C. It depends mainly on the sugar source used in the medium and the temperature in which cells are growing, YP supplemented with glucose at 24 °C will require around 1.5 h, while YP supplemented with raffinose or galactose at 24 °C will require time closer to 2 h. Growth rate will depend on the strain of interest too.
5. The main drawback of using alpha factor is that budding yeast cells secrete proteases (mainly Bar1) which degrade alpha factor in the medium, so the longer cells are incubated with mating pheromone, the less alpha factor will be left in the medium. This means that cells will not arrest properly and will leak past

the G1 block unless regular amounts of alpha factor are added. At later stages of the alpha factor arrest, look for any sign of tiny emerging buds, as that would indicate that cells have leaked from the arrest.

6. When yeast cultures are grown at 24 °C alpha factor arrest will normally take around 3 h in YP medium containing glucose or three and a half hours in YP medium supplemented with raffinose or galactose.
7. We routinely use longer G1 arrest before the release into early S phase arrest for three and a half hours in YPD medium to allow all cells to grow enough to be above the minimum size needed for START, so that they will enter S phase more synchronously.
8. Samples stored in growth medium can be kept at 4 °C for up to 1 h. For longer storage, spin cells down and resuspend in PBS.
9. Samples can be stored at this stage at 4 °C.
10. The pellet formed at this stage will not be readily visible.
11. This step is required to digest RNA content of the sample. At this stage you might extend the incubation to a few hours without disturbing the staining of samples.
12. As pepsin powder is difficult to resuspend, we recommend placing suspension at 37 °C to help porcine pepsin go into solution.
13. It is essential to keep this incubation time to 30 min precisely, as extending incubation time with pepsin will result in damage to the samples.
14. If samples are not going to be passed through flow cytometer immediately, they can be wrapped in aluminium foil and stored overnight at 4 °C. It is recommended to pass samples through cytometer within 48 h from the time of preparation.
15. The manual settings of the cytometer depend on the type of the flow cytometer and your experimental strain so need to be adjusted properly for each experiment. It is crucial to collect an asynchronous sample for each experimental strain and use it for setting up the machine.
16. We find that presenting outcome data as an overlay of histogram plots makes it easier to follow and presents the synchronous progression through the cell cycle in a much more visible way.

Acknowledgements

We are grateful for teaching and scientific advice from Professor Karim Labib. Methods described in this chapter were extensively used in the Labib laboratory and we would like to thank members of his group, past and present who contributed to our current understanding of methods presented here. We especially thank Dr. Ben Hodgson for comments on this chapter. ASD is a recipient of a Ramon y Cajal contract and received funding from the Cantabria International Campus and via grant BFU2011-23193 from the Spanish "Ministerio de Economia y Competitividad" (co-funded by the European Regional Development Fund).

References

1. Bertoli C, Skotheim JM, de Bruin RA (2013) Control of cell cycle transcription during G1 and S phases. Nat Rev Mol Cell Biol 14(8): 518–528. doi:10.1038/nrm3629
2. Malumbres M, Harlow E, Hunt T, Hunter T, Lahti JM, Manning G, Morgan DO, Tsai LH, Wolgemuth DJ (2009) Cyclin-dependent kinases: a family portrait. Nat Cell Biol 11(11): 1275–1276. doi:10.1038/ncb1109-1275
3. Hochegger H, Takeda S, Hunt T (2008) Cyclin-dependent kinases and cell-cycle transitions: does one fit all? Nat Rev Mol Cell Biol 9(11):910–916. doi:10.1038/nrm2510
4. Futcher B (1999) Cell cycle synchronization. Methods Cell Sci 21(2-3):79–86
5. Burke D, Dawson D, Stearns T (2000) Methods in yeast genetics. Cold Spring Harbor Laboratory Press, Cold Spring Harbor
6. Futcher AB (1990) Yeast cell cycle. Curr Opin Cell Biol 2(2):246–251
7. Johnston GC, Pringle JR, Hartwell LH (1977) Coordination of growth with cell division in the yeast Saccharomyces cerevisiae. Exp Cell Res 105(1):79–98
8. Broach JR, Volkert FC (1991) Circular DNA plasmids of yeasts. In: Broach JR, Pringle JR, Jones EW (eds) The molecular biology of the yeast Saccharomyces; genome dynamics, protein synthesis and energetics, vol 1. Cold Spring Harbor Press, Cold Spring Harbor, pp 297–332
9. Fitch I, Dahmann C, Surana U, Amon A, Nasmyth K, Goetsch L, Byers B, Futcher B (1992) Characterization of four B-type cyclin genes of the budding yeast Saccharomyces cerevisiae. Mol Biol Cell 3(7):805–818
10. Amberg DC, Burke DJ, Strathern JN (2006) Logarithmic growth. CSH Protoc 2006(1). doi:10.1101/pdb.prot4179
11. Breeden LL (1997) Alpha-factor synchronization of budding yeast. Methods Enzymol 283:332–341
12. Givan AL (2001) Principles of flow cytometry: an overview. Methods Cell Biol 63:19–50
13. Zhang H, Siede W (2004) Analysis of the budding yeast Saccharomyces cerevisiae cell cycle by morphological criteria and flow cytometry. Methods Mol Biol 241:77–91
14. Haase SB, Lew DJ (1997) Flow cytometric analysis of DNA content in budding yeast. Methods Enzymol 283:322–332

Chapter 20

Fission Yeast Cell Cycle Synchronization Methods

Marta Tormos-Pérez, Livia Pérez-Hidalgo, and Sergio Moreno

Abstract

Fission yeast cells can be synchronized by cell cycle arrest and release or by size selection. Cell cycle arrest synchronization is based on the block and release of temperature-sensitive cell cycle mutants or treatment with drugs. The most widely used approaches are *cdc10-129* for G1; hydroxyurea (HU) for early S-phase; *cdc25-22* for G2, and *nda3-KM311* for mitosis. Cells can also be synchronized by size selection using centrifugal elutriation or a lactose gradient. Here we describe the methods most commonly used to synchronize fission yeast cells.

Key words Fission yeast, *Schizosaccharomyces pombe*, Synchronization, Block and release, cdc10-129, cdc25-22, nda3-KM311, HU, Centrifugal elutriation, Lactose gradient, G1-phase, S-phase, G2-phase, Mitosis

1 Introduction

The fission yeast *Schizosaccharomyces pombe* has been widely used for the analysis of the cell cycle [1], mitosis [2] and cytokinesis [3]. A wild-type *S. pombe* cell is rod shaped, grows by elongation, and divides by medial fission. The identification of cell-cycle mutants has allowed the genetic and biochemical analysis of cell division [1]. Fission yeast cells can be synchronized in G1 using the temperature-sensitive *cdc10-129* mutant. Cdc10 is a transcription factor that forms a heterodimer with Res1 or Res2 and is activated at the end of G1 to promote the expression of S-phase genes [4]. Fission yeast cells can also be synchronized in early S-phase with hydroxyurea (HU), a powerful inhibitor of ribonucleotide reductase. This HU treatment leads to the depletion of dNTP pools [5]. They can also be synchronized in late G2, using the temperature-sensitive *cdc25-22* mutant [6], which is defective in the Cdc25 protein phosphatase that activates Cdc2/Cdc13 (Cdk1/Cyclin B) at the onset of mitosis. Finally, synchronization can be performed in mitosis using the cold-sensitive *nda3-KM311* mutant, defective in

Alberto Sanchez-Diaz and Pilar Perez (eds.), *Yeast Cytokinesis: Methods and Protocols*, Methods in Molecular Biology, vol. 1369, DOI 10.1007/978-1-4939-3145-3_20,

β-tubulin, which arrests in metaphase with condensed chromosomes without a mitotic spindle, when shifted to 20 °C [7].

Synchronization by size selection can be achieved by elutriation [8] or by using lactose gradients [9]. Both methods select the smallest cells in the culture, which are in early G2 in wild-type cells growing in rich media. Centrifugal elutriation allows the separation of the smallest cells based on the combination of two opposite forces: sedimentation of the cells in a centrifugal field, and liquid counterflow by the growth medium. Since the yeast cells are elutriated at 25 °C and there is a continuous flow of nutrients, they continue to grow in the rotor and perturbations are minimal. Although elutriation is by far the best choice, synchronization by lactose gradients is a cheaper alternative that does not require any special equipment and can be performed simultaneously with several strains. Lactose is used because it cannot be metabolized by fission yeast cells and does not affect growth. One drawback of this technique is that only small volumes of cells can be processed; such volumes may be sufficient for microscopy, but they limit the number of samples that can be taken for protein or RNA studies. In these cases, elutriation is the preferred method.

2 Materials and Equipment

2.1 Strains

1. Wild-type *h- 972*.
2. *h-cdc10-129*.
3. *h-cdc25-22*.
4. *h-nda3-KM311*.

If possible, prototrophic strains (without auxotrophic markers) isogenic to the wild type should be used to be able to compare the results.

Fission yeast cells are grown in yeast extract with supplements medium (YES) (0.5 % yeast extract; 3 % glucose; 225 mg/l adenine, histidine, leucine, uracil, and lysine hydrochloride; adjust the pH to 5.6 with HCl) or in Edinburg Minimal Medium (EMM) (3 g/l potassium hydrogen phthalate, 2.2 g/l Na_2HPO_4, 5 g/l NH_4Cl, 2 % glucose, 20 ml/l salts, 0.1 ml/l minerals, 1 ml/l vitamins, pH 5.6 with HCl) (Salts 50× stock: 52.5 g/l $MgCl_2{\cdot}6H_2O$, 0.735 mg/l $CaCl_2{\cdot}2H_2O$, 50 g/l KCl, plus 2 g/l Na_2SO_4; minerals 10,000× stock: 5 g/l boric acid, 4 g/l $MnSO_4$, 4 g/l $ZnSO_4{\cdot}7H_2O$, 2 g/l $FeCl_2{\cdot}6H_2O$, 0.4 g/l molybdic acid, 1 g/l KI, 0.4 g/l $CuSO_4{\cdot}5H_2O$, plus 10 g/l citric acid; vitamins 1000× stock: 1 g/l pantothenic acid, 10 g/l nicotinic acid, 10 g/l *myo-inositol*, plus 10 mg/l biotin). If auxotrophic strains are used it is required to add 225 mg/l of supplements to EMM [10].

2.2 Block and Release Synchronization

1. Shaking water baths at 20, 25, 32, and 36 °C.
2. Thermometer.
3. Ice.
4. Hydroxyurea (HU). Prepare 1 M stock solution by dissolving 0.076 g of HU powder in 1 ml of water.
5. Phloxin B. Prepare 1000× stock solution (5 mg/ml) by dissolving 1 g of phloxin-B powder in 200 ml of water. Filtrate the solution and store it at 4 °C wrapped in aluminium foil to protect it from the light. To prepare YES-phloxin-B plates add the phloxin B when the medium has cooled below 60 °C.
6. Centrifuge for 50 ml tubes.
7. 50 ml screw-cap tubes.
8. 100 ml and 1 l flasks.

2.3 Centrifugal Elutriation

1. Beckman centrifuge (Avanti J-20I, J-21 or J-6 M/E) equipped with a stroboscopic lamp and a viewing window on the lid.
2. Elutriator rotor (JE 5.0 or JE-6B).
3. Peristaltic pump.
4. Bubble trap.
5. Pressure gauge.
6. Two 5 l flasks.
7. Ten to fifteen 250 ml flasks.
8. Spectrophotometer.
9. Microscope.
10. Shaking water bath at 25 °C.
11. 0.2 % diethyl pyrocarbonate (DEPC). Add 1 ml of DEPC to 1 l of water. Shake well and incubate at 37 °C for 12 h before use.

2.4 Lactose Gradients

1. Lactose.
2. Centrifuges for 1.5 and 50 ml tubes.
3. 1.5 and 50 ml tubes.
4. Microscope.

2.5 DAPI-Calcofluor Staining

1. 1.5 ml tubes.
2. Microcentrifuge.
3. 70 % ethanol.
4. Phosphate-buffered saline (PBS).

5. DAPI (4′,6-diamidino-2-phenylindole). Prepare 1 mg/ml stock solution in PBS; dilute to 0.1 mg/ml, protect from the light, and store at −20 °C. Prepare a 5 mg/ml stock solution of Calcofluor in water. Dilute to 50 μg/ml in PBS and store at −20 °C.
6. Slides and cover slips.
7. Fluorescence microscope.

2.6 Flow Cytometry

1. 1.5 ml tubes.
2. Microcentrifuge.
3. 70 % ethanol.
4. FACS tubes.
5. 50 mM sodium citrate (freshly prepared from a stock solution of 0.5 M).
6. RNase A. Prepare 10 mg/ml stock solution in 10 mM Tris–HCl, pH 7.5, plus 15 mM NaCl. Boil for 10 min, aliquot, and store at −20 °C.
7. Propidium iodide (PI). Prepare stock solution of 4 mg/ml in water, protect from the light, and store at −20 °C.
8. Sonicator.
9. Flow cytometer.
10. Cell Quest software to analyze and represent the data acquired with the flow cytometer.

3 Methods

3.1 Synchronizing Fission Yeast Cells in G1 with the cdc10-129 Mutant

The *cdc10-129* mutant can be blocked in G1 for 4 h at 36 °C (Fig. 1) and then released to 25 °C to allow the cells to enter into S-phase synchronously. Samples can be taken at the block point (G1-phase) and every 15 min after release for flow cytometry, RNA and protein analyses, etc. In a typical experiment, S-phase is completed approximately 90 min after release.

1. Pick a single colony of the *h-cdc10-129* strain from a fresh (3-day-old) YES plate and inoculate 20 ml of YES liquid in a 100 ml flask. Incubate for 24–48 h at 25 °C in a shaking water bath.
2. Use this culture to inoculate 300 ml of EMM in a 1 l flask. Incubate overnight at 25 °C in a shaking water bath until the culture reaches a density of 0.5–1 × 10^7 cells/ml. Note that the generation time of *S. pombe* in EMM at 25 °C is approximately 4 h.

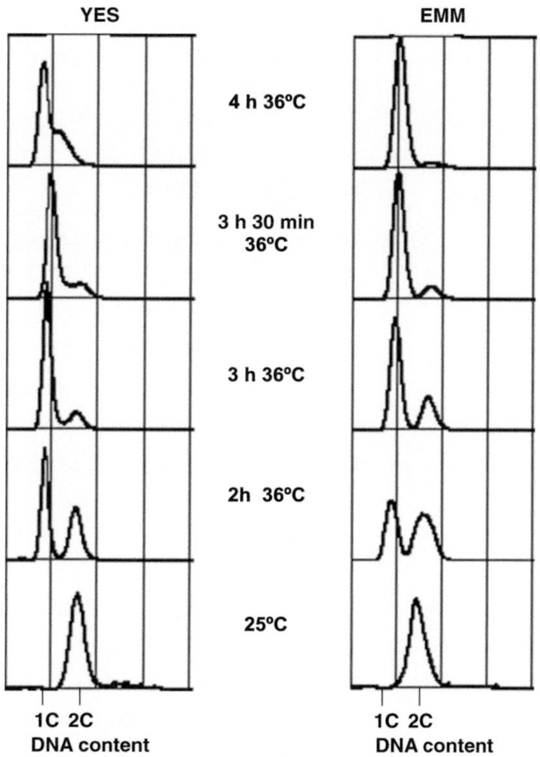

Fig. 1 Cell cycle arrest in G1 of the *cdc10-129* mutant. *cdc10-129* mutant cells were grown in two different media, YES and EMM, and shifted to 36 °C. Samples were taken at the indicated times and analysed by flow cytometry. 1C cells are in G1

3. The following day, determine the cell number and dilute the culture with fresh EMM to 4×10^6 cells/ml. Shift the 300 ml culture to 36 °C for 4 h in a shaking water bath.
4. Prepare ice-cold water in a sink. Remove the flask from the 36 °C water bath, introduce a thermometer into the liquid culture and rapidly cool the culture down by manually shaking the flask in the ice-cold water. When temperature drops to 25 °C, place the flask in the 25 °C shaking water bath (*see* **Note 1**).
5. Samples can be taken after 4 h at 36 °C (G1-arrested cells) and every 15 min after release (*see* **Notes 2–4**).
6. Take a sample for flow cytometry to determine the synchrony of cultures (Fig. 1, *see* Subheading 3.7).

3.2 Synchronizing Fission Yeast Cells in Early S-Phase with Hydroxyurea

Fission yeast cells treated with 12 mM hydroxyurea (HU) for 4 h arrest their cell cycle in early S-phase. When the drug is removed from the growth medium, cells resume the cell cycle in a synchronized manner. In comparison with *cdc10-129* and *cdc25-22* synchronization, this method has the advantage that it can be used to synchronize thermosensitive mutants at 25 °C, which, after release, can be inactivated at the restrictive temperature. Since hydroxyurea disturbs DNA replication, it might not be the most appropriate synchronization method for some DNA replication or checkpoint mutants.

1. Grow the fission yeast cells in YES or EMM to log phase (4×10^6 cells/ml) at the desired temperature in an incubator or shaking water bath.
2. Take a sample corresponding to "asynchronous cells".
3. Add the hydroxyurea to a final concentration of 12 mM from a 1 M HU stock solution (*see* **Notes 5** and **6**).
4. Grow the cells in 12 mM HU for 4 h. Depending on the temperatures and the media used, the cells need more or less time to reach S-phase and arrest, but 4 h should be sufficient at all temperatures in both YES and EMM.
5. Take a sample after 4 h in HU, corresponding to "early S-phase arrest".
6. Remove the HU from the medium to release the cells from arrest. This can be done by washing the cells twice with 40 ml of fresh pre-warmed medium without hydroxyurea. Washing can be achieved by centrifugation at $1800 \times g$, for 5 min at room temperature. Alternatively, cell filtration may be a quicker and less stressful method (*see* **Note** 7).
7. Resuspend the cell pellets in pre-warmed medium without hydroxyurea to a final concentration of 4×10^6 cells/ml. In the case of filtration, place the filter in the medium and resuspend the cells by shaking the flask manually.
8. Incubate cells at the desired temperature and take samples every 15 min to monitor cell cycle progression by flow cytometry and DAPI and Calcofluor staining (Fig. 2, *see* Subheading 3.7).

3.3 Synchronizing Fission Yeast Cells in G2 with the cdc25-22 Mutant

The *cdc25-22* mutant can be blocked at the end of G2 after incubation for 4 h at 36 °C followed by release to 25 °C to allow the cells to enter into mitosis synchronously (Fig. 3). Samples can be taken at the block point (G2/M boundary) and every 10 min after release for RNA and protein analyses, DAPI and Calcofluor staining, etc. In a typical experiment, metaphase takes place at approximately 20 min; anaphase at 30 min; and septation at 80 min after release (Fig. 3).

1. Pick a single colony of the *h-cdc25-22* strain from a fresh (3-day-old) YES plate and inoculate 20 ml of YES liquid in a 100 ml flask. Incubate for 24–48 h at 25 °C in a shaking water bath.

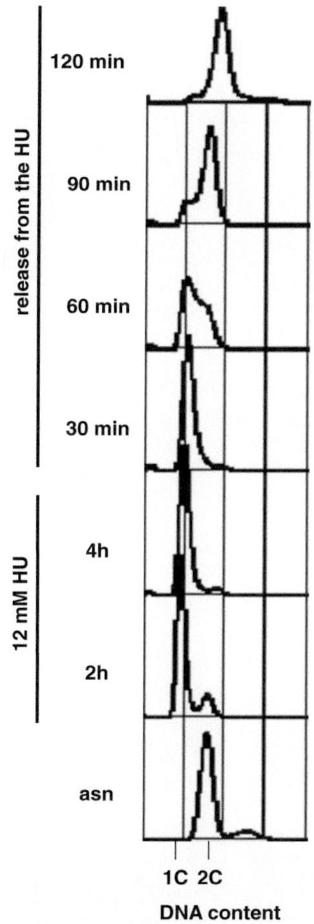

Fig. 2 Cell cycle block and release in early S-phase with HU. Wild-type cells were blocked in early S-phase in YES medium containing 12 mM of HU for 4 h and then released into fresh medium. Samples were taken at the indicated times and analyzed by flow cytometry. S-phase progression was observed at 60 min after release. *Asn* asynchronous culture

2. Use this culture to inoculate 300 ml of EMM in a 1 l flask. Incubate overnight at 25 °C in a shaking water bath until the culture reaches a density of $0.5–1 \times 10^7$ cells/ml. Note that the generation time of *S. pombe* in EMM at this temperature is approximately 4 h.
3. The following day, determine the cell concentration and dilute the culture with fresh EMM to 4×10^6 cells/ml. Shift the 300 ml culture to 36 °C for 4 h in a shaking water bath.
4. Rapidly cool down the culture by shaking the flask manually in ice-cold water. When temperature drops to 25 °C place the flask in a shaking water bath at 25 °C (*see* **Note 1**).

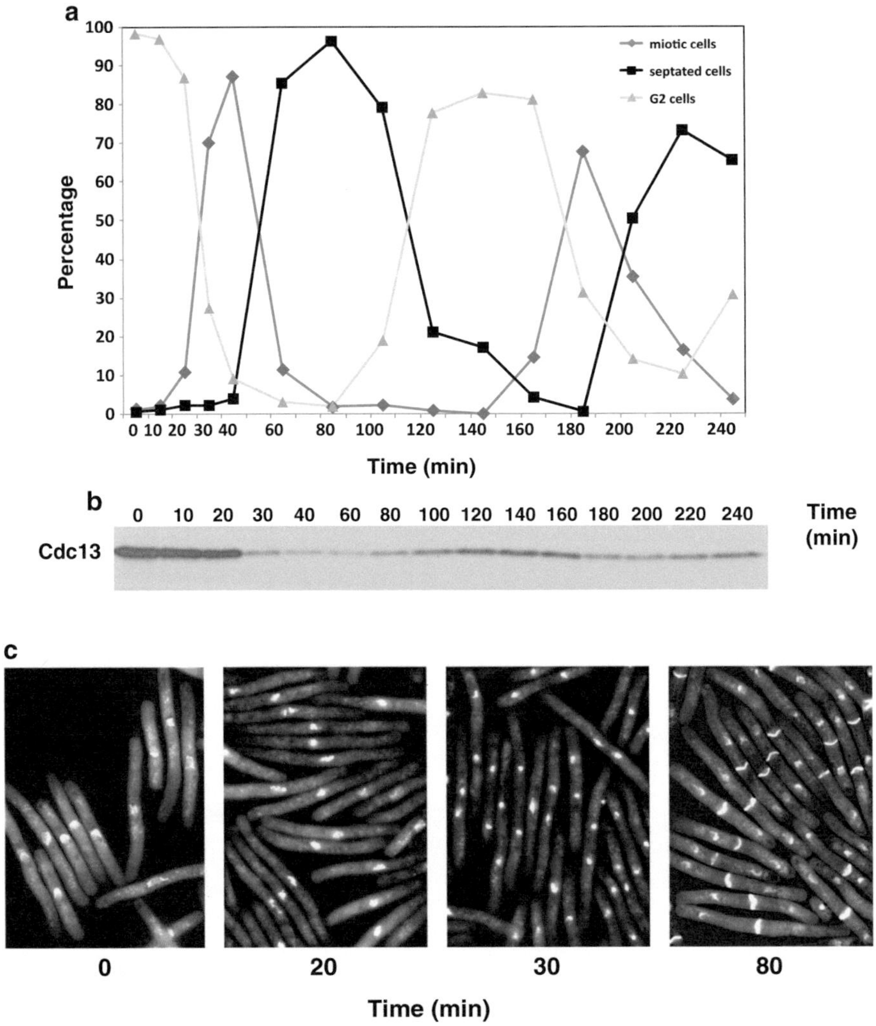

Fig. 3 Cell cycle block and release in late G2 of the *cdc25-22* mutant. (**a**) Percentage of cells in G2 (cells with a single interphase nucleus), mitosis (anaphase-cells with two nuclei), and undergoing cytokinesis (cells with one septum). Samples were taken every 10 min for the first 40 min, and then every 20 min up to 4 h after *cdc25-22* release. (**b**) Levels of Cdc13 (Cyclin B) by Western blot during the two cell cycles after *cdc25-22* release. Cdc13 becomes degraded during anaphase. (**c**) DAPI-Calcofluor staining of fixed *cdc25-22* cells at the block time (G2 phase) and at 20 min (metaphase), 30 min (anaphase), and 80 min (septa formation) after release

5. Samples can be taken after 4 h at 36 °C (G2-arrested cells) and every 10 min after release up to 40 min and then every 20 min (*see* **Notes 4** and **8**).
6. To determine the degree of synchrony, the mitotic index (percentage of cells with two nuclei) can be determined after DAPI staining and the septation index (percentage of cells with one

complete septum) can be determined after Calcofluor staining (Fig. 3) (*see* Subheading 3.7, *see* **Note 9**).

3.4 Synchronization in Mitosis with the nda3-KM311 Mutant

When *nda3-KM311* cells are incubated at 20 °C they arrest in metaphase with condensed chromosomes but no mitotic spindle [7]. The *nda3-KM311* strain has a tendency to diploidize spontaneously at 32 °C. Therefore, before starting the experiment it is important to check the ploidy of the strain by growing the cells on YES containing phloxin B and to refresh the colonies every 2 days in YES.

1. Pick a single light-pink colony from a fresh (2–3-day-old) YES-phloxin-B plate and inoculate 20 ml of YES liquid in a 100-ml flask. Incubate at 32 °C overnight in a shaking water bath.
2. Use this culture to inoculate 200 ml of YES medium in a 0.5 l flask. Incubate overnight at 32 °C in a shaking water bath until the culture reaches $0.5–1 \times 10^7$ cells/ml. Note that the generation time of *S. pombe* at 32 °C in YES is approximately 2.5 h.
3. The following day, determine the cell concentration and dilute the culture with fresh YES to 4×10^6 cells/ml. Shift the 200 ml culture to 20 °C for 8 h. After this time 80–90 % of the cells should show condensed chromosomes after DAPI staining. We normally place the water bath in the cold room and adjust the temperature to 20 °C.
4. To release the cells from the block, shift the culture to 32 °C in a shaking water bath.
5. Samples can be taken at the block (metaphase) and every 5 min after release.
6. To determine the degree of synchrony, the mitotic index (percentage of cells with two nuclei) can be determined after DAPI staining and the septation index (percentage of cells with one complete septum) can be estimated after Calcofluor staining (*see* Subheading 3.7).

3.5 Synchronization in Early G2 Using Centrifugal Elutriation (See Note 10)

1. Pick a single wild-type colony from a fresh (3-day-old) YES plate and inoculate 20 ml of YES liquid in a 100 ml flask. Incubate for 24–48 h at 32 °C in a shaking water bath.
2. Use this culture to inoculate 300 ml of EMM in a 1 l flask. Incubate overnight at 32 °C in a shaking water bath until the culture reaches a density of 10^7 cells/ml.
3. The following day, inoculate 4 l of EMM using the overnight culture. Divide the 4 l of inoculated EMM in two 5 l flasks, each containing 2 l of EMM, and incubate overnight at 25 °C. Note that the generation time of *S. pombe* at this temperature is approximately 4 h.
4. Install the elutriator rotor in the centrifuge the day before the elutriation experiment. Figure 4 shows a picture of the elutria-

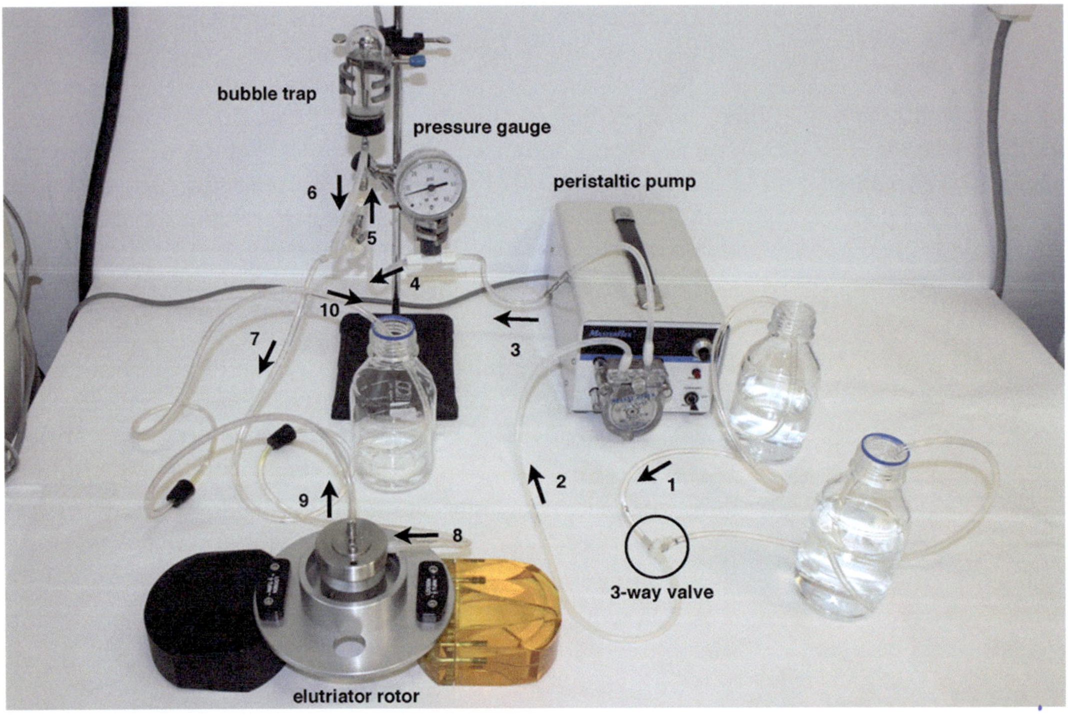

Fig. 4 Elutriation setup. Components of the elutriation system. In this case, sterile distilled water from a glass flask was pumped (1) through a 3-way valve using a peristaltic pump (2) passing through a pressure gauge (3, 4) and a bubble trap (5, 6) and then into the elutriator cell (7, 8). The water filled the chamber (*yellow* part) and exited from the top of the rotor (9) into another flask (10). Once the rotor had been mounted in the centrifuge and checked for the absence of leaks in the system, the system was filled with medium and then with the fission yeast culture

tion setup. The peristaltic pump can be inserted in the inflow line or in the outflow line. Creanor and Mitchison [8] have reported better results when the pump is inserted in the outflow line because it causes less turbulence in the rotor cell. The pump is set at 125 ml/min. The rotor is loaded at 25 °C at 4200 rpm (maximum speed) with water to ensure that there are no leaks in the system. A bubble trap and a pressure gauge are inserted in the inflow line. When the rotor is at full speed the stroboscopic lamp is switched on and adjusted to the centrifuge speed to be able to see into the transparent rotor chamber (Fig. 4, yellow chamber).

5. The following day, the centrifuge is set at 25 °C and the rotor cell is loaded first with water and spun at 4200 rpm at 125 ml/min, then with pre-warmed medium, and finally with the 4 l of exponentially growing fission yeast cells (8×10^6 cells/ml), either in EMM or in YES at 25 °C. It takes about 30 min to introduce all the culture into the elutriator chamber with a flow of 125 ml/min. Under these conditions, wild-type size

fission yeast cells should remain in the elutriator chamber. The stroboscopic lamp allows the filling of the chamber to be observed. To check that the cells are being retained in the rotor, the liquid coming out of the centrifuge should be monitored using a spectrophotometer and a microscope. This clear medium should be collected in one of the two 5 l flasks. If small cells start to leave the rotor, reduce the pump rate slightly, keeping the centrifuge speed at 4200 rpm.

6. Once all the cells are in the rotor cell, decrease the pump rate (by 2 ml/min, from 125 to 123 ml/min) and then the centrifuge speed (by 100 rpm, from 4200 to 4100 rpm). Repeat the process by consecutively decreasing the pump rate by 2 ml/min to 117 ml/min and the centrifuge speed to 3800 rpm. To pump out the cells, increase the pump rate very slowly (1 ml/min each time) and monitor the OD_{595} using a spectrophotometer until the OD_{595} starts to increase to 0.1–0.2 (equivalent to 10^6–2×10^6 cells/ml); discard the first 100 ml and start collecting the cells in the 250 ml flasks. Fill each flask with 240 ml of eluted cells and monitor the medium coming out of the rotor using the spectrophotometer. In an optimal elutriation experiment, the OD_{595} should remain between 0.1 and 0.2 and pressure should be less than 10 psi. When the OD_{595} drops below 0.1 (equivalent to 10^6 cells/ml), slowly increase the pump rate (1 ml/min each time) and wait for about 1 min until the OD_{595} increases to 0.1–0.2. Collect 10–14 aliquots of 240 ml each. Monitor synchrony in each aliquot by measuring the OD_{595} and the size of the cells under the microscope. The size of these cells should be small, and in the case of the wild-type they correspond to cells in late S-phase or early G2 (Fig. 5a). Pool the samples containing homogeneous small cells. In a typical synchronous culture, approximately 2 l at 1.5×10^6 cells/ml can be obtained, which represents approximately 10 % of the initial asynchronous culture.

7. Then incubate the selected cells in a 5 l flask and allow them to recover for 30 min. Following this, collect samples every 20 min for protein, RNA, and flow cytometry analyses, and for determination of the septation index, etc. (Fig. 5b, c, *see* Subheading 3.7).

8. The rotor is cleared of cells by stopping the centrifuge, while maintaining the flowing medium. Then, flush with distilled water and disinfect the rotor by filling the system with 0.2 % diethyl pyrocarbonate in water and leave overnight. The following day this solution is pumped out and the system is washed with 1 l of distilled water. The system can be used again for another experiment or disassembled according to the elutriator rotor instruction manual.

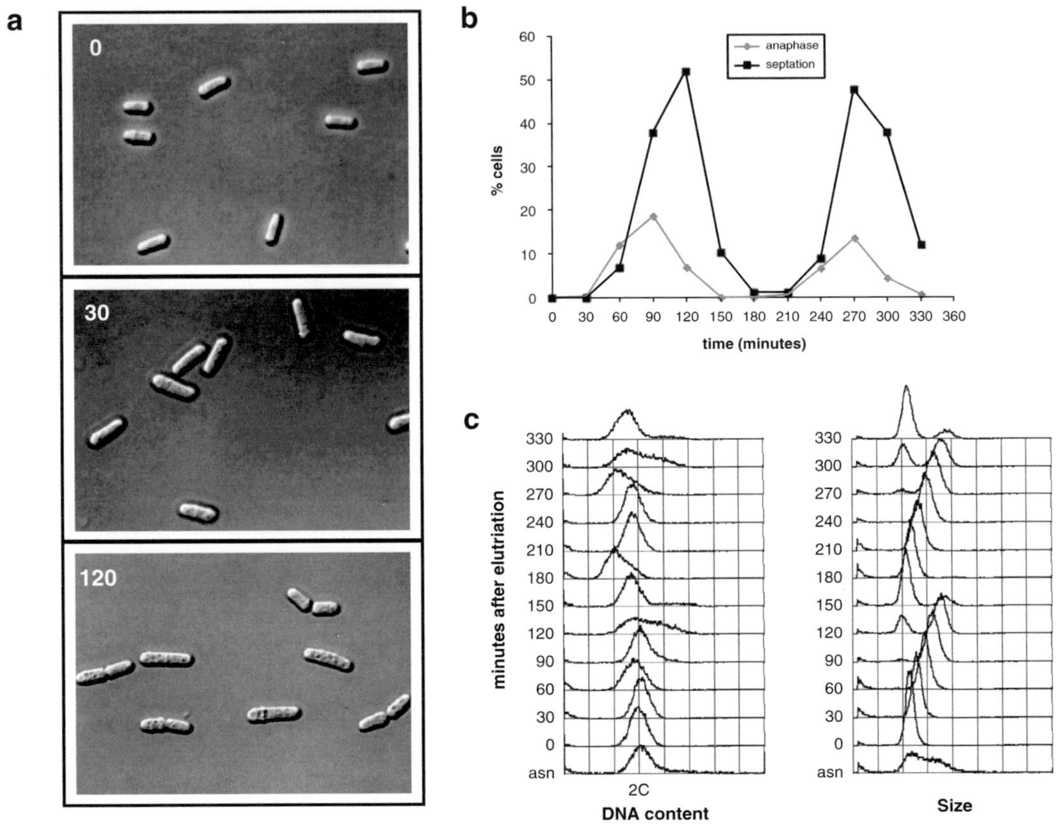

Fig. 5 Synchronous culture of fission yeast elutriated cells. Fission yeast cells were synchronised using an elutriator rotor. Samples were taken every 30 min for DAPI-Calcofluor staining and for flow cytometry. (**a**) Pictures of cells synchronised by elutriation at times 0 min (cells in late S-phase or early G2-phase), 30 min (cells in late G2-phase), and 120 min (cells undergoing cell separation) after being eluted from the rotor. (**b**) Percentage of cells in mitosis (cells in anaphase with two nuclei) and undergoing septation (cells with one septum). (**c**) DNA content and cell size determined by flow cytometry. *Asn* asynchronous culture

3.6 Synchronization in Early G2-Phase Using Lactose Gradients

1. To prepare the lactose gradient, freeze 45 ml of a 20 % lactose solution made in the desired medium and sterilized by filtration in a 50 ml tube at −80 °C for at least 4 h, and then allow it to thaw at 30 °C for 3 h. This creates a 10–30 % lactose gradient. Alternatively, a gradient maker can be used.
2. Pick a single wild-type colony from a fresh (3-day-old) YES plate and inoculate 20 ml of YES liquid in a 100 ml flask. Incubate for 24–48 h at 32 °C in a shaking water bath.
3. Use this culture to inoculate 300 ml of YES or EMM in a 1 l flask. Incubate overnight at 32 °C in a shaking water bath, until the culture reaches a concentration of 0.5×10^7 cells/ml (*see* **Note 11**).

4. Harvest 1–2 × 10^9 cells of a mid-exponential culture at 1800 × *g* for 3–5 min at room temperature.
5. Resuspend the pellets in 3–4 ml of medium (*see* **Note 12**).
6. Layer the concentrated cells on the top of the gradient.
7. Centrifuge at 500 × *g* for 8 min at room temperature.
8. Take 1 ml fractions from the top of the gradient with a Gilson pipet. This step should be done carefully and quickly, because bubbles from the bottom may disrupt the gradient.
9. Check the fractions under the microscope, and select those enriched in small cells (in early G2). Discard fractions containing septated cells, since these are in G1/S-phase, and those containing large cells, which are in late G2.
10. To remove the lactose, wash the selected cells with 1 ml of fresh medium by centrifugation at 800 × *g* for 5 min at room temperature. At this point, cells can be layered onto a second gradient to improve synchrony.
11. Resuspend the cells in fresh pre-warmed medium at the desired temperature.
12. Take a sample corresponding to "time 0".
13. Incubate and take samples every 15–30 min to monitor cell cycle progression by flow cytometry, and DAPI-Calcofluor staining (*see* Subheading 3.7).

3.7 Monitoring Fission Yeast Cell Cycle Synchrony

3.7.1 DAPI-Calcofluor Staining

For DAPI-Calcofluor staining, use cells fixed in 70 % ethanol. DAPI stains the nucleus and Calcofluor stains the septum (Fig. 3c).

1. Spin down 10^7 cells. Wash them once with 1 ml of sterile water and resuspend them in 1 ml of ice-cold 70 % ethanol. Samples fixed in 70 % ethanol may be stored indefinitely at 4 °C.
2. Pellet 60-100 μl of fixed cells. Rehydrate by adding 200 μl of PBS and vortexing.
3. Pellet the cells and resuspend in a mixture of DAPI-Calcofluor (2 μl of 0.1 mg/ml DAPI and 2 μl of 50 μg/ml Calcofluor) (*see* **Notes 13** and **14**).
4. Pipette 2.5–3 μl of cell suspension onto a clean microscope slide. Seal the slide using nail polish to prevent the samples from drying out.
5. Examine by epifluorescence microscopy (*see* **Note 15**).

3.7.2 Flow Cytometry

Cells fixed in 70 % ethanol are also used to analyze the cell cycle by flow cytometry.

1. Spin down 10^7 cells. Wash them once with 1 ml of sterile water and resuspend them in 1 ml of ice-cold 70 % ethanol. Samples fixed in 70 % ethanol may be stored indefinitely at 4 °C.
2. Pellet 3×10^6 fixed cells (300 μl) and resuspend them by vortexing with 1 ml of 50 mM sodium citrate, prepared freshly from a 0.5 M stock solution.
3. Centrifuge at full speed in a microcentrifuge for 5 min at room temperature. Be careful not to lose the cells, since they do not pellet well in sodium citrate solution.
4. Add 500 μl of 50 mM sodium citrate containing 0.1 mg/ml RNase to the cells and vortex.
5. Incubate for 2 h to overnight at 37 °C.
6. Transfer the samples to flow cytometry tubes and add 500 μl of 50 mM sodium citrate with 4 μg/ml propidium iodide (2 μg/ml, final concentration, *see* **Note 16**). Protect the samples from light.
7. Sonicate at 50 % amplitude for 10 s to separate divided cells. Sonication parameters may vary among the different sonicators.
8. The flow cytometer settings must be adjusted for each experiment using a wild-type strain growing in rich medium (2C peak), a wild-type strain blocked in MM lacking nitrogen for 12 h (1C peak), and a diploid strain growing in rich medium (4C peak). In block and release experiments, it is also important to analyze the sample corresponding to the asynchronous culture (before cell cycle arrest) as a control (*see* **Note 17**). Arrested elongated *cdc10-129* or *cdc25-22* cells usually display a DNA peak shifted to the right due to an increase in mitochondrial DNA.
9. Vortex the samples before mounting the tube in the flow cytometer (Becton Dickinson FACSCalibur™ Cell Analyser) and pass the samples at low to medium speed (500–1000 cells/s).
10. We use the CellQuest software to collect and to analyze the data.
11. The data can be represented as an overlay of histogram plots, where the number of cells is plotted against of the amount of DNA represented by the fluorescent signal of the propidium iodide (Figs. 1, 2 and 5c).

4 Notes

1. Increase and decrease the temperature as fast as possible. Be careful not to over-cool the culture.
2. The *cdc10-129* mutant is leaky, especially in YES (Fig. 1).
3. Avoid blocking the *cdc10-129* mutant for longer than 4 h to prevent cells escaping from arrest.

4. We normally take two time 0 samples, one before release when the cells are still at 36 °C and the other immediately after release at 25 °C.
5. Use freshly made 1 M hydroxyurea stock solution. If necessary, this stock solution can be stored at 4 °C for a few hours.
6. Hydroxyurea is a mutagenic compound and must be handled with caution. Follow the procedures for proper handling and disposal.
7. A vacuum filtration system may be used instead of centrifugation to wash the HU more gently. In this case, wash the cells twice with 20 ml of fresh pre-warmed medium using 0.45 μm sterile filters.
8. In contrast to the block and release of the *cdc10-129* mutant (*see* **Note 2**), in the *cdc25-22* mutant no significant differences arise, regardless of whether the experiment is done in EMM or in YES.
9. For effective temperature-driven block and release of a *cdc25-22* mutant the maximum percentage of septated cells in the population should reach 60–80 %. These cells should maintain synchrony for two complete cell cycles, although synchrony in the second cell cycle is reduced.
10. Here we describe the protocol using the Beckman J-6M centrifuge and the JE-5.0 elutriator rotor equipped with a 40 ml elutriation chamber.
11. If the cells show a tendency to clump, add 2 % peptone to the YES medium or use YPD (1 % yeast extract, 2 % peptone, 2 % glucose, pH 6.5) instead.
12. A smaller gradient in a 15-ml tube can be used if fewer cells are required. In this case, harvest 50 ml of culture and resuspend in 0.5 ml of medium.
13. Store the 0.1 mg/ml DAPI and the 50 μg/ml Calcofluor solutions at -20 °C and protect them from the light. DAPI stocks may form a yellow precipitate. Centrifuge the DAPI solution immediately before use to avoid pipetting the yellow pellet. Sometimes, the concentration of DAPI and Calcofluor needs to be adjusted, e.g., in elongated cells, such as the *cdc25-22* mutant arrested in G2, which are poorly stained with DAPI. In this case, reduce the amount of Calcofluor in the mix.
14. DAPI staining only works well in fixed cells. In contrast, Calcofluor also works in live cells, and in this case it stains cell walls as well as septa.
15. Examine the samples under the microscope no more than 2 h after the staining because the signal decreases after time, especially the Calcofluor signal.

16. Propidium iodide is toxic. Read its safety data sheet and place the waste in a suitable, labeled container for waste disposal.
17. Wild-type *S. pombe* cells exponentially growing in rich media (YES or EMM) do not present a 1C peak in flow cytometry because under these conditions G1 is very short and cytokinesis and cell separation take place when the cells are replicating their DNA.

Acknowledgements

We thank N. Skinner for corrections to the manuscript. This work was funded by grants BFU2011-28274 from the Spanish Ministry of Economy and Competitiveness MINECO and CSI151U13 from la Junta de Castilla y León. M.T.-P. is a recipient of a CSIC JAE-predoctoral fellowship.

References

1. Nurse P, Thuriaux P, Nasmyth K (1976) Genetic control of cell division cycle in the fission yeast *Schizosaccharomyces pombe*. Mol Gen Genet 146:167–178
2. Toda T, Umesono K, Hirata A, Yanagida M (1983) Cold-sensitive nuclear division arrest mutants of the fission yeast *Schizosaccharomyces pombe*. J Mol Biol 168:251–270
3. Marks J, Fankhauser C, Simanis V (1992) Genetic interactions in the control of septation in the fission yeast *Schizosaccharomyces pombe*. J Cell Sci 101:801–808
4. Lowndes N, Mclnerny L, Johnson A, Fantes P, Johnston L (1992) Control of DNA synthesis genes in fission yeast by the cell-cycle gene *cdc10*. Nature 355:449–452
5. Slater ML (1973) Effect of reversible inhibition of deoxyribonucleic acid synthesis on the yeast cell cycle. J Bacteriol 113:263–270
6. Moreno S, Hayles J, Nurse P (1989) Regulation of p34^{cdc2} protein kinase during the traverse of mitosis. Cell 58:361–372
7. Hiraoka Y, Toda T, Yanagida T (1984) The *nda3* gene of fission yeast encodes β-tubulin: a cold-sensitive *nda3* mutation reversibly blocks spindle formation and chromosome movement in mitosis. Cell 39:349–358
8. Creanor J, Mitchison JM (1979) Reduction of perturbations in leucine incorporation in synchronous cultures of *Schizosaccharomyces pombe* made by elutriation. J Gen Microbiol 112:385–388
9. Carr AM, Murray JM (1996) DNA repair and checkpoint controls in fission yeast: a practical guide. In: Adolph KW (ed) Microbial genome methods. CRC, Boca Raton, pp 133–147
10. Moreno S, Klar A, Nurse P (1991) Molecular genetic analysis of fission yeast *Schizosaccharomyces pombe*. Methods Enzymol 194:793–823

Chapter 21

A Review of Fluorescent Proteins for Use in Yeast

Maja Bialecka-Fornal, Tatyana Makushok, and Susanne M. Rafelski

Abstract

The field of fluorescent proteins (FPs) is constantly developing. The use of FPs changed the field of life sciences completely, starting a new era of direct observation and quantification of cellular processes. The broad spectrum of FPs (*see* Fig. 1) with a wide range of characteristics allows their use in many different experiments. This review discusses the use of FPs for imaging in budding yeast (Saccharomyces cerevisiae) and fission yeast Schizosaccharomyces pombe). The information included in this review is relevant for both species unless stated otherwise.

Key words Fluorescent proteins, Yeast, *Saccharomyces cerevisiae*, *Schizosaccharomyces pombe*, Protein tagging, Fluorescence microscopy

1 What Are the Applications of FPs?

Fluorescent proteins (FPs) are the core component of numerous techniques that range from single molecule studies to labeling entire organisms and include imaging of both live and fixed cells. The broad spectrum of FPs (*see* Fig. 1) with a wide range of characteristics allows their use in many different experiments. The timeline of major breakthroughs and the phylogenetic tree of FPs are illustrated in [1, 2], respectively.

Protein localization labeling. Tagging a protein of interest allows for the accurate determination of its localization, as well as observing its expression pattern, abundance, interaction with other molecules, turnover, activity, and transport in real time.

Protein-protein interactions. Direct observation of protein-protein interactions leads to a deeper understanding of the function of molecular complexes in living cells by identifying and confirming stable or dynamic interactions between their elements. The techniques used to study these interactions are Förster resonance energy transfer (FRET) between two FP molecules [3, 4], fluorescence correlation spectroscopy (FCS) detecting the average concentrations and diffusion rates of molecules [5, 6], and the bimolecular

Alberto Sanchez-Diaz and Pilar Perez (eds.), *Yeast Cytokinesis: Methods and Protocols*, Methods in Molecular Biology, vol. 1369, DOI 10.1007/978-1-4939-3145-3_21,

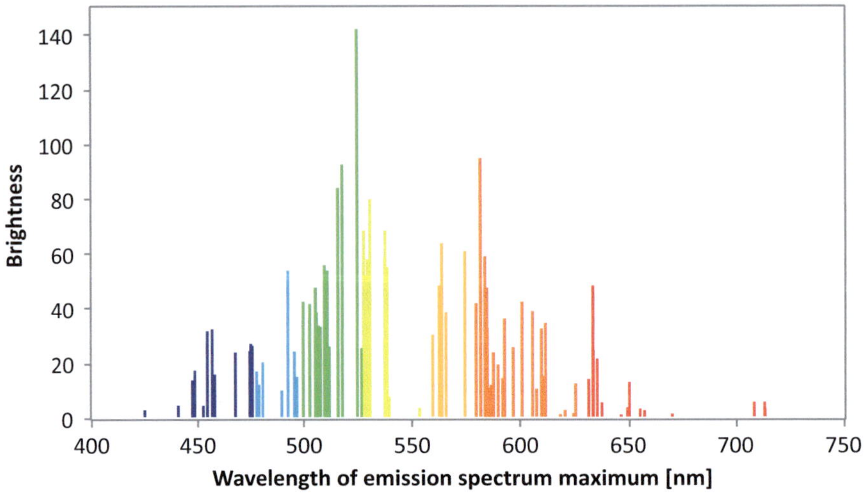

Fig. 1 Spectral diversity of FPs listed in Table 6. The emission maxima of available FPs span across almost the entire visible spectrum. Most of the brightest FPs emit in the range of 500–550 nm. Brightness values were calculated as a product of published molar extinction coefficient and fluorescence quantum yield, and divided by 1000. Molar extinction coefficient is a measure of how strongly a substance absorbs light at a given wavelength. Fluorescence quantum yield is a measure of how many times an event occurs (in this case photon emission) per photon absorbed by the system

fluorescence complementation (BiFC) assay using an FP split into two non-overlapping N- and C-terminal fragments [7, 8].

Organelle labeling. Targeting an FP to a specific organelle allows visualization and studies of local cellular physiology, structure, and dynamics. Targeting sequences are used to localize FPs to specific organelles and cellular compartments, including but not limited to the nucleus [9, 10], the plasma membrane [10, 11], peroxisomes [12], cytosol [13], mitochondria [10, 14], ER lumen [15, 16], and Golgi [10, 17].

Cell and tissue labeling. Expressing an FP in a specific cell type helps in studies of cell cycle progression (e.g., Fucci [18]), cell morphology, cell motility, and interactions within a cell culture, a tissue, or a whole animal.

Gene expression. Visualization of mRNAs and transcription factors allows for quantitative studies of their production, localization, lifetime, and dynamics in cells [19]. Expression of an FP under the control of a specific promoter shows promoter activity [20, 21] under desired conditions (activation or inactivation), in space and time, or in response to a stress factor. These measurements are crucial for studying the mechanisms underlying gene regulatory networks.

Timers. The use of a special class of FPs (timer FPs) that change their spectral properties with time [22] gives access to both the temporal and the spatial components of dynamic processes. Some examples include the age of cells, organelles, and molecules, the dynamics of gene expression, and the timing of promoter activation. These timers can be designed to measure time spans ranging from about 10 min to many hours.

Diffusionand mobility. Photobleaching techniques [23] like fluorescence recovery after photobleaching (FRAP), and FCS allow for detailed studies concerning molecule diffusion or mobility in a living cell and how these processes are affected by external factors.

Sensors. A specialized class of FPs has been developed in which an FP is combined with a detector protein to measure the changes in concentration of specific molecular species. These sensors can be used to study the variation in the levels of pH [24, 25], Ca^{2+} [26], H_2O_2 [27], redox potential [28, 29], and membrane potential [30]. For a recent review on biosensors *see* [31, 32].

2 Creating Yeast Strains Expressing FPs

All the applications mentioned above rely on expressing FPs or FP fusion proteins, created by fusing a protein of interest to an FP. Genes encoding for FPs or FP fusions (both called FP fusions in this section for simplicity) are introduced into yeast cells either transiently or stably. In the first, transient case, the FP fusion is expressed via an appropriate plasmid vector, which is introduced into the cell. To prevent the host cell from losing the plasmid, it is necessary to use a selection marker. Selection markers and types of plasmids used in yeast are described below. In the second, stable case, the FP fusion is integrated into the genome. Nearly all DNA recombination in yeast occur through homologous recombination, meaning they occur between identical or nearly identical sequences. This arrangement allows targeting any DNA sequence to a desired location in the genome with high fidelity. Such an integration can be achieved by transforming cells with a PCR product containing an FP gene (to be fused to protein of interest) and a selection marker (Fig. 2a), by transforming a linear, enzyme-digested plasmid carrying an FP fusion and a selection marker (Fig. 2b), or by transforming a linear enzyme-digested plasmid carrying a selection marker that can be removed after successful selection (Fig. 2c).

Very often more than one insertion needs to be performed in a cell. If this is the case, multiple genes can be inserted sequentially with different selection markers. Alternatively, the selection marker can be removed from the genome after the first insertion and reused for the subsequent ones (Fig. 2c). For a list of removable and non-removable selection markers *see* Tables I and III in [33], respectively.

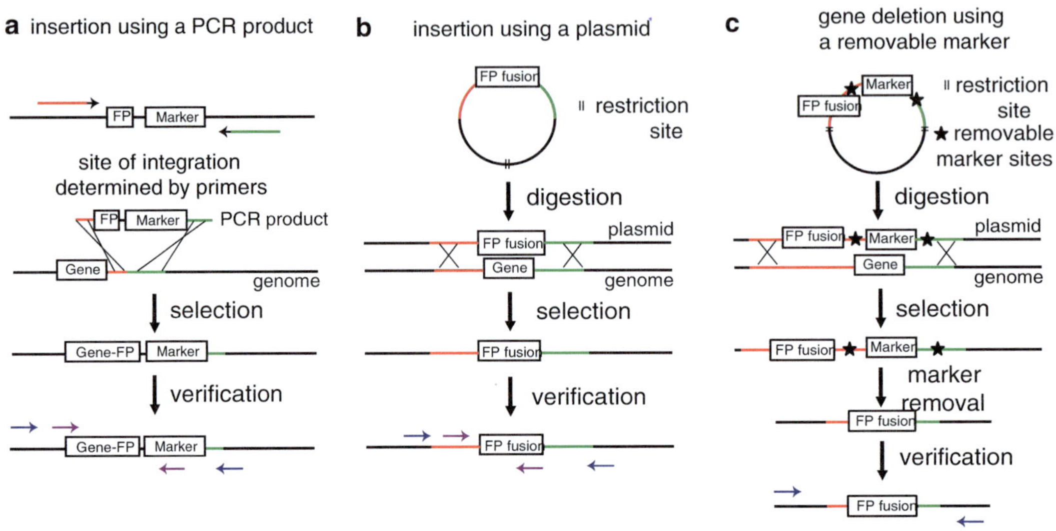

Fig. 2 Molecular biology strategies for FP tagging and insertion into the genome. (**a**) FP tagging using a PCR product. Site of integration is determined by the primers used for PCR. (**b**) FP fusion insertion using a linearized, enzyme-digested plasmid. (**c**) Gene deletion using a removable selection marker. After integration and selection, the marker gene is removed by expressing Cre or Flp recombinase [33]. In all cases, the genetic material is being inserted by homologous recombination. *Blue* and *purple arrows* indicate example primers used for verification of introduced modifications. *Red* and *green lines* represent sequences homologous to the target site in the genome

Gene disruption in yeast is based on the same strategy. The inserted DNA, however, contains only a marker gene that allows for selection.

2.1 What Types of Plasmids Are Used in Yeast?

There are three main types of plasmids used for fluorescence tagging in budding yeast: centromeric, episomal, and integrating plasmids. The properties of these different plasmid types are listed in Table 1, and the decision which one to use should be made based on how these properties match specific experimental requirements (Fig. 3a). Other types of budding yeast plasmids include yeast replicating plasmids (YRp) and yeast linear plasmids (YLp, linearized version of YRp plasmids); however, they are used very rarely due to their instability [34]. There are no centromeric plasmids for fission yeast due to the large size of its centromere [35]; however, episomal and integrating plasmids can be used [36]. Fission yeast are more prone to loosing plasmids than budding yeast [37]. Thus, integrating plasmids are preferable for fission yeast.

Almost all yeast vectors are shuttle vectors. This means that they can replicate and be selected for in two different species, in this case *Escherichia coli* (prokaryote) and yeast (eukaryote). The advantage of such an arrangement is that the plasmid can be modified and/or replicated in a rapidly dividing organism in which molecular biology techniques are most optimized (*E. coli*) and then used

Table 1
Properties of centromeric, episomal, and integrating plasmids

Centromeric	Episomal	Integrating
Distribution		
Cell-to-cell variability even in a clonal population.	Cell-to-cell variability even in a clonal population.	Homogenous distribution.
Replication		
Contain autonomously replicating sequence.	Autonomously replicating with 2μ origin of replication and the partitioning locus.	Lack the sequence for autonomous replication.
Stability		
Segregated by spindle during mitosis and meiosis. High fidelity of segregation. Plasmid-free cells are sometimes generated.	Segregate similarly to centromeric plasmids during mitosis. Less stable than integrating plasmids. High fidelity of segregation depends on the presence of the endogenous or introduced 2μ element.	Stable transformants.
Copy number		
Low copy number: 1–2/cell.	High copy number: 20–50/cell.	Single copy.
Other		
Used for plasmid shuffling, for gene cloning by complementation, for gap repair.	Used for high copy suppressor screens. High transformation efficiency.	Used for genomic integration. Low transformation efficiency. Only one or a few gene copies per cell may be integrated.
High transformation efficiency. Easy to use. Reference: [114].	Reference: [115].	Reference: [116].

in an organism of interest (yeast). The yeast shuttle vectors include a bacterial origin of replication and a bacterial selectable marker, which is usually a bacterial gene conferring resistance to an antibiotic (e.g., ampicillin or kanamycin). The yeast components are a replicating sequence, a centromere (CEN), a yeast selectable marker, and an insert (Fig. 3b). The replicating sequences are either chromosomal DNA fragments called autonomously replicating sequence (ARS in budding yeast and *ars1* in fission yeast [38]) or the replicating sequence derived from the endogenous yeast 2 μ plasmid. The biggest difference between fission yeast and budding yeast shuttle vectors is that the fission yeast ARS has a dramatically different structure and mechanism of action [39].

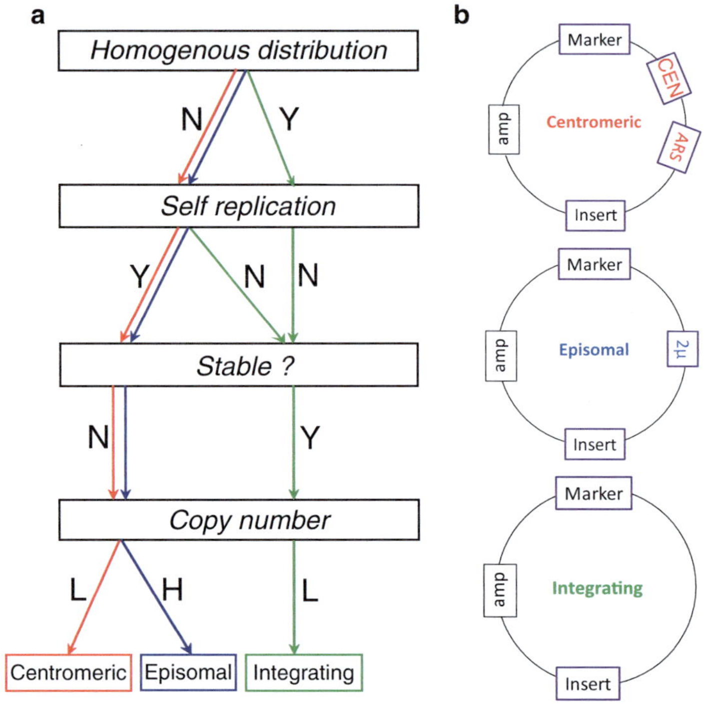

Fig. 3 A summary of plasmid properties. (**a**) The diagram is based on Table 1. The colors represent properties of centromeric (*red*), episomal (*blue*), and integrating (*green*) plasmids. Y stands for yes, N stands for no, L stands for low, and H stands for high. (**b**) The components of integrating, centromeric, and episomal shuttle vectors. Bacterial elements (plasmid backbone and selection marker ampicillin (amp)) are marked in black; yeast elements are marked in violet (yeast selectable marker, insert, a centromere CEN, autonomously replicating sequence (ARS), and 2μ element)

2.2 What Selection Markers Are Available for Yeast?

In yeast molecular genetics, three main classes of markers are available: auxotrophic markers and markers conferring resistance to antibiotics and other types of drugs. A strain is auxotrophic if it carries a mutation that prevents it from synthesizing a compound essential for growth. In this case, the strain will grow only if the compounds taken up from the environment complement its specific auxotrophy. The list of these auxotrophic markers in budding yeast includes *ade1*, *ade2*, *ade8*, *his2*, *his3*, *met15*, *leu2* (complements *S. pombeleu1*), *lys2*, *trp1*, *trp5*, *ura3* (weakly complements *S. pombe ura4*), and *ura4* [40, 41]. The auxotrophic markers used in *S. pombe* are *ade1+*, *ade6+*, *arg3+*, *his3+*, *his7+*, *leu1+*, *ura4+*, and *sup3-5* [42, 43]. One should remember that the use of auxotrophic markers requires the use of appropriate strains (auxotrophic genes deleted) that may have growth defects due to changes in the media that cells are grown on [44, 45].

There are five commonly used antibiotic resistance genes: kan (conferring resistance to G418), ble (conferring resistance to phleomycin), hph (conferring resistance to hygromycin B), nat (conferring resistance to nourseothricin), and pat (conferring resistance to bialaphos) [46–50]. Drug resistance can also be used to screen for a lack of a given gene. In this case, growth in the presence of the drug causes toxicity or death when the gene is present. The most commonly used drug of this type is 5-fluoro-orotic acid (5-FOA) used for screening for *ura3⁻* or *ura5⁻* mutants [51]. Other examples include 5-fluoroanthranilic acid 5-FAA (for *trp1⁻* mutants) [52], α-aminoadipic acid α-AAA (for *lys2⁻* or *lys5⁻* mutants) [53], and canavanine (for *can1⁻* mutant) [54]. In some cases, the canavanine resistance arises in the presence of *can1* gene, which is not fully understood [55, 56].

2.3 How to Amplify a DNA Fragment of Interest?

Most often a desired DNA fragment will be amplified from a plasmid. In order to do so, one has to target specific sites on the plasmid, before and after the fragment of interest, using primers. Primers are either designed specifically for the experiment of interest or are standard ones and can be ordered from repositories such as [57]. More information on the design of primers can be found in [58, 59]. Here we describe a standard way to design primers specifically for direct FP tagging within the yeast genome. These primers consist of two parts: one part that is homologous to the template plasmid and the other part that is homologous to the target sequence in the genome (e.g., in case of chromosomal integration, the FP coding gene on the plasmid and the region that the FP is targeted to, respectively). Generally, for *S. cerevisiae*, about 20 bp are used for the part of the primer that is homologous to the template and 30–50 bp for the part of the primer that is homologous to the target site in the genome. For *S. pombe*, the homologous sequence is recommended to be about 80 bp long [59]. For the relationship between homologous sequence length and the integration success *see* Table 1 in [60].

2.4 How to Introduce DNA into Cells?

In order to perform any genetic modifications one has to be able to introduce exogenous DNA into the cell. The DNA that is being introduced can be either an amplified DNA fragment or a plasmid. There are three main methods used to introduce DNA into yeast: the lithium acetate (LiAc) method, the spheroplast method, and electroporation. Lithium acetate transformation is fast and simple; it has, however, low efficiency. Spheroplast transformation is complicated and time consuming, it has, however, higher efficiency. Electroporation combines features of both methods: it is easy and has the highest efficiency [61, 62].

After introducing DNA into the cell, the transformants are plated on a selective medium. Only the cells that integrated the

DNA fragment or a plasmid successfully will form colonies on the dish. In the case of chromosomal integration, the correct DNA insertion should be confirmed by PCR. The size of the PCR product is measured by gel electrophoresis. PCR samples are being run alongside with DNA fragments of known size (ladder and controls) for comparison. Sequencing of the fragment of interest can also be performed as the final confirmation step.

3 What Are the Criteria When Choosing an FP Insert?

Choosing an optimal FP as an insert is critical for the success of an experiment. The criteria listed below take into consideration various properties of the FP. However, they should only be treated as a starting point: every experiment is unique and needs some degree of optimization and testing when choosing a proper fluorescent protein! Very often more than one FP should be tested for a particular use and different criteria will be important depending on the experimental design. For example, protein stability and slow photobleaching will be more important than brightness for long-duration experiments. For short experiments, the situation will be the opposite. When looking at a signal cascade or translation, fast maturation will be more important than brightness or stability. For a summary of the recommended FPs, *see* Table 2.

3.1 Brightness

One of the most important parameters when choosing a fluorescent protein is brightness. Fusion constructs with brighter FPs can be used in lower amounts (no overexpression required, minimal disturbance to cells) and they require lower intensity of excitation light, thus limiting phototoxic effects. The intrinsic brightness of a protein is determined by its molar extinction coefficient (EC) and fluorescence quantum yield (QY). The reported values are measured in vitro. One has to remember that when expressed in a cell, the brightness is affected by the transcription and translation efficiency, folding efficiency, and protein maturation speed (see below). It is also important that the FP is bright enough relative to the background signal so that it can be reliably detected. This can be achieved by choosing an FP that fluoresces in the region of the spectrum with low autofluorescence for a given organism. Unfortunately, there are many molecules contributing to high autofluorescence including flavins, lipofuscins, and metabolic intermediates in adenine biosynthesis [63–65]. Lower autofluorescence can be achieved by using illumination with longer wavelengths as they penetrate deeper into biological tissue. The optimal "optical window" for imaging in yeast is considered to be 650–700 nm [63]. Unfortunately, only very few FPs can be used for imaging in this range (Fig. 1).

Table 2
FPs recommended for imaging experiments

Category	Fluorescent proteins
UV-excitable green	**T-Sapphire**
Blue	**mTagBFP2**
Cyan	Cerulean (low photostability), CyPet, **ECFP**
Green	Emerald, **Clover**, **EGFP**, **mWasabi**
Yellow	**mCitrine**, Venus, YPet
Orange	mOrange (low photostabillity, pH sensitive), mKO
Red	mCherry, tdTomato (big), mStrawberry, **mRuby2**, **TagRFP-T**
Far-red	mPlum, AQ143 (tetrameric)
Photoactivatable	PA-GFP, PA-mRFP
Photoconvertible	KikGR, mEos2, PS-CFP
Photoswitchable	Dronpa, KFP
FRET	**Clover and mRuby2** CyPet and YPet
Acid sensitive	mOrange, GFPs, YFPs
3-color imaging	**EGFP/GFPγ, TagRFP-T /mRuby2, mTagBFP2** Cerulean/CyPet, mOrange/mKO, mPlum
4-color imaging	**T-Sapphire, mTagBFP2, mWasabi, mRuby2** **T-Sapphire, ECFP, Citrine, tdimer2** Cerulean/CyPet, YFP, mOrange/mKO, mCherry/mPlum

The FPs in bold are particularly recommended for imaging in budding yeast. The table is based on [74, 82, 88]

3.2 Maturation Time

Chromophore maturation is a sequence of modifications necessary for the protein to become fluorescent. It depends on the temperature and the presence of molecular oxygen and its concentration. Depending on the FP, the maturation can take any time between a few minutes (e.g., fast-maturing yellow FPs) and a few hours (e.g., timer FPs). One has to remember that the reported maturation times may not be accurate for every experiment, as they are measured for specific conditions (e.g., temperature), and thus may vary significantly depending on the cell type or even the cell compartment. Very often, the environmental conditions cause a delay between the end of an FP protein translation and its maturation [66].

3.3 Environmental Sensitivity

Ideally, the FP should be relatively insensitive to environmental effects (e.g., chloride concentration or acidity), especially for quantitative measurements. pH-sensitive FPs will perform badly in acidic compartments of the cell such as lysosomes. This is why the pKa of the protein (the pH at which the protein reaches 50 % of its maximal brightness at optimal pH) should be adjusted to the target compartment of the cell. pH levels above 10 usually lead to protein denaturation. Protein sensitivity to pH changes can be also used to monitor pH fluctuations in the cell.

3.4 Optical Properties of the Imaging Setup

The wavelength used for illumination as well as the light intensity have an impact on the detected signal from the FP. Using optimal filter sets is equally critical for obtaining the best fluorescence image for a particular FP. Good camera sensitivity enables imaging of dimmer FPs without increasing illumination intensity.

3.5 Photostability and Photobleaching Rate

The FP should be photostable throughout the duration of the experiment, especially for long time-lapse imaging and for quantitative measurements. In case of an experiment requiring repetitive illumination of the same region of a sample, high photostability is preferred over high brightness to maintain sufficient signal-to-noise ratio over the entire duration of imaging. The main factors influencing the photobehavior of a fluorescent protein are the intensity of illumination light, the frequency of light pulses, the duration of each pulse, the excitation wavelength, and the properties of the light source [67]. This is why the duration, and, in particular, the intensity of illumination should be minimal. Other factors contributing to FP photobehavior are composition of the medium [68], physiological state of the cell, the properties of the protein that the FP is fused to, and the cell compartment that the FP is targeted to. Due to the complexity of the factors affecting protein photostability, it is impossible to assign an absolute value describing photostability, and thus the proteins are simply described as having high or low photostability. The reported photobehavior and photostability of the protein should be treated carefully as they can be used as a general guide only for the similar imaging conditions. One should also remember that in addition to irreversible photobleaching, some FPs undergo photoconversion and change their excitation/emission spectra.

3.6 Toxicity

Most monomeric proteins are nontoxic to cells. However, multimeric proteins may be indirectly toxic by causing aggregation in the cell (especially when overexpressed), atypical localization, disruption of protein's native function [69], or interference with protein kinetics. To indirectly test for FP toxicity one should measure the growth rate of cells expressing the fusion protein. For a more elaborate assays, *see* [69, 70]. Some proteins have been reported to accumulate in lysosomes, forming large aggregates [71]. It is often helpful to create two tags for the protein of interest: C-terminal and N-terminal, since one of the constructs might lead to a non-functional fusion protein. Additionally, the length and the rigidity of the linker connecting the FP to the protein of interest may influence the stability and the abundance of the fusion protein [72].

3.7 Multicolor Imaging

Particularly careful planning is required when designing a multicolor imaging experiment. The major parameter to consider in this case is the amount of overlap between individual excitation and emission channels, as it leads to crosstalk artifacts. It has been reported that the variety of available FPs allows for six color imaging schemes [73]. The crosstalk measurement between four FPs (Sapphire, YFP, CFP, and tdimer2) is illustrated in Table 5 in [74]. FPs with a large Stokes shift (increased distance between excitation and emission maxima) are particularly useful for multicolor imaging, as they can allow for imaging of two FPs with different emission but same excitation wavelengths, reducing illumination time, and therefore phototoxicity. Alternatively, they could permit the use of two FPs with different excitation but same emission wavelengths. Since optimizing the choice of FP used for multicolor imaging does not eliminate crosstalk artifacts entirely in most cases, linear spectral unmixing techniques may be useful to separate the signals obtained with each of the FPs [75]. For the list of proteins particularly suited for multicolor imaging, *see* Table 2.

3.8 Image Analysis

Experiments involving intensity measurements and/or automated image analysis require choosing FPS particularly carefully and optimizing the imaging conditions. This maximizes the quality of the resulting image and minimizes artifacts [76–78].

4 Are There FPs Optimized for Use in Yeast?

Many fluorescent proteins have been codon-optimized for budding yeast expression. Codon optimization is a mutation of a gene sequence that switches the codons without changing the amino acid sequence of the encoded protein. Most amino acids are encoded by multiple codons. This means that there are multiple tRNAs having different anti-codon loops but carrying same amino acid. The abundance of these various tRNAs for each amino acid depends on the organism. This is why codon optimization is very important for the efficiency of heterologous gene expression (when a gene from one organism is expressed in another organism): rare codons are replaced with abundant ones in the host organism. It has been shown that codon optimization for the expression host improves FP expression significantly, up to two-fold for *S. cerevisiae* [74, 79–81].

5 Are There "Ready-to-Use" Vectors for the FP of Choice?

There exists a wide variety of tagging vectors with budding yeast codon-optimized fluorescent proteins available. However, it is not clear which FPs are optimal to use. Published vectors with FPs codon-optimized for *S. cerevisiae* are listed in Table 3 [74, 82]. This table does not list plasmids pCY 3040-05 and pOM42 with the marker Leu2.

Some of these FPs were systematically tested for brightness, photostability, and toxicity [82]. It has been shown that the theoretical brightness (a product of QY and EM) is only weakly correlated with the measured brightness, showing the importance of codon optimization, as the optimized tags are up to twice as bright as the unoptimized tags [82]. Since codon usage in fission yeast is generally similar to that in budding yeast [83], the FPs that were codon-optimized for budding yeast could be useful in fission yeast as well. The other vectors used in budding yeast but not codon-optimized are listed in Table 4. The vectors that are used for *S. pombe* are listed in Table 5.

6 What If the Vector with FP of Choice Has Not Been Created Yet?

Apart from using the published and ready to use plasmids listed above (Tables 3, 4, and 5), there are other ways of obtaining the desired plasmid. Most of the existing plasmids can be modified (e.g., see marker swap plasmids below). There are also expression cassettes ready to be used with a protein of interest. These cassettes can be used to express any available FP (*see* Table 6), also photoconvertible FPs (*see* Table 7). For example, Funk et al. [84] developed a collection of expression vectors for use in *S. cerevisiae*. This collection is based on the pRS series (see below) of centromeric or 2μ plasmids carrying the ADE2, HIS3, KanMX, LEU2, MET15, TRP1, and URA3 selection markers and various promoters (for the detailed description and a schematic map of the vectors constructed *see* Figs. 1 and 2 in [84]).

Sikorski and Hieter created a set of pRS series vectors by converting the *E. coli* pBLUESCRIPT vector to YIp pRS400 and pRS300 series vectors (using yeast DNA segments encoding HIS3, TRP1, LEU2, and URA3) and then to the YCp pRS410 and pRS310 series vectors (by addition of yeast ARS and CEN sequences). pRS vectors are uniform in structure (except for the selectable marker, *see* Fig. 2 and Table 2 in [85]) and allow almost all yeast DNA manipulations to be performed in *E. coli* using the same plasmid that is later introduced into yeast cells.

Table 3
Plasmid vectors used for tagging with codon-optimized FPs in budding yeast and their selection markers

Fluorescent protein	His3	Ura3	Trp1	KanMX	HphMX	NatMX
EBFP			pYM34	pYM33		
mTagBFP	pFA6a-link-yomTagBFP-SpHis5	pFA6a-link-yomTagBFP-CaUra3		pFA6a-link-yomTagBFP-KanR		
mTagBFP2	pFA6a-link-yomTagBFP2-SpHis5	pFA6a-link-yomTagBFP2-CaUra3		pFA6a-link-yomTagBFP2-KanR		
Sapphire	pKT0149			pTK150		
Cerulean		pFA6a-link-yECerulean-CaUra3				
mCFP	pKT0210	pKT0212			pFA6a-link-yEmCFP-Hygro; pCY 3030-02	
ECFP	pKT0101; pYM31	pKT0174	pYM32	pKT102; pYM30	pFA6a-link-yECFP-Hygro; pCY 3020-02	
EGFP	pKT0128; pFA6a-link-yoEGFP-SpHis5; pYM28	pKT0209; pFA6a-link-yoEGFP-CaUra3; pOM42	pYM29; pCY 3040-04	pKT0127; pFA6a-link-yoEGFP-Kan; pYM12, pYM27; pCY 3040-01; pOM40		
Emerald	pFA6a-link-yoEmerald-SpHis5	pFA6a-link-yoEmerald-CaUra3		pFA6a-link-yoEmerald-KanR		
mWasabi	pFA6a-link-yomWasabi-SpHis5	pFA6a-link-yomWasabi-CaUra3		pFA6a-link-yomWasabi-KanR		
GFPγ	pFA6a-link-yoGFPγ-SpHis5	pFA6a-link-yoGFPγ-CaUra3		pFA6a-link-yoGFP-KanR		
Superfolder GFP	pFA6a-link-yoSuperfolderGFP-SpHis5	pFA6a-link-yoSuperfolderGFP-CaUra3		pFA6a-link-yoSuperfolderGFP-KanR		

(continued)

Table 3 (continued)

Fluorescent protein	His3	Ura3	Trp1	KanMX	HphMX	NatMX
Clover	pFA6a-link-yoClover-SpHis5	pFA6a-link-yoClover-CaUra3		pFA6a-link-yoClover-KanR		
EYFP	pYM41			pYM39	pYM40	
Venus	pKT0090			pKT0103; pCY 3060-01		
Citrine	pKT0139	pKT0175		pKT0140; pCY 3070-01	pFA6a-link-yECitrine-Hygro	
mCitrine	pKT0211	pCY 3080-06	pCY 3080-04	pCY 3080-01	pFA6a-link-yEmCitrine-Hygro	
mKO2	pFA6a-link-yomKO2-SpHis5	pFA6a-link-yomKO2-CaUra3		pFA6a-link-yomKO2-KanR		
DsRed[a]				pYM37		
RedStar				pYM38		pYM42
RedStar2						pYM43
TagRFP-T	pFA6a-link-yoTagRFP-T-SpHis5	pFA6a-link-yoTagRFPT-CaUra3		pFA6a-link-yoTagRFPT-KanR		
mApple	pFA6a-link-yomApple-SpHis5	pFA6a-link-yomApple-CaUra3		pFA6a-link-yomApple-KanR		
mRuby	pFA6a-link-yomRuby-SpHis5	pFA6a-link-yomRuby-CaUra3		pFA6a-link-yomRuby-KanR		
mRuby2	pFA6a-link-yomRuby2-SpHis5	pFA6a-link-yomRuby2-CaUra3		pFA6a-link-yomRuby2-KanR		

LSS-mKate2	pFA6a-link-yoLSSmKate2-SpHis5	pFA6a-link-yoLSS-mKate2-CaUra3	pFA6a-link-yoLSS-mKate2-KanR	
mCherry	pFA6a-link-yomCherry-SpHis5	pFA6a-link-yomCherry-CaUra3	pFA6a-link-yomCherry-KanR	
eqFP611			pYM51	
mKeima	pFA6a-link-yomKeima-SpHis5	pFA6a-link-yomKeima-CaUra3	pFA6a-link-yomKeima-KanR	
mKate2	pFA6a-link-yomKate2-SpHis5	pFA6a-link-yomKate2-CaUra3	pFA6a-link-yomKate2-KanR	
TagRFP657	pFA6a-link-yoTagRFP657-SpHis5	pFA6a-link-yoTagRFP657-CaUra3	pFA6a-link-yoTagRFP657-KanR	
mEos2	pFA6a-link-yomEos2-SpHis5	pFA6a-link-yomEos2-CaUra3	pFA6a-link-yomEos2-KanR; pCY 3110-01	
PA-mCherry	pFA6a-link-yoPA-mCherry-SpHis5	pFA6a-link-yoPAmCherry-CaUra3	pFA6a-link-yoPA-mCherry-KanR	
PA-TagRFP	pFA6a-link-yoPA-TagRFP-SpHis5	pFA6a-link-yoPATagRFP-CaUra3	pFA6a-link-yoPATagRFP-KanR	
PSmOrange	pFA6a-link-yoPSmOrange-SpHis5	pFA6a-link-yoPSmOrange-CaUra3	pFA6a-link-yoPSmOrange-KanR	
PS-CFP2	pFA6a-link-yoPS-CFP2-SpHis5	pFA6a-link-yoPSCFP2-CaUra3	pFA6a-link-yoPS-CFP2-KanR	
PA-GFP[a]		pFA6a-link-yEPAGFP-CaUra3		pYM48
FlAsH[a,b]			pCY 3100-01	pYM47

Table was assembled based on [74, 82, 89–92]. All plasmics are available from EUROSCARF or addgene.com. For the plasmid maps refer to these websites

[a]The red fluorescent protein DsRed is not recommended for use in yeast since its maturation time is very long (~24 h); FlAsH and PA-GFP are not recommended because they are not bright enough

[b]The FlAsH tag consists of a small peptide that is recognized by specific di-arsenic compounds, which, upon binding, become fluorescent

Table 4
Plasmid vectors used for tagging with FPs in budding yeast

FP	Reference	FP	Reference	FP	Reference
T-Sapphire	[117]	Clover	[114]	tdTomato	[117]
mTurquoise2	[118]	PhiYellow	[117]	mCherry	[91, 117]
Cerulean	[91, 119]	Venus	[91, 118]	mRuby2	[118]
CFP	[120, 121]	YFP	[120, 121]	tdimer2	[74]
ECFP	[119]	EYFP	[119]	PA-GFP	[118, 122]
GFP	[84, 104, 123–125]	mKO1	[117]	tdPA-GFP	[122]
EGFP	[119, 126]	mDsRed	[117, 119]	mEos2	[91, 118]

The FPs listed in this table are not codon-optimized for yeast, but are still used widely

They can be also used to construct other yeast vectors. Christianson et al. [86] expanded this collection by creating the high copy number YEp pRS420 series vectors carrying HIS3, TRP1, LEU2, and URA3 markers (Fig. 2 in [86]). They also studied the stability and segregation rates of these plasmids (*see* Table 1 in [86]).This collection was further expanded and modified by Chee and Haase (Fig. 1 and Table 1 in [46]). Their pRSII plasmid collection consists of integrating, centromeric, and episomal plasmids carrying ADE1, ADE2, HIS2, HIS3, LEU2, TRP1, and URA3 biosynthetic markers and phleomycin, hygromycin B, nourseothricin, and bialaphos drug selection markers.

Marker swap plasmids allow for changing markers in *S. cerevisiae* when marker conflicts arise (e.g., when the strain to be transformed and the plasmid have the same selection marker, Fig. 4, Table 8). They become especially useful when deleting or modifying multiple genes, when one needs to cross two strains carrying the same marker, working with the yeast knockout collection, or transforming a plasmid into a strain carrying the same marker. Methods for switching markers also exist for *S. pombe* [42, 49, 87].

This review does not discuss the availability of tags other than FPs (e.g., epitope tags). Epitope tags that can be recognized by antibodies are often used for immunolabeling or other biochemical experiments. However, in some cases a separate tag for immunolabeling may not be necessary as many companies offer antibodies against the most commonly used FPs. This allows for using an FP as a tag for both imaging and biochemical studies.

Table 5
Plasmid vectors used for fluorescent protein tagging in fission yeast

Fluorescent protein	ura4	leu2	kanMX6	natMX6	bleMX6	hphMX6
GFP	pFA6a-GFP(S65T)-ura4MX6[d(C)] REP42GFP-C[h(C)] pSGP572[j(C)] pSGP573[j(N)] REP42GFP-N[h(N)] pINTK1GFP[k(N)] pINTK41GFP[k(N)] pINTK81GFP[k(N)]	REP41GFP-Nh[(N)] REP41GFP-Ch[(C)] pSGA[g(N)]	pFA6a-GFP(S65T)-kanMX6[a(C)] pFA6a-kanMX6-P3nmt1-GFP[a(N)] pFA6a-kanMX6-P41nmt1-GFP[a(N)] pFA6a-kanMX6-P81nmt1-GFP[a(N)] pFA6a-kanMX6-Purg1-GFP[i(N)]	pFA6a-GFP(S65T)-natMX6[f(C)] pFA6a-natMX6-P3nmt1-GFP[f(N)] pFA6a-natMX6-P41nmt1-GFPf[(N)] pFA6a-natMX6-P81nmt1-GFPf[(N)]	pFA6a-GFP(S65T)-bleMX6[d(C)]	pFA6a-GFP(S65T)-hphMX6[e(C)]
EGFP	REP42EGFP-C[h(C)] pINTL41EGFPC[k(C)] REP42EGFP-N[h(N)] pINTL41EGFPN[k(N)]	REP41EGFP-Ch[(C)] REP41EGFP-Nh[(N)]	pREP4K-EGFPCk[(C)] pREP4K-EGFPN[k(N)]	pINTH41EGFPC[k(C)] pREP4N-EGFPC[k(C)] pINTH41EGFPN[k(N)] pREP4N-EGFPN[k(N)]		
GFPmut2 (brighter GFP)		pYZ3N-GFPg[(N)]				
yEGFP3–Cln2PEST (unstable GFP)				pFA6a-yEGFP3-CLN2$_{PEST}$-natMX6f[(C)]		
mCherry			pFA6a-mCherry-kanMX6b[(C)] pFA6a-kanMX6-P3nmt1-mCherry[b(N)] pFA6a-kanMX6-P41nmt1-mCherry[b(N)] pFA6a-kanMX6-P81nmt1-mCherry[b(N)]	pFA6a-mCherry-natMX6b[(C)]		
tdTomato			pFA6a-tdTomato-kanMX6b[(C)] pFA6a-kanMX6-P3nmt1-tdTomato[b(N)] pFA6a-kanMX6-P41nmt1-tdTomato[b(N)] pFA6a-kanMX6-P81nmt1-tdTomato[b(N)]	pFA6a-tdTomato-natMX6b[(C)]		

(continued)

Table 5
(continued)

Fluorescent protein	ura4	leu2	kanMX6	natMX6	bleMX6	hphMX6
mRFP	pFA6a-mRFP-ura4MX6d[(C)]		pFA6a-mRFP-kanMX6e[(C)]	pFA6a-mRFP-natMX6e[(C)]		pFA6a-mRFP-hphMX6e[(C)]
TagRFP-T			pFA6a-TRT-kanMX6c[(C)] pFA6a-nmt1-TRT-kanMX6[c(N)] pFA6a-nmt41-TRT-kanMX6[c(N)] pFA6a-nmt81-TRT-kanMX6[c(N)]	pFA6a-TRT-natMX6c[(C)] pFA6a-nmt1-TRT-natMX6c[(N)] pFA6a-nmt41-TRT-natMX6[c(N)] pFA6a-nmt81-TRT-natMX6[c(N)]		
mKate			pFA6a-mO2-kanMX6c[(C)] pFA6a-nmt1- mK-kanMX6[c(N)] pFA6a-nmt41-mK-kanMX6[c(N)] pFA6a-nmt81-mK-kanMX6[c(N)]			
mOrange2			pFA6a-mO2-kanMX6c[(C)] pFA6a-nmt1- mO2-kanMX6[c(N)] pFA6a-nmt41-mO2-kanMX6[c(N)] pFA6a-nmt81-mO2-kanMX6[c(N)]	pFA6a- mO2-natMX6c[(C)] pFA6a-nmt1- mO2-natMX6c[(N)] pFA6a-nmt41-mO2-natMX6c[(N)] pFA6a-nmt81-mO2-natMX6c[(N)]		
YFP	pINTK1YFP[k(N)] pINTK41YFP[k(N)] pINTK81YFP[k(N)]					
EYFP				pFA6a-EYFP-natMXf[(C)]		
CFP	pINTK1CFP[k(N)] pINTK41CFP[k(N)] pINTK81CFP[k(N)]					
ECFP				pFA6a-ECFP-natMXf[(C)]		

[(C)]C-terminal tagging. [(N)]N-terminal tagging. References: [a][93], [b][94], [c][69], [d][95], [e][50], [f][96], [g][97], [h][98], [i][99], [j][100], [k][101]

Table 6
Available fluorescent proteins

Protein	Excitation peak (nm)	Emission peak (nm)	QY	EM (M^{-1} cm^{-1})	Brightness	p*K*a	Oligomerization	Reference
Sirius	355	424	0.24	15,000	3.6	<3	Monomer	[127]
EBFP	380	440	0.17	31,000	5.3	6.3	Monomer	[128]
Azurite	383	447	0.55	26,200	14.4	5	Monomer	[129]
EBFP2	383	448	0.56	32,000	18	5.3	Monomer	[130]
mBlueberry1	398	452	0.48	11,000	5.3			[130]
mTagBFP2	399	454	0.64	50,600	32.4	2.7	Monomer	[131]
mTagBFP	399	456	0.63	52,000	32.8	2.7	Monomer	[132]
mKalama1	385	456	0.45	36,000	16.2	5.5	Monomer	[130]
mBlueberry2	402	467	0.48	51,000	24.5	<2.5		[130]
SCFP	433	474	0.56	30,000	16.8	<4.5	Monomer	[133]
mTurquoise	434	474	0.84	30,000	25.2	4.5	Monomer	[134]
mTurquoise2	434	474	0.93	30,000	27.9	3.1	Monomer	[135]
Cerulean	433	475	0.62	43,000	26.7	4.7	Weak dimer	[136]
mCFP	433	475	0.4	32,500	13	4.7	Monomer	[137]
CyPet	435	477	0.51	35,000	17.9	5	Weak dimer	[138]
ECFP	434	477	0.4	32,500	13	4.7	Monomer	[2]
TagCFP	458	480	0.57	37,000	21.1	4.7	Monomer	[139]
AmCyan	458	489	0.24	44,000	10.6		Tetramer	[140]

(continued)

Table 6 (continued)

Protein	Excitation peak (nm)	Emission peak (nm)	QY	EM (M^{-1} cm^{-1})	Brightness	pKa	Oligomerization	Reference
mTFP1	462	492	0.85	64,000	54.4	4.3	Monomer	[141]
Midoriishi Cyan (MiCy)	472	495	0.9	27,300	24.6	6.6	Dimer	[142]
mMidoriishi Cyan	470	496	0.7	22,150	15.5	7	Monomer	[143]
mUKG	483	499	0.72	60,000	43.2	5.2	Monomer	[30]
UKG (Umi-Kinoko)	483	499	0.41	71,000	29.1		Dimer	[30]
TurboGFP	482	502	0.53	70,000	37.1	5.2	Weak dimer	[139]
CopGFP	482	502	0.6	70,000	42	4.3	Tetramer	[144]
TagGFP	481	505	0.6	55,000	33	5	Monomer	[145]
AzamiGreen (AG)	492	505	0.67	72,300	48.4	<5	Tetramer	[146]
mAzami-Green (mAG)	492	505	0.81	41,800	33.9	6.2	Monomer	[146]
AceGFP	480	505	0.55	50,000	27.5		Weak dimer	[147]
ZsGreen	493	505	0.91	43,000	39.1		Tetramer	[140]
mTagGFP	483	506	0.61	56,500	34.5	5	Monomer	[132]
EGFP	488	507	0.6	56,000	33.6	6	Weak dimer	[148]
Emerald	487	509	0.68	57,500	39.1	6	Monomer	[149]
GFP	395/475	509	0.77	21,000	16.2		Monomer	[150]
mWasabi	493	509	0.8	70,000	56	6.5	Monomer	[151]

Superfolder GFP	485	510	0.65	83,300	54.2		Monomer	[152]
Sapphire[a]	399	511	0.64	29,000	18.6		Monomer	[149]
T-Sapphire[a]	399	511	0.6	44,000	26.4	4.9	Monomer	[153]
Clover	505	515	0.76	111,000	84.4	6.2	Monomer	[154]
mNeonGreen	506	517	0.8	116,000	92.8	5.7	Monomer	[155]
LanYFP	513	524	0.95	150,000	142.5	3.5	Tetramer	[155]
TagYFP	508	524	0.62	50,000	31	5.5	Monomer	[139]
mYFP	514	527	0.62	79,000	49		Monomer	[137]
mAmetrine	406	526	0.58	45,000	26.1	6	Monomer	[156]
EYFP	514	527	0.61	84,000	51.2	6.5	Weak dimer	[2]
Topaz	514	527	0.6	94,500	56.7		Weak dimer	[157]
SYFP2	515	527	0.68	101,000	68.7	6	Monomer	[133]
Venus	515	528	0.57	92,200	52.6	6	Weak dimer	[158]
mCitrine	516	529	0.76	77,000	58.5	5.7	Monomer	[159]
YPet	517	530	0.77	104,000	80.1	5.6	Weak dimer	[138]
PhiYFP	525	537	0.6	115,000	69	6	Dimer	[144]
mHoneydew	487	537	0.79	75,000	59.3	4.7	Monomer	[160]
TurboYFP	525	538	0.53	105,000	55.7	5.9	Dimer	[139]
ZsYellow	529	539	0.42	20,200	8.5		Tetramer	[140]
mBanana	540	553	0.7	6000	4.2	6.7	Monomer	[160]
mKO (Kusabira Orange)	548	559	0.6	51,600	31	5	Monomer	[142]

(continued)

Table 6 (continued)

Protein	Excitation peak (nm)	Emission peak (nm)	QY	EM (M^{-1} cm^{-1})	Brightness	p*K*a	Oligomerization	Reference
mOrange	548	562	0.69	71,000	49	6.5	Monomer	[160]
mKOk	551	563	0.61	105,000	64.1	4.2	Monomer	[30]
mKO2	551	565	0.62	63,800	39.6	5.5	Monomer	[18]
mOrange2	549	565	0.6	58,000	34.8	6.5	Monomer	[161]
TurboRFP	553	574	0.67	92,000	61.6	4.4	Dimer	[162]
Dimer2	552	579	0.69	60,000	41.4	4.9	Dimer	[163]
tdimer2	552	579	0.68	120,000	81.6	4.8	Tandem dimer	[163]
tdTomato	554	581	0.69	138,000	95.2	4.7	Tandem dimer	[160]
dTomato	554	581	0.69	69,000	47.6	4.7	Dimer	[160]
DsRed	558	583	0.79	75,000	59.3	4.7	Tetramer	[140]
TagRFP	555	584	0.48	100,000	48	3.8	Monomer	[162]
TagRFP-T	555	584	0.41	81,000	33.2	4.6	Monomer	[161]
mTangerine	568	585	0.3	38,000	11.4	5.7	Monomer	[160]
DsRed-Monomer	556	586	0.1	35,000	3.5	<4.5	Monomer	[150]
DsRed-Express (T1)	554	586	0.42	30,100	12.6	4.8	Tetramer	[164]
DsRed2	561	587	0.55	43,800	24.1	4.5	Tetramer	[164]
DsRed-Max	560	589	0.41	48,000	19.7		Tetramer	[165]

DsRed-Express2	554	591	0.42	35,600	15		Tetramer	[165]
mApple	568	592	0.49	75,000	36.8	6.5	Monomer	[161]
AsRed2	576	592	0.05	56,200	2.81		Tetramer	[150]
mStrawberry	574	596	0.29	90,000	26.1	<4.5	Monomer	[160]
mRuby2	559	600	0.38	113,000	42.9	5.3	Monomer	[154]
mRuby	558	605	0.35	112,000	39.2	4.4	Monomer	[166]
LSS-mKate2	460	605	0.17	26,000	4.4	2.7	Monomer	[167]
mRFP1	584	607	0.25	44,000	11	4.5	Monomer	[163]
tdRFP611	558	609	0.47	70,000	32.9		Tandem dimer	[168]
mCherry	587	610	0.22	72,000	15.8	<4.5	Monomer	[160]
JRed	584	610	0.2	44,000	8.8	5	Dimer	[139]
eqFP611	559	611	0.45	78,000	35.1		Tetramer	[169]
mBeRFP	446	611	0.27	65,000	17.6	5.6	Monomer	[170]
HcRed1	588	618	0.015	20,000	0.3		Dimer	[171]
mKeima Red	440	620	0.24	14,400	3.5	6.5	Monomer	[73]
LSS-mKate1	463	624	0.08	31,200	2.5	3.2	Monomer	[167]
mRaspberry	598	625	0.15	86,000	12.9	<4.5	Monomer	[172]
tdRFP639	589	631	0.16	90,400	14.5		Tandem dimer	[168]
mKate2	588	633	0.4	62,500	25	5.4	Monomer	[67]
tdKatushka2	588	633	0.37	132,500	49	5.4	Tandem dimer	[67]
mKate	588	635	0.33	45,000	14.9	6.2	Monomer	[173]

(continued)

Table 6 (continued)

Protein	Excitation peak (nm)	Emission peak (nm)	QY	EM (M^{-1} cm^{-1})	Brightness	p*K*a	Oligomerization	Reference
Katushka	588	635	0.34	65,000	22.1		Dimer	[173]
HcRed-Tandem	590	637	0.04	160,000	6.4		Tandem dimer	[174]
mGrape3	608	646	0.03	40,000	1.2	7		[175]
mPlum	590	649	0.1	41,000	4.1	<4.5	Monomer	[172]
Neptune	600	650	0.18	72,000	13	5.8		[175]
mNeptune	600	650	0.2	67,000	13.4	5.4	Monomer	[175]
AQ143	595	655	0.04	90,000	3.6		Tetramer	[176]
TagRFP657	611	657	0.1	34,000	3.4	5	Monomer	[177]
NirFP	605	670	0.06	15,700	0.9	4.5	Dimer	[139]
IFP1.4	684	708	0.07	92,000	6.4	4.6	Monomer	[178]
IFP1.1	686	713	0.05	86,000	4.3		Dimer	[178]
iRFP	690	713	0.059	105,000	6.2	4	Dimer	[179]

Fluorescence quantum yield (QY) is a measure of how many times an event occurs (in this case photon emission) per photon absorbed by the system. Molar extinction coefficient (EM) is a measure of how strongly a substance absorbs light at a given wavelength. Brightness values were calculated as a product of molar extinction coefficient and fluorescence quantum yield, and divided by 1000. p*K*a is the pH at which the protein reaches 50 % of its maximal brightness at optimal pH

[a]Denotes UV-excitable FPs. Relevant references are listed, including the primary paper publishing the FP and occasionally a secondary paper of relevance

Table 7
The optical highlighter fluorescent proteins [102, 103] are known to undergo light-induced photoactivation, photoconversion, and photoswitching upon specific illumination

Protein	Excitation peak (nm)	Emission peak (nm)	QY	EM (M^{-1} cm^{-1})	Brightness	Color change	Oligomerization	Reference
Photoactivatable (*irreversible*)								
PA-GFP	400/504	517/517	0.13/0.79	20,700/17,400	2.7/13.7	None/red	Monomer	[180]
PAmCherry1	404/564	–/595	–/0.46	6500/18,000	–/8.3	None/red	Monomer	[181]
PAmCherry2	406/570	–/596	–/0.53	1900/24,000	–/12.7		Monomer	[181]
PATagRFP	–/562	–/595	–/0.38	–/66,000	–/25.1	None/red	Monomer	[182]
PA-mRFP1	–/578	–/605	–/0.08	–/10,000	–/0.8	None/red	Monomer	[183]
PS-CFP2	400/490	468/511	0.2/0.23	43,000/47,000	8.6/10.8	Cyan/green	Monomer	[184]
KFP1[a]	580	600	<0.001/0.07	123,000/59,000	0.1/4.1	None/red	Tetramer	[185]
Phamret	458/458	475/517	0.4/0.79	32,500/17,400	13/13.7	Cyan/green	Monomer	[186]
Photoconvertible (*irreversible*)								
Kaede	508/572	518/580	0.88/0.33	98,800/60,400	86.9/19.9	Green/red	Tetramer	[187]
KikGR	507/583	517/593	0.7/0.65	28,200/32,600	19.7/21.2	Green/red	Tetramer	[188]
mKikGR	505/580	515/591	0.69/0.63	49,000/28,000	33.8/17.6	Green/red	Monomer	[189]
mEos2	506/573	519/584	0.84/0.66	56,000/46,000	47/30.4	Green/red	Monomer	[190]
wtEosFP	506/571	516/581	0.7/0.55	72,000/41,000	50.4/22.6	Green/red	Tetramer	[191]
dEos	506/569	516/581	0.66/0.6	84,000/33,000	55.4/19.8	Green/red	Dimer	[191]
Dendra	486/558	505/575	0.7/0.72	21,000/20,000	14.7/14.4	Green/red	Monomer	[192]

(continued)

Table 7 (continued)

Protein	Excitation peak (nm)	Emission peak (nm)	QY	EM (M^{-1} cm^{-1})	Brightness	Color change	Oligomerization	Reference
mEos3.2	507/572	516/580	0.84/0.55	63,400/32,200	53.3/17.7	Green/red	Monomer	[193]
Cy11.5	436/436	527/475				Yellow/cyan	Monomer	[194]
PSmOrange	548/634	565/662	0.51/0.28	113,300/32,700	57.8/9.2		Monomer	[195]
Photoswitchable (reversible)								
Dronpa	392/503	517	–/0.85	–/95,000	–/80.8	None/green	Monomer	[196]
Dronpa-2	486	513	0.28	56,000	15.7		Monomer	[197]
Dronpa-3	487	514	0.33	58,000	19.1		Monomer	[197]
rsFastLime	384/496	518	0.77	39,094	30.1		Monomer	[198]
Padron	505/503	522	0.64	43,000	27.5	Green	Monomer	[199]
bsDronpa	385/460	504	0.5	45,000	22.5		Monomer	[199]
mTFP0.7	453	488	0.5	60,000	30	Cyan	Monomer	[200]
E2GFP	515	523	0.91	22,400	20.4		Monomer	[201]
rsCherry	572/572	610/610	0.009/0.02	81,000/80,000	0.7/1.6	Red	Monomer	[181]
rsCherryRev	572/572	608/608	0.0003/0.005	85,000/84,000	0.03/0.42	Red	Monomer	[181]
IrisFP	488/551	516/580	0.43/0.47	52,200/35,400	22.4/16.6	Green/red	Tetramer	[202]

Photoactivatable FPs change their emission brightness from low to high, photoconvertible FPs change their fluorescence emission bandwidth, and photoswitchable FPs can be turned "on" or "off." These optical highlighters are used for various purposes ranging from selectively labeling subpopulation of molecules and tracking their behavior to investigating extremely rapid cellular processes since the time required for photoactivation is typically very brief. They are also very important for superresolution techniques like photoactivated localization microscopy (PALM) or stochastic optical reconstruction microscopy (STORM) [104–107]. The first number in the excitation and emission column denotes pre-activation; the second number denotes post-activation

[a]A reversibly photoactivatable KFP1 is also available

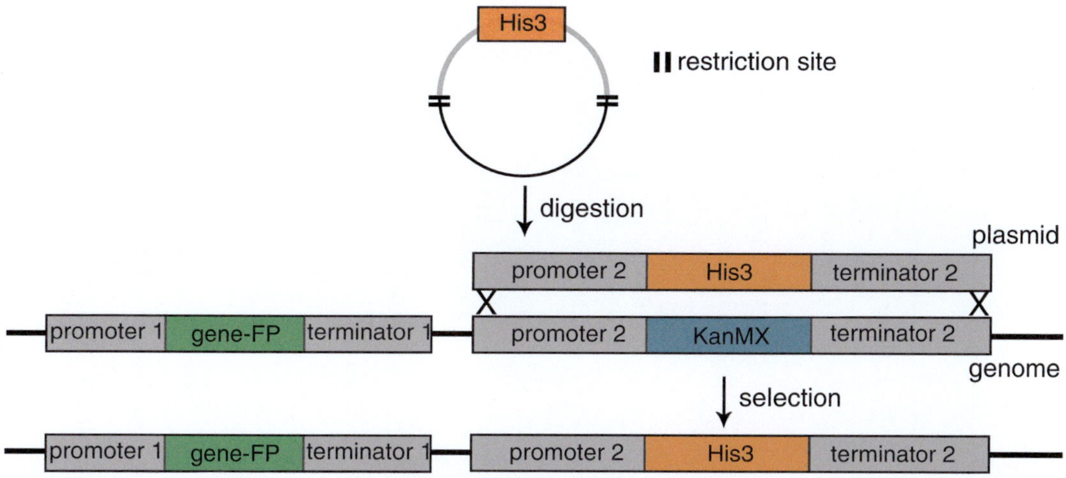

Fig. 4 A diagram illustrating marker swapping. Marker swap plasmids can be used to change selectable marker in the genome or on a plasmid. Marker conversion does not disrupt other genetic modifications, e.g., fluorescent protein fusion. To convert the old marker to a new one, a linearized, restriction enzyme digested vector fragment is used. The markers are swapped through homologous recombination. Then, the desired strain is selected by growth on an appropriate selection medium

7 Useful Websites and Databases

http://web.uni-frankfurt.de/fb15/mikro/euroscarf/index.html EUROSCARF (EUROpean Saccharomyces Cerevisiae Archive for Functional Analysis).

https://www.Addgene.org addgene distributes several ready-to-use plasmids.

http://www.addgene.org/fluorescent-proteins/davidson/ Michael Davidson fluorescent protein collection.

http://www.atcc.org/cydac/cydac.cfm American type Culture Collection ATCC.

http://depts.washington.edu/yeastrc/pages/plasmids_protocols.html Yeast resource center.

http://info.addgene.org/download-addgenes-ebook-plasmids-101-1st-edition Plasmids 101 eBook.

http://nic.ucsf.edu/FPvisualization/ Fluorescent Protein Properties.

http://nic.ucsf.edu/FPvisualization/PSFP.html Photoswitchable Fluorescent Protein Properties.

http://www.yeastgenome.org/help/analyze/design-primers Primer design tool.

Table 8
Marker swap plasmids

		Replacement marker						
		His3	KanMX	Leu2	Lys2	Trp1	Ura3	Other
Disrupted marker	His3		M3929 (KanMX3)[a]*	pHL3[b]	D1433[a]	pHT6[b]	pHU10[b]	M2371 (Ade2)[a]
	KanMX	M4754[a]; pFA6a:His3MX6 (His3MX6)[c]*		M4755[a]		M4757[a]	M4758[a] pAG60 (Ura3MX4)[d]*	M4753 (Ade2)[a]; pAG25 (NatMX4)[e]* pAG29(PatMX4)[e]* pAG32(HphMX4)[e]* M4786 (hisG-Ura3-hisG)[a]**
	Leu2	pLH7[b]	M3926 (KanMX3) [a]*		D771[a]	pLT11[b]	pLU12[b]	M3939 (Ade2)[a]; pNKY85 (hisG-Ura3-hisG)[f]**
	Lys2	D588[a]						
	Trp1	pTH4[b]	M3925 (KanMX3)[a]*	pTL7[b]			pTU10[b]	M3938 (Ade2)[a]; pNKY1009 (hisG-Ura3-hisG)[f]**
	Ura3	pUH7[b]	M3927 (KanMX3)[a]*	pUL9[b]	M2660[a]	pUT11[b]		M3499 (Ade2)[a]
	Other							peltaADE2 (Ade2 to hisG-Ura3-hisG)[g]*

[a]Denotes plasmids from [108]
[b]Denotes plasmids from [109]
[c]Denotes plasmids from [110]
[d]Denotes plasmids from [111]
[e]Denotes plasmids from [48]
[f]Denotes plasmids from [112]
[g]Denotes plasmids from [40]
[a,b,f,] and [g]available from ATCC; [c]available from Mark Longtine (mark_longtine@biochem.okstate.edu) or Peter Philippsen (peter.philippsen@unibas.ch).; [d] and [e]available from EUROSCARF
*These genes are expressed from the promoter and terminator from the *Ashbya gossypii* TEF gene [113]. These plasmids are covered by a BASF patent, but can be used without restriction by academic labs
**hisG-Ura3-hisG can be converted to hisG by growth on 5-FOA [51]

Acknowledgements

We would like to thank Rafelski lab, especially Dr. Irina Mueller, and Marshall lab for fruitful discussions.

References

1. Tomosugi W, Matsuda T, Tani T, Nemoto T, Kotera I, Saito K, Horikawa K, Nagai T (2009) An ultramarine fluorescent protein with increased photostability and pH insensitivity. Nat Methods 6(5):351–353. doi:10.1038/nmeth.1317
2. Patterson GH, Knobel SM, Sharif WD, Kain SR, Piston DW (1997) Use of the green fluorescent protein and its mutants in quantitative fluorescence microscopy. Biophys J 73(5):2782–2790. doi:10.1016/S0006-3495(97)78307-3
3. Mena MA, Treynor TP, Mayo SL, Daugherty PS (2006) Blue fluorescent proteins with enhanced brightness and photostability from a structurally targeted library. Nat Biotechnol 24(12):1569–1571. doi:10.1038/nbt1264
4. Ai HW, Shaner NC, Cheng Z, Tsien RY, Campbell RE (2007) Exploration of new chromophore structures leads to the identification of improved blue fluorescent proteins. Biochemistry 46(20):5904–5910. doi:10.1021/bi700199g
5. Subach OM, Cranfill PJ, Davidson MW, Verkhusha VV (2011) An enhanced monomeric blue fluorescent protein with the high chemical stability of the chromophore. PLoS One 6(12), e28674. doi:10.1371/journal.pone.0028674
6. Subach OM, Gundorov IS, Yoshimura M, Subach FV, Zhang J, Gruenwald D, Souslova EA, Chudakov DM, Verkhusha VV (2008) Conversion of red fluorescent protein into a bright blue probe. Chem Biol 15(10):1116–1124. doi:10.1016/j.chembiol.2008.08.006
7. Kremers GJ, Goedhart J, van Munster EB, Gadella TW Jr (2006) Cyan and yellow super fluorescent proteins with improved brightness, protein folding, and FRET Forster radius. Biochemistry 45(21):6570–6580. doi:10.1021/bi0516273
8. Goedhart J, van Weeren L, Hink MA, Vischer NO, Jalink K, Gadella TW Jr (2010) Bright cyan fluorescent protein variants identified by fluorescence lifetime screening. Nat Methods 7(2):137–139. doi:10.1038/nmeth.1415
9. Goedhart J, von Stetten D, Noirclerc-Savoye M, Lelimousin M, Joosen L, Hink MA, van Weeren L, Gadella TW Jr, Royant A (2012) Structure-guided evolution of cyan fluorescent proteins towards a quantum yield of 93%. Nat Commun 3:751. doi:10.1038/ncomms1738
10. Rizzo MA, Springer GH, Granada B, Piston DW (2004) An improved cyan fluorescent protein variant useful for FRET. Nat Biotechnol 22(4):445–449. doi:10.1038/nbt945
11. Zacharias DA, Violin JD, Newton AC, Tsien RY (2002) Partitioning of lipid-modified monomeric GFPs into membrane microdomains of live cells. Science 296(5569):913–916. doi:10.1126/science.1068539
12. Nguyen AW, Daugherty PS (2005) Evolutionary optimization of fluorescent proteins for intracellular FRET. Nat Biotechnol 23(3):355–360. doi:10.1038/nbt1066
13. Chudakov DM, Matz MV, Lukyanov S, Lukyanov KA (2010) Fluorescent proteins and their applications in imaging living cells and tissues. Physiol Rev 90(3):1103–1163. doi:10.1152/physrev.00038.2009
14. Basic fluorescent proteins. http://www.evrogen.com/products/BasicFPs-app.shtml
15. Matz MV, Fradkov AF, Labas YA, Savitsky AP, Zaraisky AG, Markelov ML, Lukyanov SA (1999) Fluorescent proteins from nonbioluminescent Anthozoa species. Nat Biotechnol 17(10):969–973. doi:10.1038/13657
16. Ai HW, Henderson JN, Remington SJ, Campbell RE (2006) Directed evolution of a monomeric, bright and photostable version of Clavularia cyan fluorescent protein: structural characterization and applications in fluorescence imaging. Biochem J 400(3):531–540. doi:10.1042/BJ20060874
17. Karasawa S, Araki T, Nagai T, Mizuno H, Miyawaki A (2004) Cyan-emitting and orange-emitting fluorescent proteins as a donor/acceptor pair for fluorescence resonance energy transfer. Biochem J 381(Pt 1):307–312. doi:10.1042/BJ20040321
18. Fluorescent Proteins. http://www.mblintl.com/research/fluorescent-proteins.aspx
19. Tsutsui H, Karasawa S, Okamura Y, Miyawaki A (2008) Improving membrane voltage measurements using FRET with new fluorescent

proteins. Nat Methods 5(8):683–685. doi:10.1038/nmeth.1235

20. Shagin DA, Barsova EV, Yanushevich YG, Fradkov AF, Lukyanov KA, Labas YA, Semenova TN, Ugalde JA, Meyers A, Nunez JM, Widder EA, Lukyanov SA, Matz MV (2004) GFP-like proteins as ubiquitous metazoan superfamily: evolution of functional features and structural complexity. Mol Biol Evol 21(5):841–850. doi:10.1093/molbev/msh079
21. Xia NS, Luo WX, Zhang J, Xie XY, Yang HJ, Li SW, Chen M, Ng MH (2002) Bioluminescence of Aequorea macrodactyla, a common jellyfish species in the East China Sea. Marine Biotechnol 4(2):155–162. doi:10.1007/s10126-001-0081-7
22. Karasawa S, Araki T, Yamamoto-Hino M, Miyawaki A (2003) A green-emitting fluorescent protein from Galaxeidae coral and its monomeric version for use in fluorescent labeling. J Biol Chem 278(36):34167–34171. doi:10.1074/jbc.M304063200
23. Gurskaya NG, Fradkov AF, Pounkova NI, Staroverov DB, Bulina ME, Yanushevich YG, Labas YA, Lukyanov S, Lukyanov KA (2003) A colourless green fluorescent protein homologue from the non-fluorescent hydromedusa Aequorea coerulescens and its fluorescent mutants. Biochem J 373(Pt 2):403–408. doi:10.1042/BJ20021966
24. Heim R, Cubitt AB, Tsien RY (1995) Improved green fluorescence. Nature 373(6516):663–664. doi:10.1038/373663b0
25. Cubitt AB, Woollenweber LA, Heim R (1999) Understanding structure-function relationships in the Aequorea victoria green fluorescent protein. Methods Cell Biol 58:19–30
26. The Fluorescent Protein Color Palette. http://www.olympusconfocal.com/applications/fpcolorpalette.html
27. Ai HW, Olenych SG, Wong P, Davidson MW, Campbell RE (2008) Hue-shifted monomeric variants of Clavularia cyan fluorescent protein: identification of the molecular determinants of color and applications in fluorescence imaging. BMC Biol 6:13. doi:10.1186/1741-7007-6-13
28. Pedelacq JD, Cabantous S, Tran T, Terwilliger TC, Waldo GS (2006) Engineering and characterization of a superfolder green fluorescent protein. Nat Biotechnol 24(1):79–88. doi:10.1038/nbt1172
29. Zapata-Hommer O, Griesbeck O (2003) Efficiently folding and circularly permuted variants of the Sapphire mutant of GFP. BMC Biotechnol 3:5. doi:10.1186/1472-6750-3-5
30. Lam AJ, St-Pierre F, Gong Y, Marshall JD, Cranfill PJ, Baird MA, McKeown MR, Wiedenmann J, Davidson MW, Schnitzer MJ, Tsien RY, Lin MZ (2012) Improving FRET dynamic range with bright green and red fluorescent proteins. Nat Methods 9(10):1005–1012. doi:10.1038/nmeth.2171
31. Shaner NC, Lambert GG, Chammas A, Ni Y, Cranfill PJ, Baird MA, Sell BR, Allen JR, Day RN, Israelsson M, Davidson MW, Wang J (2013) A bright monomeric green fluorescent protein derived from Branchiostoma lanceolatum. Nat Methods 10(5):407–409. doi:10.1038/nmeth.2413
32. Ai HW, Hazelwood KL, Davidson MW, Campbell RE (2008) Fluorescent protein FRET pairs for ratiometric imaging of dual biosensors. Nat Methods 5(5):401–403. doi:10.1038/nmeth.1207
33. Tsien RY (1998) The green fluorescent protein. Annu Rev Biochem 67:509–544. doi:10.1146/annurev.biochem.67.1.509
34. Nagai T, Ibata K, Park ES, Kubota M, Mikoshiba K, Miyawaki A (2002) A variant of yellow fluorescent protein with fast and efficient maturation for cell-biological applications. Nat Biotechnol 20(1):87–90. doi:10.1038/nbt0102-87
35. Griesbeck O, Baird GS, Campbell RE, Zacharias DA, Tsien RY (2001) Reducing the environmental sensitivity of yellow fluorescent protein. Mechanism and applications. J Biol Chem 276(31):29188–29194. doi:10.1074/jbc.M102815200
36. Shaner NC, Campbell RE, Steinbach PA, Giepmans BN, Palmer AE, Tsien RY (2004) Improved monomeric red, orange and yellow fluorescent proteins derived from Discosoma sp. red fluorescent protein. Nat Biotechnol 22(12):1567–1572. doi:10.1038/nbt1037
37. Sakaue-Sawano A, Kurokawa H, Morimura T, Hanyu A, Hama H, Osawa H, Kashiwagi S, Fukami K, Miyata T, Miyoshi H, Imamura T, Ogawa M, Masai H, Miyawaki A (2008) Visualizing spatiotemporal dynamics of multicellular cell-cycle progression. Cell 132(3):487–498. doi:10.1016/j.cell.2007.12.033
38. Shaner NC, Lin MZ, McKeown MR, Steinbach PA, Hazelwood KL, Davidson MW, Tsien RY (2008) Improving the photostability of bright monomeric orange and red fluorescent proteins. Nat Methods 5(6):545–551. doi:10.1038/nmeth.1209
39. Merzlyak EM, Goedhart J, Shcherbo D, Bulina ME, Shcheglov AS, Fradkov AF, Gaintzeva A, Lukyanov KA, Lukyanov S, Gadella TW, Chudakov DM (2007) Bright monomeric red fluorescent protein with an extended fluorescence lifetime. Nat Methods 4(7):555–557. doi:10.1038/nmeth1062

40. Campbell RE, Tour O, Palmer AE, Steinbach PA, Baird GS, Zacharias DA, Tsien RY (2002) A monomeric red fluorescent protein. Proc Natl Acad Sci U S A 99(12):7877–7882. doi:10.1073/pnas.082243699
41. Bevis BJ, Glick BS (2002) Rapidly maturing variants of the Discosoma red fluorescent protein (DsRed). Nat Biotechnol 20(1):83–87. doi:10.1038/nbt0102-83
42. Strack RL, Strongin DE, Bhattacharyya D, Tao W, Berman A, Broxmeyer HE, Keenan RJ, Glick BS (2008) A noncytotoxic DsRed variant for whole-cell labeling. Nat Methods 5(11):955–957. doi:10.1038/nmeth.1264
43. Kredel S, Oswald F, Nienhaus K, Deuschle K, Rocker C, Wolff M, Heilker R, Nienhaus GU, Wiedenmann J (2009) mRuby, a bright monomeric red fluorescent protein for labeling of subcellular structures. PLoS One 4(2), e4391. doi:10.1371/journal.pone.0004391
44. Piatkevich KD, Hulit J, Subach OM, Wu B, Abdulla A, Segall JE, Verkhusha VV (2010) Monomeric red fluorescent proteins with a large Stokes shift. Proc Natl Acad Sci U S A 107(12):5369–5374. doi:10.1073/pnas.0914365107
45. Kredel S, Nienhaus K, Oswald F, Wolff M, Ivanchenko S, Cymer F, Jeromin A, Michel FJ, Spindler KD, Heilker R, Nienhaus GU, Wiedenmann J (2008) Optimized and far-red-emitting variants of fluorescent protein eqFP611. Chem Biol 15(3):224–233. doi:10.1016/j.chembiol.2008.02.008
46. Wiedenmann J, Schenk A, Rocker C, Girod A, Spindler KD, Nienhaus GU (2002) A far-red fluorescent protein with fast maturation and reduced oligomerization tendency from Entacmaea quadricolor (Anthozoa, Actinaria). Proc Natl Acad Sci U S A 99(18):11646–11651. doi:10.1073/pnas.182157199
47. Yang J, Wang L, Yang F, Luo H, Xu L, Lu J, Zeng S, Zhang Z (2013) mBeRFP, an improved large stokes shift red fluorescent protein. PLoS One 8(6), e64849. doi:10.1371/journal.pone.0064849
48. Gurskaya NG, Fradkov AF, Terskikh A, Matz MV, Labas YA, Martynov VI, Yanushevich YG, Lukyanov KA, Lukyanov SA (2001) GFP-like chromoproteins as a source of far-red fluorescent proteins. FEBS Lett 507(1):16–20
49. Kogure T, Karasawa S, Araki T, Saito K, Kinjo M, Miyawaki A (2006) A fluorescent variant of a protein from the stony coral Montipora facilitates dual-color single-laser fluorescence cross-correlation spectroscopy. Nat Biotechnol 24(5):577–581. doi:10.1038/nbt1207
50. Wang L, Jackson WC, Steinbach PA, Tsien RY (2004) Evolution of new nonantibody proteins via iterative somatic hypermutation. Proc Natl Acad Sci U S A 101(48):16745–16749. doi:10.1073/pnas.0407752101
51. Shcherbo D, Murphy CS, Ermakova GV, Solovieva EA, Chepurnykh TV, Shcheglov AS, Verkhusha VV, Pletnev VZ, Hazelwood KL, Roche PM, Lukyanov S, Zaraisky AG, Davidson MW, Chudakov DM (2009) Far-red fluorescent tags for protein imaging in living tissues. Biochem J 418(3):567–574. doi:10.1042/BJ20081949
52. Shcherbo D, Merzlyak EM, Chepurnykh TV, Fradkov AF, Ermakova GV, Solovieva EA, Lukyanov KA, Bogdanova EA, Zaraisky AG, Lukyanov S, Chudakov DM (2007) Bright far-red fluorescent protein for wholebody imaging. Nat Methods 4(9):741–746. doi:10.1038/nmeth1083
53. Fradkov AF, Verkhusha VV, Staroverov DB, Bulina ME, Yanushevich YG, Martynov VI, Lukyanov S, Lukyanov KA (2002) Far-red fluorescent tag for protein labelling. Biochem J 368(Pt 1):17–21. doi:10.1042/BJ20021191
54. Lin MZ, McKeown MR, Ng HL, Aguilera TA, Shaner NC, Campbell RE, Adams SR, Gross LA, Ma W, Alber T, Tsien RY (2009) Autofluorescent proteins with excitation in the optical window for intravital imaging in mammals. Chem Biol 16(11):1169–1179. doi:10.1016/j.chembiol.2009.10.009
55. Shkrob MA, Yanushevich YG, Chudakov DM, Gurskaya NG, Labas YA, Poponov SY, Mudrik NN, Lukyanov S, Lukyanov KA (2005) Far-red fluorescent proteins evolved from a blue chromoprotein from Actinia equina. Biochem J 392(Pt 3):649–654. doi:10.1042/BJ20051314
56. Morozova KS, Piatkevich KD, Gould TJ, Zhang J, Bewersdorf J, Verkhusha VV (2010) Far-red fluorescent protein excitable with red lasers for flow cytometry and superresolution STED nanoscopy. Biophys J 99(2):L13–L15. doi:10.1016/j.bpj.2010.04.025
57. Shu X, Royant A, Lin MZ, Aguilera TA, Lev-Ram V, Steinbach PA, Tsien RY (2009) Mammalian expression of infrared fluorescent proteins engineered from a bacterial phytochrome. Science 324(5928):804–807. doi:10.1126/science.1168683
58. Filonov GS, Piatkevich KD, Ting LM, Zhang J, Kim K, Verkhusha VV (2011) Bright and stable nearinfrared fluorescent protein for in vivo imaging. Nat Biotechnol 29(8):757–761. doi:10.1038/nbt.1918
59. Day RN, Davidson MW (2009) The fluorescent protein palette: tools for cellular imaging. Chem Soc Rev 38(10):2887–2921. doi:10.1039/b901966a

60. Periasamy A, Day RN (1999) Visualizing protein interactions in living cells using digitized GFP imaging and FRET microscopy. Methods Cell Biol 58:293–314
61. Jares-Erijman EA, Jovin TM (2003) FRET imaging. Nat Biotechnol 21(11):1387–1395. doi:10.1038/nbt896
62. Hink MA (2015) Fluorescence correlation spectroscopy. Methods Mol Biol 1251:135–150. doi:10.1007/978-1-4939-2080-8_8
63. Kim SA, Heinze KG, Schwille P (2007) Fluorescence correlation spectroscopy in living cells. Nat Methods 4(11):963–973. doi:10.1038/nmeth1104
64. Kerppola TK (2008) Bimolecular fluorescence complementation (BiFC) analysis as a probe of protein interactions in living cells. Annu Rev Biophys 37:465–487. doi:10.1146/annurev.biophys.37.032807.125842
65. Akman G, MacNeill SA (2009) MCM-GINS and MCM-MCM interactions in vivo visualised by bimolecular fluorescence complementation in fission yeast. BMC Cell Biol 10:12. doi:10.1186/1471-2121-10-12
66. DiPilato LM, Cheng X, Zhang J (2004) Fluorescent indicators of cAMP and Epac activation reveal differential dynamics of cAMP signaling within discrete subcellular compartments. Proc Natl Acad Sci U S A 101(47):16513–16518. doi:10.1073/pnas.0405973101
67. Gallegos LL, Kunkel MT, Newton AC (2006) Targeting protein kinase C activity reporter to discrete intracellular regions reveals spatiotemporal differences in agonist-dependent signaling. J Biol Chem 281(41):30947–30956. doi:10.1074/jbc.M603741200
68. Kunkel MT, Toker A, Tsien RY, Newton AC (2007) Calcium-dependent regulation of protein kinase D revealed by a genetically encoded kinase activity reporter. J Biol Chem 282(9):6733–6742. doi:10.1074/jbc.M608086200
69. Gould SJ, Keller GA, Hosken N, Wilkinson J, Subramani S (1989) A conserved tripeptide sorts proteins to peroxisomes. J Cell Biol 108(5):1657–1664
70. Wen W, Meinkoth JL, Tsien RY, Taylor SS (1995) Identification of a signal for rapid export of proteins from the nucleus. Cell 82(3):463–473
71. Palmer AE, Giacomello M, Kortemme T, Hires SA, Lev-Ram V, Baker D, Tsien RY (2006) Ca2+ indicators based on computationally redesigned calmodulin-peptide pairs. Chem Biol 13(5):521–530. doi:10.1016/j.chembiol.2006.03.007
72. Miyawaki A, Llopis J, Heim R, McCaffery JM, Adams JA, Ikura M, Tsien RY (1997) Fluorescent indicators for Ca2+ based on green fluorescent proteins and calmodulin. Nature 388(6645):882–887. doi:10.1038/42264
73. Palmer AE, Jin C, Reed JC, Tsien RY (2004) Bcl-2-mediated alterations in endoplasmic reticulum Ca2+ analyzed with an improved genetically encoded fluorescent sensor. Proc Natl Acad Sci U S A 101(50):17404–17409. doi:10.1073/pnas.0408030101
74. Gleeson PA (1998) Targeting of proteins to the Golgi apparatus. Histochem Cell Biol 109(5-6):517–532
75. Tyagi S (2009) Imaging intracellular RNA distribution and dynamics in living cells. Nat Methods 6(5):331–338. doi:10.1038/nmeth.1321
76. Chalfie M, Tu Y, Euskirchen G, Ward WW, Prasher DC (1994) Green fluorescent protein as a marker for gene expression. Science 263(5148):802–805
77. Plautz JD, Day RN, Dailey GM, Welsh SB, Hall JC, Halpain S, Kay SA (1996) Green fluorescent protein and its derivatives as versatile markers for gene expression in living Drosophila melanogaster, plant and mammalian cells. Gene 173(1 Spec No):83–87
78. Khmelinskii A, Keller PJ, Bartosik A, Meurer M, Barry JD, Mardin BR, Kaufmann A, Trautmann S, Wachsmuth M, Pereira G, Huber W, Schiebel E, Knop M (2012) Tandem fluorescent protein timers for in vivo analysis of protein dynamics. Nat Biotechnol 30(7):708–714. doi:10.1038/nbt.2281
79. Lippincott-Schwartz J, Altan-Bonnet N, Patterson GH (2003) Photobleaching and photoactivation: following protein dynamics in living cells. Nat Cell Biol 5(Suppl):S7–S14
80. Miesenbock G, De Angelis DA, Rothman JE (1998) Visualizing secretion and synaptic transmission with pH-sensitive green fluorescent proteins. Nature 394(6689):192–195. doi:10.1038/28190
81. Johnson DE, Ai HW, Wong P, Young JD, Campbell RE, Casey JR (2009) Red fluorescent protein pH biosensor to detect concentrative nucleoside transport. J Biol Chem 284(31):20499–20511. doi:10.1074/jbc.M109.019042
82. Zhao Y, Araki S, Wu J, Teramoto T, Chang YF, Nakano M, Abdelfattah AS, Fujiwara M, Ishihara T, Nagai T, Campbell RE (2011) An expanded palette of genetically encoded Ca(2)(+) indicators. Science 333(6051):1888–1891. doi:10.1126/science.1208592

83. Belousov VV, Fradkov AF, Lukyanov KA, Staroverov DB, Shakhbazov KS, Terskikh AV, Lukyanov S (2006) Genetically encoded fluorescent indicator for intracellular hydrogen peroxide. Nat Methods 3(4):281–286. doi:10.1038/nmeth866
84. Dooley CT, Dore TM, Hanson GT, Jackson WC, Remington SJ, Tsien RY (2004) Imaging dynamic redox changes in mammalian cells with green fluorescent protein indicators. J Biol Chem 279(21):22284–22293. doi:10.1074/jbc.M312847200
85. Hanson GT, Aggeler R, Oglesbee D, Cannon M, Capaldi RA, Tsien RY, Remington SJ (2004) Investigating mitochondrial redox potential with redox-sensitive green fluorescent protein indicators. J Biol Chem 279(13):13044–13053. doi:10.1074/jbc.M312846200
86. Okumoto S, Jones A, Frommer WB (2012) Quantitative imaging with fluorescent biosensors. Annu Rev Plant Biol 63:663–706. doi:10.1146/annurev-arplant-042110-103745
87. De Michele R, Carimi F, Frommer WB (2014) Mitochondrial biosensors. Int J Biochem Cell Biol 48:39–44. doi:10.1016/j.biocel.2013.12.014
88. Johnston M, Riles L, Hegemann JH (2002) Gene disruption. Methods Enzymol 350: 290–315
89. Beggs JD (1978) Transformation of yeast by a replicating hybrid plasmid. Nature 275(5676): 104–109
90. Hinnen A, Hicks JB, Fink GR (1992) Transformation of yeast. Biotechnology 24:337–341
91. Clarke L, Carbon J (1980) Isolation of a yeast centromere and construction of functional small circular chromosomes. Nature 287(5782):504–509
92. Zhang Z, Moo-Young M, Chisti Y (1996) Plasmid stability in recombinant Saccharomyces cerevisiae. Biotechnol Adv 14(4):401–435
93. Clarke L (1998) Centromeres: proteins, protein complexes, and repeated domains at centromeres of simple eukaryotes. Curr Opin Genet Dev 8(2):212–218
94. Siam R, Dolan WP, Forsburg SL (2004) Choosing and using Schizosaccharomyces pombe plasmids. Methods 33(3):189–198. doi:10.1016/j.ymeth.2003.11.013
95. Hayles J, Nurse P (1992) Genetics of the fission yeast Schizosaccharomyces pombe. Annu Rev Genet 26:373–402. doi:10.1146/annurev.ge.26.120192.002105
96. Clyne RK, Kelly TJ (1995) Genetic analysis of an ARS element from the fission yeast Schizosaccharomyces pombe. EMBO J 14(24): 6348–6357
97. Dubey DD, Kim SM, Todorov IT, Huberman JA (1996) Large, complex modular structure of a fission yeast DNA replication origin. Curr Biol 6(4):467–473
98. Brachmann CB, Davies A, Cost GJ, Caputo E, Li J, Hieter P, Boeke JD (1998) Designer deletion strains derived from Saccharomyces cerevisiae S288C: a useful set of strains and plasmids for PCR-mediated gene disruption and other applications. Yeast 14(2):115–132. doi:10.1002/(SICI)1097-0061(19980130)14:2<115::AID-YEA204>3.0.CO;2-2
99. Commonly used auxotrophic markers. Retrieved from http://wiki.yeastgenome.org/index.php/Commonly_used_auxotrophic_markers
100. Fujita Y, Giga-Hama Y, Takegawa K (2005) Development of a genetic transformation system using new selectable markers for fission yeast Schizosaccharomyces pombe. Yeast 22(3):193–202. doi:10.1002/yea.1201
101. Ma Y, Sugiura R, Saito M, Koike A, Sio SO, Fujita Y, Takegawa K, Kuno T (2007) Six new amino acid-auxotrophic markers for targeted gene integration and disruption in fission yeast. Curr Genet 52(2):97–105. doi:10.1007/s00294-007-0142-1
102. Conrad M, Schothorst J, Kankipati HN, Van Zeebroeck G, Rubio-Texeira M, Thevelein JM (2014) Nutrient sensing and signaling in the yeast Saccharomyces cerevisiae. FEMS Microbiol Rev 38(2):254–299. doi:10.1111/1574-6976.12065
103. Pronk JT (2002) Auxotrophic yeast strains in fundamental and applied research. Appl Environ Microbiol 68(5):2095–2100
104. Chee MK, Haase SB (2012) New and redesigned pRS plasmid shuttle vectors for genetic manipulation of Saccharomyces cerevisiae. G3 2(5):515–526. doi:10.1534/g3.111.001917
105. Frazer LN, O'Keefe RT (2007) A new series of yeast shuttle vectors for the recovery and identification of multiple plasmids from Saccharomyces cerevisiae. Yeast 24(9):777–789. doi:10.1002/yea.1509
106. Goldstein AL, McCusker JH (1999) Three new dominant drug resistance cassettes for gene disruption in Saccharomyces cerevisiae. Yeast 15(14):1541–1553. doi:10.1002/(SICI)1097-0061(199910)15:14<1541::AID-YEA476>3.0.CO;2-K
107. Hentges P, Van Driessche B, Tafforeau L, Vandenhaute J, Carr AM (2005) Three novel antibiotic marker cassettes for gene disruption and marker switching in Schizosaccharomyces

pombe. Yeast 22(13):1013–1019. doi:10.1002/yea.1291
108. Sato M, Dhut S, Toda T (2005) New drug-resistant cassettes for gene disruption and epitope tagging in Schizosaccharomyces pombe. Yeast 22(7):583–591. doi:10.1002/yea.1233
109. Boeke JD, LaCroute F, Fink GR (1984) A positive selection for mutants lacking orotidine-5′-phosphate decarboxylase activity in yeast: 5-fluoro-orotic acid resistance. Mol Gen Genet 197(2):345–346
110. Toyn JH, Gunyuzlu PL, White WH, Thompson LA, Hollis GF (2000) A counterselection for the tryptophan pathway in yeast: 5-fluoroanthranilic acid resistance. Yeast 16(6):553–560. doi:10.1002/(SICI)1097-0061(200004)16:6<553::AID-YEA554>3.0.CO;2-7
111. Zaret KS, Sherman F (1985) alpha-Aminoadipate as a primary nitrogen source for Saccharomyces cerevisiae mutants. J Bacteriol 162(2):579–583
112. Urano J, Tabancay AP, Yang W, Tamanoi F (2000) The Saccharomyces cerevisiae Rheb G-protein is involved in regulating canavanine resistance and arginine uptake. J Biol Chem 275(15):11198–11206
113. Wilkie D (1970) Selective inhibition of mitochondrial synthesis in Saccharomyces cerevisiae by canavanine. J Mol Biol 47(1): 107–113
114. Lemontt JF (1977) Pathways of ultraviolet mutability in Saccharomyces cerevisiae. IV The relation between canavanine toxicity and ultraviolet mutability to canavanine resistance. Mutat Res 43(3):339–355
115. Sequencing Primers. Retrieved from http://www.addgene.org/mol-bio-reference/sequencing-primers/
116. Yofe I, Schuldiner M (2014) Primers-4-Yeast: a comprehensive web tool for planning primers for Saccharomyces cerevisiae. Yeast 31(2):77–80. doi:10.1002/yea.2998
117. Penkett CJ, Birtle ZE, Bahler J (2006) Simplified primer design for PCR-based gene targeting and microarray primer database: two web tools for fission yeast. Yeast 23(13):921–928. doi:10.1002/yea.1422
118. Manivasakam P, Weber SC, McElver J, Schiestl RH (1995) Micro-homology mediated PCR targeting in Saccharomyces cerevisiae. Nucleic Acids Res 23(14):2799–2800
119. Becker DM, Guarente L (1991) High-efficiency transformation of yeast by electroporation. Methods Enzymol 194:182–187
120. Forsburg SL (2003) Introduction of DNA into S. pombe cells. In: Frederick M. Ausubel et al. Current protocols in molecular biology. Chapter 13:Unit 13 17. doi:10.1002/0471142727.mb1317s64
121. Sheff MA, Thorn KS (2004) Optimized cassettes for fluorescent protein tagging in Saccharomyces cerevisiae. Yeast 21(8):661–670. doi:10.1002/yea.1130
122. Lee S, Lim WA, Thorn KS (2013) Improved blue, green, and red fluorescent protein tagging vectors for S. cerevisiae. PLoS One 8(7), e67902. doi:10.1371/journal.pone.0067902
123. Shaner NC, Steinbach PA, Tsien RY (2005) A guide to choosing fluorescent proteins. Nat Methods 2(12):905–909. doi:10.1038/nmeth819
124. Billinton N, Knight AW (2001) Seeing the wood through the trees: a review of techniques for distinguishing green fluorescent protein from endogenous autofluorescence. Anal Biochem 291(2):175–197. doi:10.1006/abio.2000.5006
125. Smirnov MN, Smirnov VN, Budowsky EI, Inge-Vechtomov SG, Serebrjakov NG (1967) Red pigment of adenine-deficient yeast Saccharomyces cerevisiae. Biochem Biophys Res Commun 27(3):299–304
126. Weisman LS, Bacallao R, Wickner W (1987) Multiple methods of visualizing the yeast vacuole permit evaluation of its morphology and inheritance during the cell cycle. J Cell Biol 105(4):1539–1547
127. Drepper T, Huber R, Heck A, Circolone F, Hillmer AK, Buchs J, Jaeger KE (2010) Flavin mononucleotide-based fluorescent reporter proteins outperform green fluorescent protein-like proteins as quantitative in vivo real-time reporters. Appl Environ Microbiol 76(17):5990–5994. doi:10.1128/AEM.00701-10
128. Bogdanov AM, Bogdanova EA, Chudakov DM, Gorodnicheva TV, Lukyanov S, Lukyanov KA (2009) Cell culture medium affects GFP photostability: a solution. Nat Methods 6(12):859–860. doi:10.1038/nmeth1209-859
129. Snaith HA, Anders A, Samejima I, Sawin KE (2010) New and old reagents for fluorescent protein tagging of microtubules in fission yeast; experimental and critical evaluation. Methods Cell Biol 97:147–172. doi:10.1016/S0091-679X(10)97009-X
130. Costantini LM, Fossati M, Francolini M, Snapp EL (2012) Assessing the tendency of fluorescent proteins to oligomerize under physiologic conditions. Traffic 13(5):643–649.doi:10.1111/j.1600-0854.2012.01336.x
131. Katayama H, Yamamoto A, Mizushima N, Yoshimori T, Miyawaki A (2008) GFP-like proteins stably accumulate in lysosomes. Cell Struct Funct 33(1):1–12

132. Prescott M, Nowakowski S, Nagley P, Devenish RJ (1999) The length of polypeptide linker affects the stability of green fluorescent protein fusion proteins. Anal Biochem 273(2):305–307. doi:10.1006/abio.1999.4235

133. Nadrigny F, Rivals I, Hirrlinger PG, Koulakoff A, Personnaz L, Vernet M, Allioux M, Chaumeil M, Ropert N, Giaume C, Kirchhoff F, Oheim M (2006) Detecting fluorescent protein expression and co-localisation on single secretory vesicles with linear spectral unmixing. Eur Biophy J 35(6):533–547. doi:10.1007/s00249-005-0040-8

134. Waters JC (2009) Accuracy and precision in quantitative fluorescence microscopy. J Cell Biol 185(7):1135–1148. doi:10.1083/jcb.200903097

135. Viana MP, Lim S, Rafelski SM (2015) Quantifying mitochondrial content in living cells. Methods Cell Biol 125:77–93

136. Tischer C, Brunner D, Dogterom M (2009) Force- and kinesin-8-dependent effects in the spatial regulation of fission yeast microtubule dynamics. Mol Syst Biol 5:250. doi:10.1038/msb.2009.5

137. Cormack BP, Bertram G, Egerton M, Gow NA, Falkow S, Brown AJ (1997) Yeast-enhanced green fluorescent protein (yEGFP): a reporter of gene expression in Candida albicans. Microbiology 143(Pt 2):303–311

138. Crameri A, Whitehorn EA, Tate E, Stemmer WP (1996) Improved green fluorescent protein by molecular evolution using DNA shuffling. Nat Biotechnol 14(3):315–319. doi:10.1038/nbt0396-315

139. Hiraoka Y, Kawamata K, Haraguchi T, Chikashige Y (2009) Codon usage bias is correlated with gene expression levels in the fission yeast Schizosaccharomyces pombe. Genes Cells 14(4):499–509. doi:10.1111/j.1365-2443.2009.01284.x

140. Janke C, Magiera MM, Rathfelder N, Taxis C, Reber S, Maekawa H, Moreno-Borchart A, Doenges G, Schwob E, Schiebel E, Knop M (2004) A versatile toolbox for PCR-based tagging of yeast genes: new fluorescent proteins, more markers and promoter substitution cassettes. Yeast 21(11):947–962. doi:10.1002/yea.1142

141. Knop M, Siegers K, Pereira G, Zachariae W, Winsor B, Nasmyth K, Schiebel E (1999) Epitope tagging of yeast genes using a PCR-based strategy: more tags and improved practical routines. Yeast 15(10B):963–972. doi:10.1002/(SICI)1097-0061(199907)15:10B<963::AID-YEA399>3.0.CO;2-W

142. Young CL, Raden DL, Caplan JL, Czymmek KJ, Robinson AS (2012) Cassette series designed for live-cell imaging of proteins and high-resolution techniques in yeast. Yeast 29(3-4):119–136. doi:10.1002/yea.2895

143. Gauss R, Trautwein M, Sommer T, Spang A (2005) New modules for the repeated internal and N-terminal epitope tagging of genes in Saccharomyces cerevisiae. Yeast 22(1):1–12. doi:10.1002/yea.1187

144. Forsburg SL (1994) Codon usage table for Schizosaccharomyces pombe. Yeast 10(8):1045–1047. doi:10.1002/yea.320100806

145. Kurt Thorn Lab Plasmids. http://www.addgene.org/Kurt_Thorn/

146. Wei-Lih Lee Lab Plasmids. http://www.addgene.org/browse/pi/280/

147. Alberti S, Gitler AD, Lindquist S (2007) A suite of Gateway cloning vectors for high-throughput genetic analysis in Saccharomyces cerevisiae. Yeast 24(10):913–919. doi:10.1002/yea.1502

148. Hailey DW, Davis TN, Muller EG (2002) Fluorescence resonance energy transfer using color variants of green fluorescent protein. Methods Enzymol 351:34–49

149. Reid RJ, Lisby M, Rothstein R (2002) Cloning-free genome alterations in Saccharomyces cerevisiae using adaptamer-mediated PCR. Methods Enzymol 350:258–277

150. Vorvis C, Markus SM, Lee WL (2008) Photoactivatable GFP tagging cassettes for protein-tracking studies in the budding yeast Saccharomyces cerevisiae. Yeast 25(9):651–659. doi:10.1002/yea.1611

151. Funk M, Niedenthal R, Mumberg D, Brinkmann K, Ronicke V, Henkel T (2002) Vector systems for heterologous expression of proteins in Saccharomyces cerevisiae. Methods Enzymol 350:248–257

152. Hell SW (2007) Far-field optical nanoscopy. Science 316(5828):1153–1158. doi:10.1126/science.1137395

153. Jansen G, Wu C, Schade B, Thomas DY, Whiteway M (2005) Drag&Drop cloning in yeast. Gene 344:43–51. doi:10.1016/j.gene.2004.10.016

154. Wach A, Brachat A, Alberti-Segui C, Rebischung C, Philippsen P (1997) Heterologous HIS3 marker and GFP reporter modules for PCR-targeting in Saccharomyces cerevisiae. Yeast 13(11):1065–1075. doi:10.1002/(SICI)1097-0061(19970915)13:11<1065::AID-YEA159>3.0.CO;2-K

155. Patino MM, Liu JJ, Glover JR, Lindquist S (1996) Support for the prion hypothesis for

inheritance of a phenotypic trait in yeast. Science 273(5275):622–626

156. McLellan CA, Whitesell L, King OD, Lancaster AK, Mazitschek R, Lindquist S (2012) Inhibiting GPI anchor biosynthesis in fungi stresses the endoplasmic reticulum and enhances immunogenicity. ACS Chem Biol 7(9):1520–1528. doi:10.1021/cb300235m
157. Bahler J, Wu JQ, Longtine MS, Shah NG, McKenzie A 3rd, Steever AB, Wach A, Philippsen P, Pringle JR (1998) Heterologous modules for efficient and versatile PCR-based gene targeting in Schizosaccharomyces pombe. Yeast 14(10):943–951. doi:10.1002/(SICI)1097-0061(199807)14:10 <943::AID-YEA292>3.0.CO;2-Y
158. Snaith HA, Samejima I, Sawin KE (2005) Multistep and multimode cortical anchoring of tea1p at cell tips in fission yeast. EMBO J 24(21):3690–3699. doi:10.1038/sj.emboj.7600838
159. Gadaleta MC, Iwasaki O, Noguchi C, Noma K, Noguchi E (2013) New vectors for epitope tagging and gene disruption in Schizosaccharomyces pombe. Biotechniques 55(5):257–263. doi:10.2144/000114100
160. Van Driessche B, Tafforeau L, Hentges P, Carr AM, Vandenhaute J (2005) Additional vectors for PCR-based gene tagging in Saccharomyces cerevisiae and Schizosaccharomyces pombe using nourseothricin resistance. Yeast 22(13):1061–1068. doi:10.1002/yea.1293
161. Zhao Y, Elder RT, Chen M, Cao J (1998) Fission yeast expression vectors adapted for positive identification of gene insertion and green fluorescent protein fusion. BioTechniques 25(3):438–440, 442, 444
162. Craven RA, Griffiths DJ, Sheldrick KS, Randall RE, Hagan IM, Carr AM (1998) Vectors for the expression of tagged proteins in Schizosaccharomyces pombe. Gene 221(1):59–68
163. Watt S, Mata J, Lopez-Maury L, Marguerat S, Burns G, Bahler J (2008) urg1: a uracil-regulatable promoter system for fission yeast with short induction and repression times. PLoS One 3(1):e1428. doi:10.1371/journal.pone.0001428
164. Pasion SG, Forsburg SL (1999) Nuclear localization of Schizosaccharomyces pombe Mcm2/Cdc19p requires MCM complex assembly. Mol Biol Cell 10(12):4043–4057
165. Fennessy D, Grallert A, Krapp A, Cokoja A, Bridge AJ, Petersen J, Patel A, Tallada VA, Boke E, Hodgson B, Simanis V, Hagan IM (2014) Extending the Schizosaccharomyces pombe molecular genetic toolbox. PLoS One 9(5), e97683. doi:10.1371/journal.pone.0097683
166. Habuchi S, Ando R, Dedecker P, Verheijen W, Mizuno H, Miyawaki A, Hofkens J (2005) Reversible single-molecule photoswitching in the GFP-like fluorescent protein Dronpa. Proc Natl Acad Sci U S A 102(27):9511–9516. doi:10.1073/pnas.0500489102
167. Habuchi S, Dedecker P, Hotta J, Flors C, Ando R, Mizuno H, Miyawaki A, Hofkens J (2006) Photo-induced protonation/deprotonation in the GFP-like fluorescent protein Dronpa: mechanism responsible for the reversible photoswitching. Photochem Photobiol Sci 5(6):567–576. doi:10.1039/b516339k
168. Patterson GH, Lippincott-Schwartz J (2002) A photoactivatable GFP for selective photolabeling of proteins and cells. Science 297(5588):1873–1877. doi:10.1126/science.1074952
169. Subach FV, Patterson GH, Manley S, Gillette JM, Lippincott-Schwartz J, Verkhusha VV (2009) Photoactivatable mCherry for high-resolution two-color fluorescence microscopy. Nat Methods 6(2):153–159. doi:10.1038/nmeth.1298
170. Subach FV, Patterson GH, Renz M, Lippincott-Schwartz J, Verkhusha VV (2010) Bright monomeric photoactivatable red fluorescent protein for two-color super-resolution spt-PALM of live cells. J Am Chem Soc 132(18):6481–6491. doi:10.1021/ja100906g
171. Verkhusha VV, Sorkin A (2005) Conversion of the monomeric red fluorescent protein into a photoactivatable probe. Chem Biol 12(3):279–285. doi:10.1016/j.chembiol.2005.01.005
172. Chudakov DM, Verkhusha VV, Staroverov DB, Souslova EA, Lukyanov S, Lukyanov KA (2004) Photoswitchable cyan fluorescent protein for protein tracking. Nat Biotechnol 22(11):1435–1439. doi:10.1038/nbt1025
173. Chudakov DM, Belousov VV, Zaraisky AG, Novoselov VV, Staroverov DB, Zorov DB, Lukyanov S, Lukyanov KA (2003) Kindling fluorescent proteins for precise in vivo photolabeling. Nat Biotechnol 21(2):191–194. doi:10.1038/nbt778
174. Matsuda T, Miyawaki A, Nagai T (2008) Direct measurement of protein dynamics inside cells using a rationally designed photoconvertible protein. Nat Methods 5(4):339–345. doi:10.1038/nmeth.1193
175. Ando R, Hama H, Yamamoto-Hino M, Mizuno H, Miyawaki A (2002) An optical

marker based on the UVinduced green-to-red photoconversion of a fluorescent protein. Proc Natl Acad Sci U S A 99(20):12651–12656. doi:10.1073/pnas.202320599

176. Tsutsui H, Karasawa S, Shimizu H, Nukina N, Miyawaki A (2005) Semi-rational engineering of a coral fluorescent protein into an efficient highlighter. EMBO Rep 6(3):233–238. doi:10.1038/sj.embor.7400361

177. Habuchi S, Tsutsui H, Kochaniak AB, Miyawaki A, van Oijen AM (2008) mKikGR, a monomeric photoswitchable fluorescent protein. PLoS One 3(12), e3944. doi:10.1371/journal.pone.0003944

178. McKinney SA, Murphy CS, Hazelwood KL, Davidson MW, Looger LL (2009) A bright and photostable photoconvertible fluorescent protein. Nat Methods 6(2):131–133. doi:10.1038/nmeth.1296

179. Wiedenmann J, Ivanchenko S, Oswald F, Schmitt F, Rocker C, Salih A, Spindler KD, Nienhaus GU (2004) EosFP, a fluorescent marker protein with UV-inducible green-to-red fluorescence conversion. Proc Natl Acad Sci U S A 101(45):15905–15910. doi:10.1073/pnas.0403668101

180. Gurskaya NG, Verkhusha VV, Shcheglov AS, Staroverov DB, Chepurnykh TV, Fradkov AF, Lukyanov S, Lukyanov KA (2006) Engineering of a monomeric green-to-red photoactivatable fluorescent protein induced by blue light. Nat Biotechnol 24(4):461–465. doi:10.1038/nbt1191

181. Zhang M, Chang H, Zhang Y, Yu J, Wu L, Ji W, Chen J, Liu B, Lu J, Liu Y, Zhang J, Xu P, Xu T (2012) Rational design of true monomeric and bright photoactivatable fluorescent proteins. Nat Methods 9(7):727–729. doi:10.1038/nmeth.2021

182. Shimozono S, Hosoi H, Mizuno H, Fukano T, Tahara T, Miyawaki A (2006) Concatenation of cyan and yellow fluorescent proteins for efficient resonance energy transfer. Biochemistry 45(20):6267–6271. doi:10.1021/bi060093i

183. Subach OM, Patterson GH, Ting LM, Wang Y, Condeelis JS, Verkhusha VV (2011) A photoswitchable orange-to-far-red fluorescent protein, PSmOrange. Nat Methods 8(9):771–777. doi:10.1038/nmeth.1664

184. Ando R, Mizuno H, Miyawaki A (2004) Regulated fast nucleocytoplasmic shuttling observed by reversible protein highlighting. Science 306(5700):1370–1373. doi:10.1126/science.1102506

185. Flors C, Hotta J, Uji-i H, Dedecker P, Ando R, Mizuno H, Miyawaki A, Hofkens J (2007) A stroboscopic approach for fast photoactivation-localization microscopy with Dronpa mutants. J Am Chem Soc 129(45):13970–13977. doi:10.1021/ja074704l

186. Stiel AC, Trowitzsch S, Weber G, Andresen M, Eggeling C, Hell SW, Jakobs S, Wahl MC (2007) 1.8 A bright-state structure of the reversibly switchable fluorescent protein Dronpa guides the generation of fast switching variants. Biochem J 402(1):35–42. doi:10.1042/BJ20061401

187. Andresen M, Stiel AC, Folling J, Wenzel D, Schonle A, Egner A, Eggeling C, Hell SW, Jakobs S (2008) Photoswitchable fluorescent proteins enable monochromatic multilabel imaging and dual color fluorescence nanoscopy. Nat Biotechnol 26(9):1035–1040. doi:10.1038/nbt.1493

188. Henderson JN, Ai HW, Campbell RE, Remington SJ (2007) Structural basis for reversible photobleaching of a green fluorescent protein homologue. Proc Natl Acad Sci U S A 104(16):6672–6677. doi:10.1073/pnas.0700059104

189. Bizzarri R, Arcangeli C, Arosio D, Ricci F, Faraci P, Cardarelli F, Beltram F (2006) Development of a novel GFP-based ratiometric excitation and emission pH indicator for intracellular studies. Biophys J 90(9):3300–3314. doi:10.1529/biophysj.105.074708

190. Adam V, Lelimousin M, Boehme S, Desfonds G, Nienhaus K, Field MJ, Wiedenmann J, McSweeney S, Nienhaus GU, Bourgeois D (2008) Structural characterization of IrisFP, an optical highlighter undergoing multiple photo-induced transformations. Proc Natl Acad Sci U S A 105(47):18343–18348. doi:10.1073/pnas.0805949105

191. Betzig E, Patterson GH, Sougrat R, Lindwasser OW, Olenych S, Bonifacino JS, Davidson MW, Lippincott-Schwartz J, Hess HF (2006) Imaging intracellular fluorescent proteins at nanometer resolution. Science 313(5793):1642–1645. doi:10.1126/science.1127344

192. Egner A, Geisler C, von Middendorff C, Bock H, Wenzel D, Medda R, Andresen M, Stiel AC, Jakobs S, Eggeling C, Schonle A, Hell SW (2007) Fluorescence nanoscopy in whole cells by asynchronous localization of photoswitching emitters. Biophys J 93(9):3285–3290. doi:10.1529/biophysj.107.112201

193. Hess ST, Girirajan TP, Mason MD (2006) Ultra-high resolution imaging by fluorescence photoactivation localization microscopy. Biophys J 91(11):4258–4272. doi:10.1529/biophysj.106.091116

194. Sikorski RS, Hieter P (1989) A system of shuttle vectors and yeast host strains designed for efficient manipulation of DNA in Saccharomyces cerevisiae. Genetics 122(1):19–27

195. Christianson TW, Sikorski RS, Dante M, Shero JH, Hieter P (1992) Multifunctional yeast high-copy-number shuttle vectors. Gene 110(1):119–122

196. Voth WP, Jiang YW, Stillman DJ (2003) New 'marker swap' plasmids for converting selectable markers on budding yeast gene disruptions and plasmids. Yeast 20(11):985–993. doi:10.1002/yea.1018

197. Cross FR (1997) 'Marker swap' plasmids: convenient tools for budding yeast molecular genetics. Yeast 13(7):647–653. doi:10.1002/(SICI)1097-0061(19970615)13:7<647::AID-YEA115>3.0.CO;2-#

198. Longtine MS, McKenzie A III, Demarini DJ, Shah NG, Wach A, Brachat A, Philippsen P, Pringle JR (1998) Additional modules for versatile and economical PCR-based gene deletion and modification in Saccharomyces cerevisiae. Yeast 14(10):953–961. doi:10.1002/(SICI)1097-0061(199807)14:10<953::AID-YEA293>3.0.CO;2-U

199. Goldstein AL, Pan X, McCusker JH (1999) Heterologous URA3MX cassettes for gene replacement in Saccharomyces cerevisiae. Yeast 15(6):507–511. doi:10.1002/(SICI)1097-0061(199904)15:6<507::AID-YEA369>3.0.CO;2-P

200. Alani E, Cao L, Kleckner N (1987) A method for gene disruption that allows repeated use of URA3 selection in the construction of multiply disrupted yeast strains. Genetics 116(4):541–545

201. Wach A, Brachat A, Pohlmann R, Philippsen P (1994) New heterologous modules for classical or PCR-based gene disruptions in Saccharomyces cerevisiae. Yeast 10(13):1793–1808

202. Chino A, Watanabe K, Moriya H (2010) Plasmid construction using recombination activity in the fission yeast Schizosaccharomyces pombe. PLoS One 5(3), e9652. doi:10.1371/journal.pone.0009652

Chapter 22

Visualization and Image Analysis of Yeast Cells

Steve Bagley

Abstract

When converting real-life data via visualization to numbers and then onto statistics the whole system needs to be considered so that conversion from the analogue to the digital is accurate and repeatable. Here we describe the points to consider when approaching yeast cell analysis visualization, processing, and analysis of a population by screening techniques.

Key words Microscopy, Visualization, Processing, Screening, Analysis

1 Choosing the Correct Hardware

When considering the analysis of fixed or live biological structures through a microscope, the nature of the system, the quality of the optics, the light source, illumination and detection technique, and the filters are of utmost importance and are often overlooked by the researcher. Consequently it is important to understand the fundamental design of the light microscope in order to achieve suitable signal but also in a standardized and repeatable manner.

The choice of fluorescence visualisation platform whether wide-field, confocal (and its variants), high-throughput/content screening or super-resolution is critical and will have impact on the study as a whole. Subsequently it is customary to employ multiple imaging platforms to describe biological structures using the advantages and circumventing the inherent disadvantages of each system.

1.1 Wide-Field Microscopy

A UV wide-field microscope presents advantages to the researcher such as a wide range of possible labels, sensitivity, and the ability to readily replace equipment when new technological advances present themselves. As UV imaging is filter based, the excitation light which passes from the bulb through a filter is often a range of wavelengths, commonly in the range of 30–100 nm (for example standard green fluorescent protein filter pass light from 470 to 494 nm);

Alberto Sanchez-Diaz and Pilar Perez (eds.), *Yeast Cytokinesis: Methods and Protocols*, Methods in Molecular Biology, vol. 1369, DOI 10.1007/978-1-4939-3145-3_22,

hence contamination of the final signal via autofluorescence or cross contamination of other fluorophores, know as bleed-through, is common and must be accounted for.

Light sources for wide-field imaging traditional utilized halogen (also tungsten-halogen), xenon, and mercury (also known as an HBO or a UV burner) bulbs which had limited life-span and a nonuniform spectral response. Commonplace in wide-field microscope is the metal-halide, with the advantage of high light output, or light-emitting diode (LED) illumination that present to the cell discrete wavelengths.

1.2 Confocal Microscopy

Confocal methods produce three-dimensional data sets by the use of a highly focused light source, the laser, and either scanning the laser across the sample in a raster fashion or to break up the light into discrete points via a spinning disk. Details about the development of the confocal microscope can be found in an informative memoir by Minsky [1]. Laser scanning confocal microscopy (LSCM) illumination is achieved by the manipulation of XY galvo-mirrors to produce a raster beam so to produce illumination across the field of view. The advantage of the confocal methods, depth discrimination by the rejection of signal at a point distant from the plane of focus is achieved by the pinhole, which in LSCM can be variable. Detection historically was via filters and photomultiplier tubes; however modern designs use spectrophotometric detection so to be able to select a user-defined proportion of the spectrum so to construct the output. Spinning Disk Confocal Microscopy (SDCM) illuminates and detects through a series of pinholes in two spinning disks. Consequently the laser light, unlike LSCM which is to a manipulated single spot, is employed to illuminate a field of view by creating a series of 356 discrete spots produced by the spinning disk. SDCM achieves an image with less of a light demand than LSCM and is also more rapid at image formation so more suited to live cell imaging.

1.3 High-Content/ Throughput Screening

High-content screening (HCS) methods commonly use UV wide-field optics; however improvements in design are now available which integrate a spinning disk confocal and automation so to allow faster screening of multi-well plates for multiple individual mutants, and differing biochemical environments. Screening techniques have been available since 1996 when the technologies were started to be used in drug discovery [2]. Whereas traditional HCS utilized low-magnification air lenses, systems are available which permit high-resolution, high-numerical-aperture water immersion lenses and consequently HCS is becoming a suitable technique for automated population analysis. As a consequence it is possible to screen a 384-well plate with multiple fields of view per well in less than 20 min under conditions that allow the elucidation of single cell events. To the researcher HCS allows for standardization, repeatability and automation so more complex investigations can be imaged and numerated.

1.4 Super-Resolution Methods

Super-resolution microscopy methods whether they are achieved by deterministic (STED, GSD, RESOLFT and SSIM) or stochastic super-resolution (all single-molecule localisation methods such as the variants of PALM and STORM) are beginning to allow for methods to visualize events at the molecular level. As super-resolution microscopy requires high magnification, and high electronic or optical zoom to achieve the required capture resolution, the technique lends itself to single cell rather than population analysis. Where these techniques are proving interesting is for achieve molecular levels of resolution for interaction studies and when combined with other imaging technologies in a correlative manner such as electron microscopy.

1.5 Transmitted Light Microscopy

When visualizing over time a variety of transmitted light techniques are employed such as differential interference contrast (DIC) or phase microscopy in conjunction with fluorescence imaging. Capture software can employ non-sequential imaging, for example utilizing transmitted light at every time point and fluorescence at every "nth" time point, thus omitting some time points from both transmitted and fluorescence visualization. This is advantageous to medium- and long-term studies as this methodology reduces the possibility of light induced damage or photo-bleaching of the fluorophore. Both DIC and phase microscopy suffer with light efficiency with DIC (due to the increase of optical components in the light path) or a halo artifact around the cells in the case of phase. Digital phase microscopy (DPM) is being to be employed as a more light efficient technique which presents transmitted light image that is reversed to what is normally produced as an output (i.e. white cells on a black background). A major advantage of DPM is producing an image that does not have the bright halo effect around the cells such as that often produced under phase microscopy, but also produces an image more suitable for tracking and hereditary studies.

1.6 Choice of Objective Lens

Whatever system imaging is employed in the laboratory, and quite often several methods are employed to describe biological phenomena, the optics are imperative. The objective lens that sits between the sample and the microscope is required to demonstrate resolution, faithfulness in both spatial and chromatic space, across a wide range of spectral frequencies, with both transmitted and fluorescence light, a flat field so peripheral distortion is not an issue, light efficient and inexpensive, unfortunately a lens that demonstrates all of these parameters is not available. So when choosing an, or a range of, objective lenses compromises may have to be made. For example the fewer the components within an objective lens the better the light efficiency but to achieve a greater resolution a heavily corrected lens is required which in turn effects light efficiency. Higher resolution lenses may achieve a signal with greater clarity and signal-to-noise ratio with a shallow depth of focus however there may be limits to the spectral range or greater peripheral distortion.

1.7 Detection

With any form of fluorescence microscopy the excitation light burden causes photo-bleaching of the fluorophore however if a decrease in single intensity is observed then photo-induced stress to the examined live cells has occurred. The light path of the microscope from sample through to the imaging device must be of high quality optics to allow maximum transmissivity of the required signal. When imaging live cells imaging an increase in sensitivity allows for a reduction of the excitation light intensity thus a reduction in the possibility of photo-induced stress. When selecting an appropriate camera for fluorescence wide-field, spinning disk confocal and super resolution imaging there are often compromises between capture speed, resolution, sensitivity, and cost.

1.7.1 CCD Camera

The charge-coupled device (CCD) has been available to the general scientific community since 1975 and is available in both grayscale and color versions. The incoming light interacts with a photo-active region which causes a capacitor to accumulate an electrical charge. Once the signal has been accumulated each capacitor passes its charge to the neighbor and so on, eventually dumping the charge to an amplifier which then converts the accumulated charge into a voltage. These systems are in expensive and suitable for transmitted light and general fluorescence applications but certainly not suitable for single cell investigations due to the low sensitivity and high background signal (noise).

1.7.2 Electron-Multiplying CCD Camera

The electron-multiplying CCD camera was introduced in the early 2000 which allowed high sensitivity so to be able to detect single photons. By the addition to the CCD technology of an electron multiplying register amplification of the signal without adding measurable noise to the final image. These cameras are commonly high sensitivity, with a frame rate of 10 images per second and a resolution of up to up 1024 × 1024 pixels

1.7.3 sCMOS Camera

The scientific-grade complementary metal–oxide–semiconductor (sCMOS) camera is based upon pipeline technologies used since the 1960s, the name of which refers to the circuitry on the integrated chips that is utilised. These devices are typified by low noise, high resolution (typically 2048 × 2048), rapid frame rates (100 frames per second), and large field of view.

1.7.4 Spectrophotometric Imaging

As the requirement for multiple labels within a single cell, their detection, separation, and quantification become paramount; hence the use of spectrophotometric detectors is becoming an attractive tool alongside standard camera technologies. As discussed previously the band-pass tolerances of UV filter-based systems lead to contamination of signal as labels cross into other imaging channels, known as bleed-through. Typically these spectrophotometric devices allow scanning through a spectral range from 400 to 700 nm at 10 nm intervals hence with software control only part of the total spectral output can be made to contribute towards the final image. Although image formation is slower than traditional camera systems and more suitable to fixed cell imaging, certainly data capture, separation and quantification of a high number of fluorophores becomes attainable.

1.7.5 Comparison of Detection Technologies

Although CCD has been included for completeness this technology is unsuitable for single-cell, low fluorescence signal investigations.

Spectrophotometric detection is more suited for fixed cell imaging under wide-field modes, due to a present lack of suitable software.

However when implemented well in confocal systems allows for rapid time-lapse investigations.

When comparing the detection efficiency of cameras the quantum efficiency (QE) is utilized, which is a measure of the efficacy of photons converted to an electronic signal. Also used as a camera of camera suitability is the Signal to Noise ratio (SNR), which compares the level of a signal to the level of background noise. With EMCCD and sCMOS technologies the typical QE is 90–95 % and 70–75 % conversely, whilst the SNR is higher with EMCCD technologies. However the improved frame rate and resolution of sCMOS is more suitable to screening and rapid time-lapse applications.

2 Visualization

Whatever the choice of equipment, optics, and camera there are a variety of conditions that are controlled by the user and will impact whether the conversion of the analogue biological data to digital is true or representative. Image intensity and contrast of the digital data are controlled by the amount of energy (light) being put into the system, the concentration of the labeling at the biology (which is controlled by the availability of the object being labeled, the quantity of object, and the techniques used at the bench for labeling), the efficiency of the hardware (microscope, optics, filters, as detailed above) and the user defined parameters set for image capture.

With almost every sample brought to the microscope it is quite simple to produce a data set that may be passable to the casual reviewer however when using a microscope choices need to be made that influence how faithful the final output is to biology and represents the real-life object. For the researcher to make these decisions experience must be gained in the wide range of phenotypes that may present themselves; consequently this can be summed up with, there is a whole range of minor phenotypic differences that can be described as "normal" cells. Capturing multiple fields of view to express the difference in the population is essential to fully characterise a biological difference or event. The device which is being used as the data recording device whether this be a camera, photo-multiplier, spectrophotometer, etc. must be set similarly across all samples, so that a response can be measured, sometimes this will result in sacrificing a good image for one that is representative of the range of cells being imaged. Similarly in addition to cell handling, preparation and labeling the conditions for visualisation must be recorded with every captured image. As a consequence data capture via high content screening which employs the same conditions across different cell mutants or treatments and records these imaging conditions in addition to the data is rapidly becoming the technique of choice. It is becoming a matter of course for data as well as imaging parameters to be included in preparation of a publication. The raw data, processed data, and comparative images are required so as to further proofs when preparing a publication for reviewers.

3 Live Cell Three-Dimensional Imaging

Due to the shallow depth of focus of high-resolution objective lenses, the ability to image over a discrete volume so as to capture all of the data and perhaps achieve this over time has become commonplace in the field. Three-dimensional imaging is achieved by manipulating the place of focus in relation to the sample, by moving the sample (via the stage) or the objective lens. However with live cell imaging in a three dimensional space over time the effects of photo-damage become cumulative which result in loss of signal or, more importantly in live cells, damage to the functioning cell. Reduction of the harmful effect of UV illuminating light for fluorescence microscopy can be lessened by

- The utilising of high-end light-efficient cameras so the light input burden can be decreased.
- A camera with rapid frame rates so to enable high-speed imaging.
- Automated and rapid light shuttering so to reduce needless light input.

- High-quality, high-efficiency fluorescence filters within the UV light path.
- If the camera allows binning can increase the sensitivity of the camera so to enable high frame rates with less required light. Binning is the process where the signal from a group of adjacent pixels on the image sensor are combined and assigned to a single pixel value.

As a consequence of many of the considerations list above sCMOS cameras (higher frame rates, larger sensor size) have become popular over EMCCD (higher sensitivity, but lower frame rates) in UV wide-field and SDCM applications.

4 Image Processing

Image processing is an offshoot from signal processing. With microscopy a real-life object is converted into a digitized model by conversion of a photon to an electronic signal, the strength of signal converted into a grayscale image (the contrast of which is governed by good laboratory and microscopy technique). Suitable image processing requires prior knowledge, so the conversion of the visualized image (the raw data) reflects the inherent cellular biology that has been observed (the model or processed data).

4.1 Image Processing via Deconvolution

When combining a stack of images flaws inherent within the optical system and how the biological sample is presented become evident when examining the final output. When imaging a volume of data as a cross section objects that should be discrete appear to be elongated, a consequence of a transform known as the point spread function (PSF). The PSF was first studied by Airy in the ninetieth century and then later in the 1930s by Zernike and Nijboer, who describes a transform that distorts and masks the data contained within an image. Suitable sampling of the data is critical which is controlled by the microscope optics, software control, and sample preparation.

By the application of image restoration techniques, such as image deconvolution, the data can be recovered from the visualised data, improves resolution, removes noise and increases the contrast (the signal to noise ratio). Modeling of the sample and optical system allows for image correction to be applied, so remove the mathematically described function which distorts the data, and the convolution; however the data must then be considered as a model of the data rather than the true measure of the data [3].

Methods such as image deconvolution correction are applied to correct for the PSF but due to the contrast enhancements in the processing algorithms the data intensities must not be compared across the same or between different datasets. Over the last 10 years image deconvolution embedded within data capture software has become commonplace however in a multi-microscope environment it is important for data consistency that the same algorithms be applied to all data sets whatever the microscope source; hence stand-alone server-based software is beneficial in a high-throughput laboratory that requires standardization across multiple microscope systems. It must be noted that the original and processed data should be stored for publication purposes, as image deconvolution alters intensity and reassigns spatial location.

A suggested workflow for suitable sample preparation, microscopy, and processing is outlined below

1. For sample preparation ensure that all cover slips are optical grade and are No. 1.5 grade (0.16–0.19 mm thick).
2. If the sample preparation has been in an environment of a temperature differential to the microscope, ensure that condensation has not resulted.
3. Ensure that all microscopes optics are clean and dust free.
4. Measure the refractive index of a sample of the mounting media that surrounds the sample (media, antifade, etc.) with a simple hand held refractometer.
5. Obtain the lens medium refractive index (if using immersion oil often the manufacturer will quote on the bottle, which is usually measured at 23 °C). As refractive index is directly related to temperature, if visualising cells at a different temperature than stated by the manufacturer it is advisable to measure using a refractometer.
6. Collection of the data at a suitable sampling frequency. When imaging Green Fluorescent Protein (GFP) using a ×100 objective lens with a NA of 1.4 as a guide please consult Table 1.
7. Software is available that will calculate the convolution, how the data has been transformed due to the change in refractive

Table 1
Suitable sample frequencies when imaging Green fluorescent Protein, calculation from the Scientific Volume Imaging (see useful websites)

	Sampling frequency (nm)	
Microscopy	**Lateral (*xy*)**	**Axial (*z*)**
Wide field	90	270
Laser scanning confocal	40	100
Spinning disk confocal	43	130
gSTED	19	100

index and optics, from the original data. However for it may be advisable to check using fluorescent beads that can be considered as a point source and of know geometry (less than 175 nm in size).

8. Once the convolution is calculated this is then applied to the original data resulting in a model of the original data (*see* Fig. 1).

4.2 General Image Processing

Imaging processing can be described as a defined process, which may make features that may be masked or hidden in the original data set discernable. As detailed previously image deconvolution

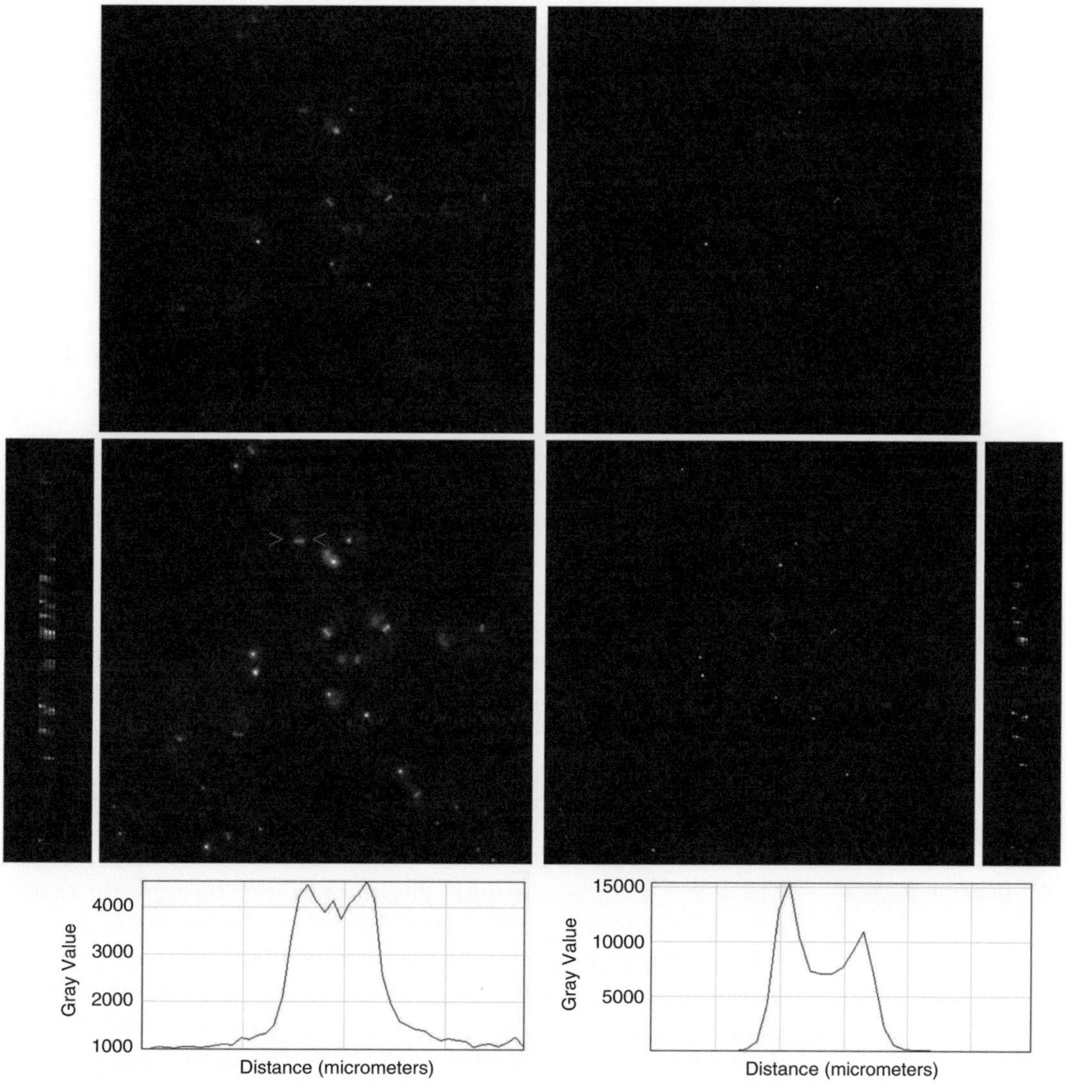

Fig. 1 Image deconvolution demonstrating the blurring function or convolution that has been applied during data collection and the data's deconvolution illustrating the reassigning of pixel location in three-dimensional space (original microscopy data supplied by Karim Labib and Alberto Sanchez-Diaz)

is a mathematical based form of image processing. When employing any form of image processing the raw data and final data must be archived along with the procedures utilised to achieve the final image.

The processing must employ mathematically defined tools rather than arbitrary sliders, not remove or mask data from the final data set but only improve clarity. For example a macro can be written in a popular graphics program so that the shadows, midtones, and highlights could be manipulated so as to remove faint structures from an image. Although an argument could be formed that if this processing is carried out across all data sets then it could be considered fair use, the processing has removed data, be it unwanted data, from the original visualised data set hence the process should be considered unsuitable for scientific data. All image processing must be repeatable, mathematically and experimentally justifiable, applied across all data sets, and applied to the same extent.

5 Image Reconstruction

Many of the methods for image reconstruction were derived when processing hardware was costly and computer time long; thus shortcuts were conceived to cut down on the hardware and time requirement. Projection methods such as the commonly used maximum projection reassign intensity so as to give an impression of depth. Consequently methods of reassigning pixels although can be useful should not be utilized for analysis, only visual inspection of the data. If volume data is to be analyzed then software can process in 3- and 4D space such as the open-source Fiji software and commercial software (see useful websites). However reconstruction of the data although visually rewarding and aiding interpretation is not essential using 2D software and quantifying each plane of the 3D data space in a batch processing mode will often yield a more suitable and accurate result.

6 Image Analysis

6.1 General Analysis

The term of image analysis describes where meaningful extraction of numerical data from a global or scene parameters, or measuring values for each individual feature that is present by digital image processing techniques. It must be remembered that the digital image is a pictorial methods of representing numerical data; however the extraction of meaningful data which can then be pipelined through to statistical analysis requires image processing and a large number of cells to extract data so that the statistical analysis becomes meaningful. The strength of applying image analysis over several fields of view, treatment or cell type, is that the population and the individual cell can be mined to examine variations in population and also to statistically validate the perceived biological phenomena.

To achieve object extraction and subsequently analysis some form of user based discrimination is to be applied which requires prior knowledge by the researcher [4], hence with any form of visualization and analysis getting to know the range of the biology present in the study is essential. Sampling a small part of the overall population requires experience and knowledge of the biological principles in action across all treatments and cell types; thus if it is possible to utilize high-content screening where many thousands of cells are imaged and assessed rather than sampling the population more satisfactory statistics result.

Discrimination of objects can be achieved via a suitable thresholding technique, which depends upon labelling intensity, suitable illumination across the field of view and image capture technique (wide field, confocal, etc.). Automated thresholding which utilizes a local area contrast scheme or discriminates above a certainly level from the background is applicable such as a Gaussian function or to assume that the difference between object and background can be classified above a certain point. Certainly if the illumination uniformity is not optimal, this will have a severe impact on the suitability of automated feature extraction.

6.2 Screening

High-content analysis (HCA) where processing, analysis, and statistical modeling has been perceived as a large-scale drug discovery endeavor using commercial software, has been traditionally directed towards mammalian cells. Techniques such as nuclear and cytoplasmic detection, translocation, presence or absence of labels, population dynamics, and live cell assessment are carried out on large-scale mammalian screens; however these procedures can be applied to a wide variety of cell types including yeast cells. Utilizing optical grade multi-well plates, an automated XYZ stage, and control over illumination and detection high-throughput/content phenotypic screening can be achieved. Certainly with free open source software such as CellProfiler or macros within software such as ImageJ and Fiji there is a wide range of 2D and 3D tools for screening populations.

A method of correcting illumination across a field, separation of objects and analysis can be straightforward depending upon the sample preparation, microscopy technique and image quality (resolution and contrast). Outlined below is a simple workflow that can be utilized for analysis and classification of a cell population.

1. Illumination correction, to reduce background fluorescence or to improve clarity.
2. Separation of objects of interest from the background via image threshold techniques, which converts the data into a

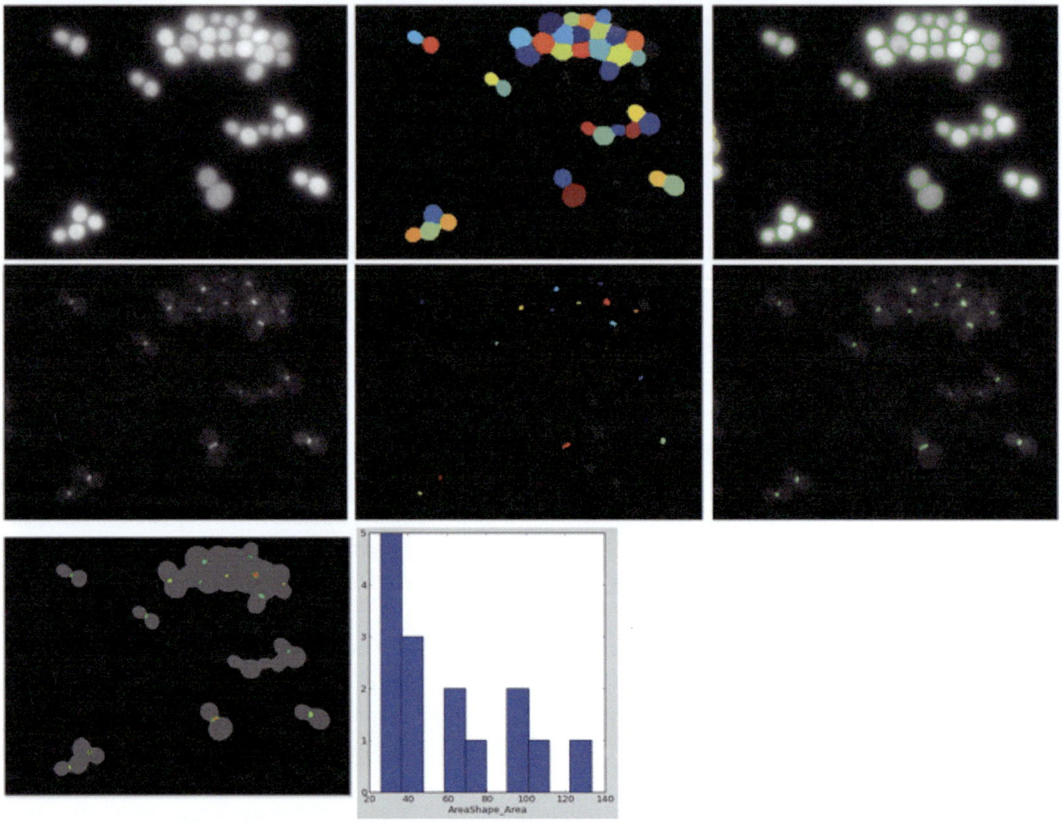

Fig. 2 Screening demonstrating the automated detection of the cell and the actomyosin ring (AMR). Using the cells as a mask the lower image shows the cell with associated AMR and profiles each cell. All data processing and analysis carried out within the CellProfiler software (original microscopy data supplied by Karim Labib and Alberto Sanchez-Diaz)

binary image. When using multiple labels the discrimination step of thresholding can be applied several times so as to extract different features from the original image.

3. Exclusion of cells that touch the edges of the image, so only whole cells are assessed.
4. Measure the fluorescence of a population of cells, singularly. Typically this step performs multiple measures of each discrete object of interest.
5. Creation of subpopulation depending upon one or several measurable characteristics or phenotype.
6. Score each image of separated features per cell.
7. Iterative machine learning, develop the scoring procedure across batches of cells so as to check the extraction technique and validity.
8. Automated scoring per cell, well, and treatment (*see* Figs. 2 and 3).

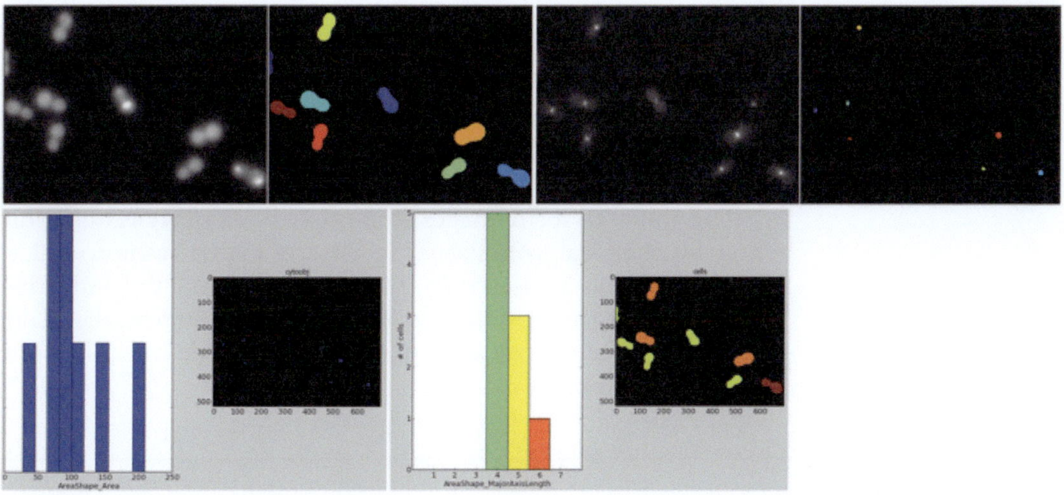

Fig. 3 Screening demonstrating the automated detection of the cell and the actomyosin ring (AMR) and numerating size of the AMR (area) and cell size (major axis length). All data processing and analysis carried out within the CellProfiler software (original microscopy data supplied by Karim Labib and Alberto Sanchez-Diaz)

7 Image Management

As the number of images increase so to describe a biological event or phenomena which require data processing, analysis, classification and statistical testing the requirement for data storage increases. As data storage requirement increases so does the necessity for data management. In the mid-1990s it was common for core facility laboratories or research laboratories that use a large amount of imaging to use optical disks that would contain 1 GB of data which in turn would contain a whole 3-year project. Commonly due to multi-plane, labeled, field, and treatment data a research project of 3 years can be contained in 500 Gb to 1 Tb; consequently data solutions in core facilities (storage, backup, and archive) require Peta bytes of data to contain not just the imaging data but from integrated molecular biology, mass spectrophotometry, and bioinformatics data streams.

Solutions such as Omero (Open Microscopy Environment) allow for the management of data streams, which is rapidly becoming essential for publication due to the necessity of showing the workflow from the raw data to the final draft of a paper.

8 Conclusion

The field of forming data using visual techniques such as microscopy is rapidly changing, however standardized cell preparation, good practice at the microscope, and well-defined image processing and analysis routines will always be the basis of data extraction and classification. Whereas capturing a few cells (under a hundred in some studies) and describing a biological phenomena or response to an external stimulus was often carried out, this is rapidly becoming unsuitable for the detection of rare or temporally fleeting events.

The ability to form images at multiple fields of view, under differing biochemical conditions, in an automated fashion and to apply algorithms to describe the population and the individual response is becoming essential so to produce statistically viable data as researchers move away from the single slide towards multi-chambers and multi-well plates.

9 Useful Websites

9.1 Free-Open-Source Software

CellProfiler

http://www.cellprofiler.org

Carpenter AE, Jones TR, Lamprecht MR, Clarke C, Kang IH, Friman O, Guertin DA, Chang JH, Lindquist RA, Moffat J, Golland P, Sabatini DM (2006). "CellProfiler: image analysis software for identifying and quantifying cell phenotypes". *Genome Biology* 7(10): R100

Fiji

http://fiji.sc/Fiji

Schindelin J, Arganda-Carreras I, Frise E et al. (2012). "Fiji: an open-source platform for biological-image analysis". *Nature methods* **9**(7): 676–682

ImageJ

http://imagej.nih.gov/ij/

Schneider CA, Rasband WS, Eliceiri KW (2012). "NIH Image to ImageJ: 25 years of image analysis". *Nature Methods* **9**(7): 671–675

Open Microscopy Environment

https://www.openmicroscopy.org/

Goldberg IG, Allan C, Burel J-M, Creager D, Falconi A, Hochheiser H, Johnston J, Mellen J, Sorger PK, Swedlow JR (2005). "The Open Microscopy Environment (OME) Data Model and XML File: open tools for informatics and quantitative analysis in biological imaging". *Genome Biology* **6**:R47

9.2 Commercial Software

Bitplane, Imaris

http://www.bitplane.com

Media Cybernetics, Image Pro Plus

http://www.mediacy.com

PerkinElmer, Volocity

http://www.perkinelmer.co.uk/catalog/category/id/imaging%20analysis%20software

Perkin Elmer Columbus
http://www.perkinelmer.co.uk/pages/020/cellularimaging/products/columbus.xhtml
Scientific Volume Imaging, Huygens
http://www.svi.nl/

Acknowledgment

This work was supported by Cancer Research UK.

References

1. Minsky M (1998) Memoir on inventing the confocal scanning microscope. Scanning 10: 128–138
2. Taylor DL (2010) A personal perspective on high-content screening (HCS): from the beginning. J Biomol Screen 15(7):720–725. doi:10.1177/1087057110374995
3. Booth MJ, Neil MA, Juskaitis R, Wilson T (2002) Adaptive aberration correction in a confocal microscope. Proc Natl Acad Sci U S A 99(9): 5788–5792. doi:10.1073/pnas.082544799
4. Wolf G (1991) Usage of global information and a priori knowledge for object isolation. Proc 8th Int Cong. Stereol. Irvine, CA, 56

Chapter 23

Toolbox for Protein Structure Prediction

Daniel Barry Roche and Liam James McGuffin

Abstract

Protein tertiary structure prediction algorithms aim to predict, from amino acid sequence, the tertiary structure of a protein. In silico protein structure prediction methods have become extremely important, as in vitro-based structural elucidation is unable to keep pace with the current growth of sequence databases due to high-throughput next-generation sequencing, which has exacerbated the gaps in our knowledge between sequences and structures.

Here we briefly discuss protein tertiary structure prediction, the biennial competition for the Critical Assessment of Techniques for Protein Structure Prediction (CASP) and its role in shaping the field. We also discuss, in detail, our cutting-edge web-server method IntFOLD2-TS for tertiary structure prediction. Furthermore, we provide a step-by-step guide on using the IntFOLD2-TS web server, along with some real world examples, where the IntFOLD server can and has been used to improve protein tertiary structure prediction and aid in functional elucidation.

Key words Protein tertiary structure prediction, Protein structure, Fold recognition, Template-based modeling, Template-free modeling, Critical Assessment of Techniques for Protein Structure Prediction (CASP), Bioinformatics web servers, Model quality assessment methods, Continuous Automated Model EvaluatiOn (CAMEO), Protein Model Portal (PMP), Protein Structure Initiative (PSI)

1 Introduction

Proteins are essential molecules in all organisms, playing key roles in maintaining the function and structural integrity of all living cells [1–3]. Thus, the determination of the tertiary structure of proteins within a system is an important step, which can aid in the elucidation of key cellular mechanisms and molecular functionality.

Current experimental techniques for structural determination, which include X-ray crystallography and nuclear magnetic resonance (NMR), have numerous limitations. The cloning, expression, and purification of a protein in its correctly folded state, and in the case of X-ray crystallography, followed by the production of diffraction quality crystals, are time consuming and costly. Conversely, computational methods for protein tertiary structure prediction are easily automated, fast, and cheap; they help in

Alberto Sanchez-Diaz and Pilar Perez (eds.), *Yeast Cytokinesis: Methods and Protocols*, Methods in Molecular Biology, vol. 1369, DOI 10.1007/978-1-4939-3145-3_23, © Springer Science+Business Media New York 2016

functional inference and can help guide in silico, in vitro and in vivo experiments [2]. Furthermore, with the advent of high-throughput next-generation sequencing technologies, the gap between protein sequence and experimentally elucidated structures is increasing at an exponential rate; only ~1 % of Uniprot sequences are structurally elucidated [4]. Thus, computational methods to predict protein tertiary structure, which include the template-based and template-free modeling methods detailed in this chapter, need to be utilized.

1.1 A Brief Introduction to 3D Structure Prediction

Protein tertiary structure prediction methods can be subdivided into two categories: the template-based modeling methods (TBM) and template-free modeling methods (FM). Essentially, if it is possible to locate a template, for the target protein, in the PDB [5], then TBM methods, such as fold recognition and homology modeling are utilized [2]. TBM is based on three key concepts: (1) similar sequences have similar structural folds; (2) many unrelated sequences also fold into similar structures; and (3) there are a finite number of unique folds, when compared to the number of naturally occurring proteins, almost all of fold space has already been structurally determined, so the likelihood of a protein having a new fold is small [1, 2, 6].

However, in the rare circumstances, when it is not possible to find a template for your protein of interest in the PDB, then template-free modeling algorithms are the only option. Template-free modeling (FM) methods are often divided into knowledge-based methods and physics-based methods. FM methods are also referred to as ab initio modeling, de novo modeling, or modeling from first principles. Generally, FM of a 3D protein structure is performed using only sequence information and without the use of a template structure. FM algorithms use designed energy functions to carry out conformational searches that usually results in the building of a number of structural decoys based on likely conformations, which are used for the selection of the final model. The energy functions that are utilized in FM can usually be divided into two categories; physics-based energy functions and knowledge-based energy functions. Knowledge-based energy functions rely on statistics derived from experimentally elucidated structures [1, 2, 6, 7].

1.2 A Brief Introduction to CASP

The major goal of the biennial Critical Assessment of Techniques for Protein Structure Prediction (CASP) experiments is to drive methods development for the prediction of protein 3D structure from sequence. CASP utilizes blind prediction experiment to provide objective testing for cutting edge methods. The CASP experiment takes the form of a competition which can be thought of as the "World Championships" for structure prediction methods. Over the years the competition has been divided into several prediction categories, which has included tertiary structure prediction—template-based and template-free modeling; disorder prediction; contact prediction; model quality assessment; binding site prediction; protein-protein interactions; oligomerization state; and protein model refinement [1, 2, 8, 9].

The advent of the CASP competition in 1994 [10] has resulted in major improvements in the predictive power of algorithms, with each successive CASP competition. Now methods have reached a level of accuracy and consistency, with the most recent CASP competitions showing that fully automated protein structure prediction servers can often produce models fairly close in quality to the very best human expert modelers [11, 12].

1.3 The IntFOLD2-TS Method for 3D Structure Prediction

The IntFOLD2 server has been developed to be easy to use by non-experts and experts biologists alike, by the provision of an intuitive interface, allowing users to effortlessly predict the 3D structure of their proteins of interest [8, 13, 14]. Additionally, for the more expert user, PDB files containing per-residue errors of the top five models can be downloaded, for more detailed examination of the raw data. The IntFOLD2-TS (Tertiary Structure) method [13] integrates numerous profile-based fold recognition algorithms including: SP3 and SPARKS2 [15], HHsearch [16, 17], COMA [18, 19], and several algorithms from the LOMETS [20] package which produce up to 90 alternative 3D models from bespoke non-redundant template libraries. This set of models is subsequently analysed using ModFOLDclust2 [21], which carries out both global and local model quality assessment, ranking the models accordingly. The per-residue error prediction scores are additionally added to the B-factor column in the PDB files, which can be utilized by the user to determine which parts of the model they can trust and which parts they cannot trust, which is indispensable information for future *in silico, in vitro* and *in vivo* studies.

The IntFOLD2-TS methods also includes a novel algorithm for multi-template modelling [13]. Firstly, taking the per-residue errors of the models from ModFOLDclust2 [21], to determine which areas of the model have low quality. Secondly, the method finds new templates for low confidence areas, in order to improve on the model-template alignments. Finally, these model-template alignments are used to build a new set of models with both higher predicted local and global model quality [13].

The IntFOLD2-TS method has been benchmarked at CASP9 and CASP10 and was identified by the assessors as one of the better performing independent automated methods [14] and the best method overall for providing information about local model accuracy [22]. In addition to CASP, the method is continuously benchmarked by CAMEO (Continuous Automated Model EvaluatiOn—http://www.cameo3d.org). The models constructed by IntFOLD2-TS are further used in our other methods, along with ModFOLDclust2 model quality results for: disorder prediction (DISOclust [23]); domain boundary prediction (DomFOLD) and predicting protein-ligand interactions (FunFOLD [24]) (*see* **Note 4** for more details). These component methods of the IntFOLD2 server are now integrated into the Protein Model Portal (PMP) [25] of the Protein Structure Initiative (PSI) structural genomics knowledgebase [26, 27]. The latest implementation of the server (IntFOLD3) includes updated versions of these methods, which were recently also benchmarked in the CASP11 experiment.

The IntFOLD2-TS method and its previous implementations have been utilized in numerous large scale projects to predict the 3D structure of selected proteins within genomes and proteomes [28, 29]. These projects have included, the study and structure based functional annotation of Secreted Effector Proteins from the genome of the pathogen barley powdery mildew *Blumeria graminis f.* sp. *hordei* [28, 29]. The fold recognition approach implemented in the IntFOLD-TS methods were necessary in these studies due to the lack of information that could be gleaned from PSI-BLAST [30] searches alone.

In summary, the use of computational methods for protein structure prediction is essential in the era of high-throughput next-generation sequencing, as experimental methods are unable to provide structurally information at the same pace as next-generation genomic sequencing. The prediction of protein structures along with the structural annotation of proteomes enables the interpretation of a proteins general function and prediction of binding sites (*see* **Notes 1–4**). These predictions can subsequently be exploited in future in silico, in vitro and in vivo studies, for the discovery and design of novel proteins, which will impact on twenty-first-century issues such as food and energy security, health and disease, and living with environmental change.

2 Materials and Systems Requirements

1. The user requires a computer with an internet connection and a web browser, which is Java Script enabled. The IntFOLD2-TS server [14] has been extensively tested on Google Chrome and Firefox, which are recommended for proper use. The server also works on Internet Explorer, Safari, and Opera, but has not been tested as extensively.

   The IntFOLD2-TS server is available at the following link: http://www.reading.ac.uk/bioinf/IntFOLD/IntFOLD2_form.html.

2. To run your predictions on the IntFOLD2-TS server you require an amino acid sequence for your protein of interest, in single-letter code format. If the length of the target amino acid

sequence is much longer than ~500 amino acids, it is best to divide the target sequence into domains (e.g., using PFAM [31] or SMART [32]) and submit each domain sequence separately (dividing sequences into domains is useful as predictions for longer sequences increase prediction time, and finding a single template for a long sequence can be difficult, thus not all of the sequence may necessarily be modeled). Additionally, a short name can be given to the submitted sequence and an email address can be included to alert the user when predictions are complete. For a more detailed explanation and common problems that may be encountered *see* **Note 1**.

3 Methods

In this section, we present a step-by-step guide on using the IntFOLD2-TS server, to generate 3D structural models for the user's sequence of interest. We also describe some interesting use cases of the IntFOLD2-TS server.

3.1 The Submission Process

1. Navigate to the IntFOLD2-TS server submission form: http://www.reading.ac.uk/bioinf/IntFOLD/IntFOLD2_form.html
2. Paste the full single-letter format amino acid sequence of the target protein into the textbox on the server submission page (*see* Fig. 1). In addition the user has the option to upload additional models, obtained from other servers/methods, which will be assessed along with the IntFOLD2-TS models.
3. Optionally, a short name can be given to the target sequence.
4. The user may choose to give their e-mail address in order to be alerted once the prediction is complete.
5. Once all the required information boxes, on the submission page, have been complete, the user needs to click on the submit button to submit their prediction.
6. Currently, submissions are limited to one per IP address, to maintain prediction speed and server capacity. Once the user's prediction is complete their IP address is unlocked and they can then submit their next target for prediction. *See* **Note 1** for common problems encountered at the submission step.

3.2 How to Interpret the Results

1. Once the user's job is complete an email is sent containing a link to the prediction results of their target protein. Clicking on the link brings you to the results page for your submitted target. *See* Fig. 2 for an example results page.
2. The results page contains links to the graphical results for the target along with links to downloadable machine-readable

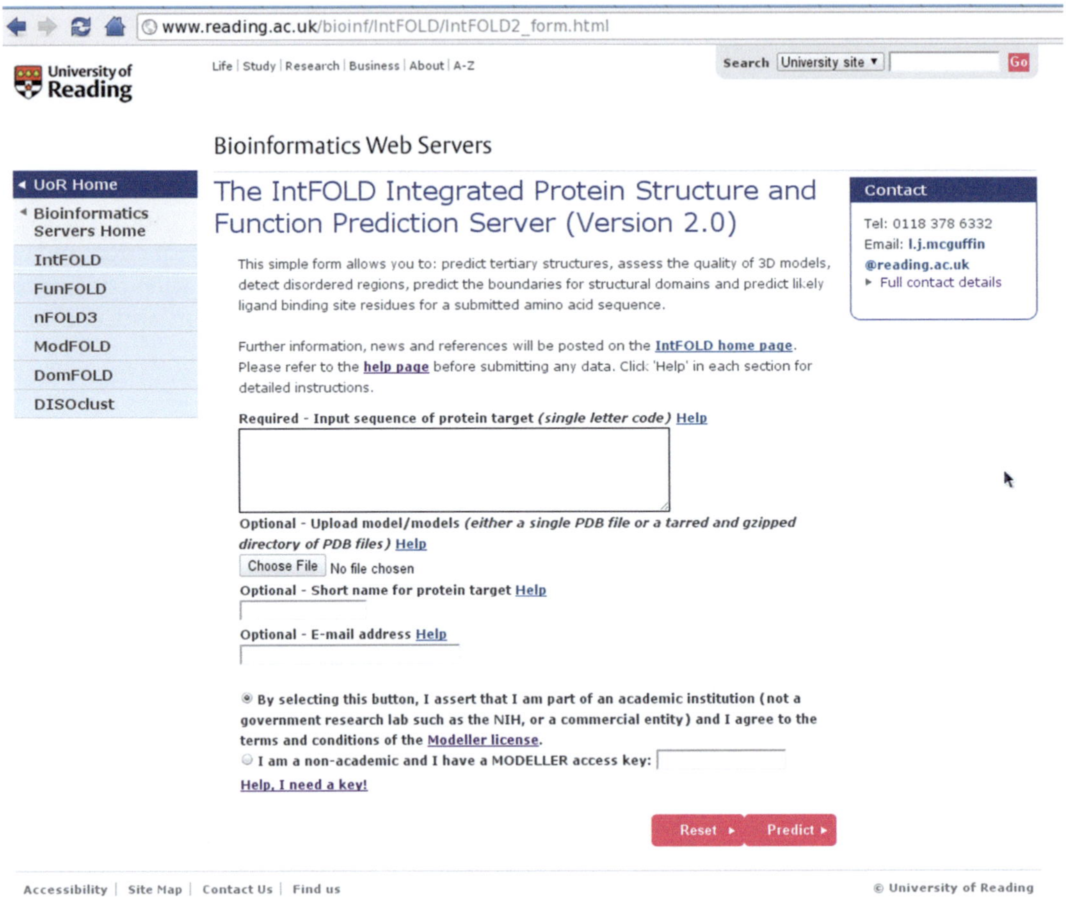

Fig. 1 Screenshot showing the IntFOLD submission page

results in CASP format. Firstly, a graphical representation of the top five models ranked by ModFOLDclust2 [21] global model quality score (between 0 and 1—bad to good) is shown, along with an associated *p*-value colored in relation to the confidence of each prediction, basically the likelihood of the model having the correct fold, $p < 0.001$ ("certain," colored blue) to $p < 0.01$ ("high," colored green) to $p > 0.1$ ("poor" confidence, colored red) (*see* Fig. 2, Table 1 and **Note 2**). A plot of the per-residue error for each model is also shown, which can be clicked on to display a larger per-residue error plot that is downloadable in postscript format. Additionally, each model coloured in reverse rainbow from blue to red (based on the per-residue model quality scores) is shown. Secondly, graphical results for disorder prediction using DISOclust [23] are shown. Thirdly, a graphical representation of predicted domain boundaries is also included. Fourthly, the FunFOLD2 [3, 23, 33] protein-ligand interaction prediction, performed using the top

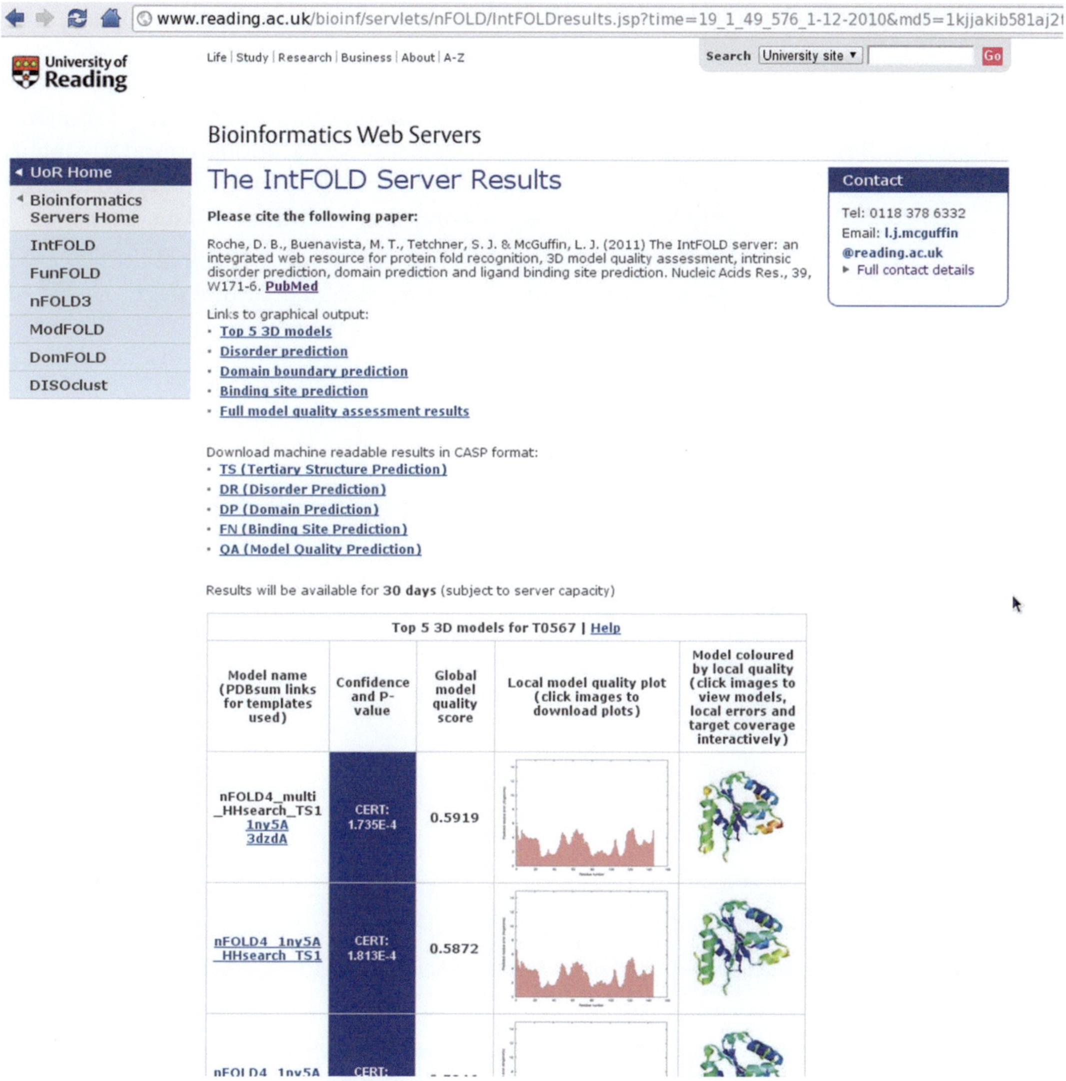

Fig. 2 Screenshot highlighting the IntFOLD-TS results page for CASP9 target T0567. Machine readable results files are available for download at the *top* of the page

Table 1
Confidence assigned to *p*-values associated to IntFOLD2-TS structural predictions

P-value cutoff	Confidence	Description
$p < 0.001$	*CERT*	Less than a 1/1000 chance that the model is incorrect.
$p < 0.01$	*HIGH*	Less than a 1/100 chance that the model is incorrect.
$p < 0.05$	*MEDIUM*	Less than a 1/20 chance that the model is incorrect.
$p < 0.1$	*LOW*	Less than a 1/10 chance that the model is incorrect.
$p > 0.1$	*POOR*	Likely to be a poor model with little or no similarity to the native structure.

The global model quality score allows for the calculation of a *p*-value which represents the probability that each model is incorrect. Each model is also assigned a color-coded confidence level depending on the *p*-value from red (POOR) to blue (CERT)

model, can also be seen. Finally, at the end of the results page, there is a graphical list of all models built for the target sequence ranked by ModFOLDclust2 [21] global model quality. Clicking on each of the graphics links the user to a more detailed results page containing quantitative data.

3. Clicking on the graphical representation of any of the 3D models brings the user to a more detailed results page, *see* Fig. 3.

Fig. 3 Screenshot showing the results page for the top model from Fig. 2 for CASP9 target T0567. This page shows a large graphical representation of the top model, which can be downloaded in PDB format containing per-residue errors in the B-factor column. Furthermore, a Jmol application enables users to inspect the model in 3D space

The detailed model results page contains a large representation of the model, coloured in reverse rainbow (blue to red—good to bad), the model can be downloaded in PDB format with the per-residue errors included in the B-factor column. Additionally, the results page contains a Jmol plug-in [34] allowing users to interrogate the model in 3D space. Furthermore, structural superpositions of the model to the top templates used to build the model are shown. Model-template superpositions are carried out using TM-align [35]. Furthermore, a Jmol plug-in, again allowing users to examine the model-template superposition in 3D space (*see* Fig. 4 and **Note 2**).

3.3 Server Fair Usage Policy

Users have the ability to submit one job for each IP address. Once the first job is complete, then notification of results is made via e-mail, if an e-mail address is provided. If a user does not wish to provide their e-mail address, then a link to the result page is provided from bookmarking and later viewing. Upon job completion, the users IP address will be unlocked and the server is again ready for the user to submit a new job. The results of each completed job are saved on the server for 30 days.

3.4 Case Studies

The IntFOLD2 server has been used in numerous applied studies which have led to interesting biological findings (*see* **Note 4** for more details on the IntFOLD2 server components). Our studies have focused on a diverse range of subjects, for example, investigating neurodevelopmental disorder proteins [36], cardiovascular disease proteins [37, 38], *Drosophila* proteins [39] and plant fungal pathogens [28, 29]. We will focus on two related studies that were carried out in collaboration with the McGuffin group, on the plant fungal pathogen barley powdery mildew (*Blumeria graminis f.* sp. *Hordei*) [28, 29]—as they are more directly relevant to the readers of this book.

The first study combined proteogenomic and in silico structural and functional annotation, to investigate the proteome of the barley powdery mildew pathogen [28]. Genome-wide fold recognition was carried out using the IntFOLD server [14]. The models were subsequently assessed for both local and global model quality using ModFOLD3 [1, 21]. The results lead to a number of interesting conclusions about the structural diversity of the proteome. Firstly, the low model quality scores for a large number of models of fungal effectors lend to the idea that many of the proteins have novel folds or are evolutionarily very distantly homologous to known protein structures. Secondly, for six proteins good-quality models were constructed (model quality score greater than 0.4). These proteins were confidently assigned putative functions—glycosyl hydrolase activity—using the IntFOLD models (an earlier implementation of the server) and FunFOLD [33]. Additionally, the functionality of these putative glycosyl

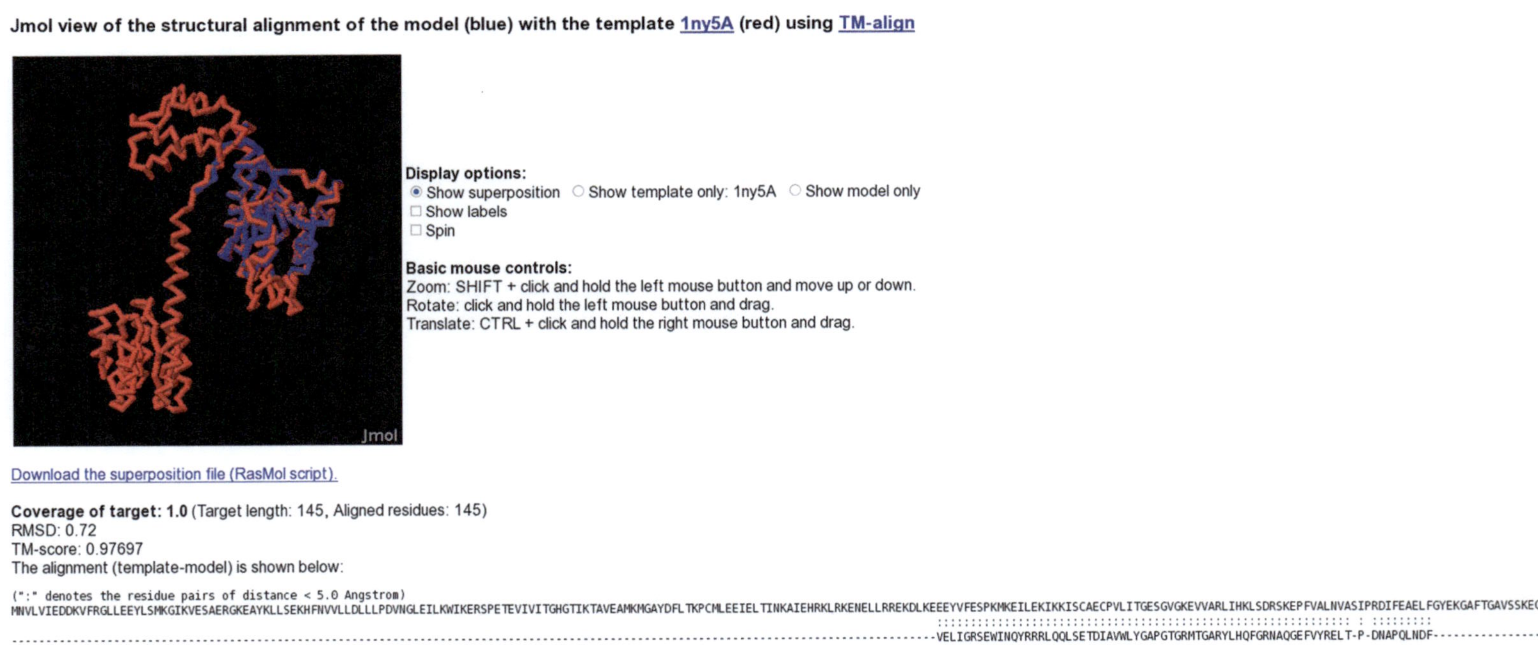

Fig 4. Screenshot showing the structural alignment of the top model from Fig. 3 (*blue*) with the template (*red*) using TM-align. The structural alignment is shown on the results page (Fig. 3) for each individual model

hydrolases was experimentally verified. In conclusion, the IntFOLD server was able to build high quality models for six novel proteins and confidently assign their functions. Furthermore, the fold diversity encoded by the *Blumeria graminis* genome was highlighted [28] (*see* **Note 3** for an analysis of model quality/ confidence scores).

The second follow-up study also investigated the barley powdery mildew, but expanded on the initial strategy, focusing on the study of Candidates for Secreted Effector Proteins (CSEPs) [29]. A combination of genomics, transcriptomics, and proteomics, in addition to in silico protein structure prediction was utilized to analyze the evolutionary relationship of the CSEPs superfamily in *Blumeria graminis f.* sp. *hordei*. This study revealed that 491 genes (7 % of the genome) encoded for CSEPs, grouped into 72 families, predominantly located in the haustoria. Studying the models built for the effector proteins, highlighted their similarity to known fungal effectors, but unexpectedly also linked the entire effector superfamily proteins structurally to ribonucleases. This structural evidence along with experimental evidence and predictive phylogenetic evidence, allowed the authors to hypothesize that the associated effector genes originated from an ancestral secreted ribonuclease gene. It was also hypothesized that this gene, which was duplicated, subsequently underwent functional diversification over the course of the evolution of the grass and cereal powdery mildew lineages [29].

4 Notes

1. In order to be successful in constructing a model for your target sequence using the IntFOLD2-TS server [13, 14], it is important to take several ideas into consideration. The target protein sequence should be supplied in the single-letter amino acid format and pasted into the text box labeled "Input sequence of protein target." Here errors can occur if the single letter format of the target amino acid sequence is not submitted or if the sequence is submitted in and alternative format. Additionally, it is recommended not to submit sequences over 500 amino acids as a single prediction. Firstly, because longer sequences increase both the load on the server and the runtime of the users prediction. Secondly, for longer sequences it is not always possible to locate a good template for all of the multiple domains within the protein; thus some domains may not be modeled well. Hence, it is advisable to partition your sequence into separate domains and submit a number of separate jobs. Furthermore, if the server is unable to build a model for one of the domains of the target protein, again it is a good idea to

partition this protein sequences using the domain boundary predictions from the results page and resubmit the split sequences to the server.

The next text box which needs to be completed is the short name for your protein target; this is useful to keep track of your predictions by assigning them a meaningful descriptor. The set of characters you can utilize for the short descriptor of your target protein are restricted to letters A–Z (either case), the numbers 0–9 and the following other characters: .~_- . The protein descriptor name specified by the user will be included in the subject line of the e-mail messages sent to the user, which contains a link to the user's results. Again, the user may optionally provide an e-mail address. This will enable a link, to the graphical machine readable results to be sent to the user, when the prediction process is completed. Furthermore, for non-academic users a MODELLER [40] access key is required to comply with the terms and conditions of the modeler user guidelines, there is a help button that can be clicked, which redirects users to the MODELLER [40] registration page where a license key can be obtained.

2. Once the IntFOLD2-TS [13, 14] server has completed the prediction process, an e-mail is returned to the user, containing a link to the results page. Clicking on the link or pasting the link into a web browser, brings the user to the results page of their target protein (*see* Fig. 2 for an example results page). Previously described in Subheading 3.2—How to interpret the results. Problems that can be encountered with the results include, models having bad model quality scores (global and local) and or not having one (several) domain(s) constructed for a particular target. The main reason is that a good template or set of templates was not located in the non-redundant template databases to enable model construction. Thus, the domain or protein not built may be completely disordered or have a novel fold (template-free methods will need to be used to build a model).
3. It is extremely important to assess both the global and local quality of a model. Model quality assessment is the cornerstone of all our methods including IntFOLD2-TS [13, 14]. For a more detailed review of model quality assessment methods *see*—Roche et al. [1].

   The IntFOLD2-TS server integrates model quality scores from the ModFOLDclust2 [21] method to rank the constructed models according to their global model quality score from 0 to 1 with scores above 0.4 being more trustworthy, *See* Table 1. The color scheme gives users at a glance an idea of the global model quality. Moreover, a plot of the per-residue model quality score is also shown, with the amino acid sequence from N-ter to C-ter along the *x*-axis and the per-residue error score in Å on the *y*-axis (from 1 to 15 Å)(downloadable in postscript format). This plot allows the user to assess which sections of the model they can trust and use in future in silico, in vitro and in vivo studies. Furthermore, the per-residue error is also included in the B-factor column of all of the models. Finally, the per-residue errors from ModFOLDclust2 are utilized in the IntFOLD2-TS multi-template modeling process to help determine the "good" and "bad" parts of the model, find new templates for the "bad" parts and building a subsequent round of models, with better global and local model quality scores.

4. Once a model has been constructed for a target protein it can be utilized for numerous in silico, in vitro and in vivo studies in numerous fields of application. These applications include drug discovery, cardiovascular disease, neurodegenerative disease research, biofuels research, and plant pathogen research [26], to name a few, as outlined in the case studies presented in Subheading 3.4 [28, 29, 36–39].

   The IntFOLD server utilizes the models built by IntFOLD2-TS additionally for protein-ligand interaction predictions (FunFOLD2 method [23]), and disorder and domain boundary predictions. The FunFOLD2 algorithm [23] combines FunFOLD [33] and FunFOLDQA [3] to predict protein-ligand interactions and binding site quality scores. Briefly, FunFOLD carries out model-template superpositions of the top-ranked model from IntFOLD2-TS and the templates used to build the model, which have biologically relevant bound ligands. In order to identify putative binding site residues, a novel agglomerative hierarchical clustering algorithm is used, along with a voting system. Additionally, the FunFOLDQA algorithm subsequently assesses the quality of the predicted ligand binding site, producing five feature-based scores, which are combined via a neural network to predict two global binding site quality scores, predicted MCC [41] and BDT [42] scores.

Acknowledgements

DBR is a recipient of a Young Investigator Fellowship from the Institut de Biologie Computationnelle, Université de Montpellier (ANR Investissements D'Avenir Bio-informatique: projet IBC). This research leading to these results has received funding from the European Union Seventh Framework Programme (FP7/2007–2013) under grant agreement No. 246556 [to D.B.R.].

References

1. Roche DB, Buenavista MT, McGuffin LJ (2014) Assessing the quality of modelled 3D protein structures using the ModFOLD server. Methods Mol Biol 1137:83–103. doi:10.1007/978-1-4939-0366-5_7
2. Roche DB, Buenavista MT, McGuffin LJ (2012) Predicting protein structures and structural annotation of proteomes. In: Roberts GCK (ed) Encyclopedia of biophysics, vol 1. Springer, Berlin
3. Roche DB, Buenavista MT, McGuffin LJ (2012) FunFOLDQA: a quality assessment tool for protein-ligand binding site residue predictions. PLoS One 7(5):e38219. doi:10.1371/journal.pone.0038219
4. Kajan L, Hopf TA, Kalas M, Marks DS, Rost B (2014) FreeContact: fast and free software for protein contact prediction from residue co-evolution. BMC Bioinformatics 15:85. doi:10.1186/1471-2105-15-85
5. Berman HM, Westbrook J, Feng Z, Gilliland G, Bhat TN, Weissig H, Shindyalov IN, Bourne PE (2000) The Protein Data Bank. Nucleic Acids Res 28(1):235–242, doi:gkd090 [pii]
6. McGuffin LJ (2008) Protein fold recognition and threading. Computational structural biology. World Scientific, London, pp 37–60
7. Lee J, Wu S, Zhang Y (2009) Ab initio protein structure prediction. From protein structure to function with bioinformatics. Springer, London, pp 1–26
8. McGuffin LJ, Roche DB (2011) Automated tertiary structure prediction with accurate local model quality assessment using the IntFOLD-TS method. Proteins 79(Suppl 10):137–146. doi:10.1002/prot.23120
9. Moult J, Fidelis K, Kryshtafovych A, Rost B, Tramontano A (2009) Critical assessment of methods of protein structure prediction—round VIII. Proteins 77(Suppl 9):1–4. doi:10.1002/prot.22589
10. Moult J, Pedersen JT, Judson R, Fidelis K (1995) A large-scale experiment to assess protein structure prediction methods. Proteins 23(3):ii–v. doi:10.1002/prot.340230303
11. Kryshtafovych A, Fidelis K, Moult J (2014) CASP10 results compared to those of previous CASP experiments. Proteins 82(Suppl 2):164–174. doi:10.1002/prot.24448
12. Kryshtafovych A, Krysko O, Daniluk P, Dmytriv Z, Fidelis K (2009) Protein structure prediction center in CASP8. Proteins 77(Suppl 9):5–9. doi:10.1002/prot.22517
13. Buenavista MT, Roche DB, McGuffin LJ (2012) Improvement of 3D protein models using multiple templates guided by single-template model quality assessment. Bioinformatics 28(14):1851–1857. doi:10.1093/bioinformatics/bts292
14. Roche DB, Buenavista MT, Tetchner SJ, McGuffin LJ (2011) The IntFOLD server: an integrated web resource for protein fold recognition, 3D model quality assessment, intrinsic disorder prediction, domain prediction and ligand binding site prediction. Nucleic acids research 39(Web Server issue):W171–W176. doi:10.1093/nar/gkr184
15. Zhou H, Zhou Y (2005) SPARKS 2 and SP3 servers in CASP6. Proteins 61(Suppl 7):152–156. doi:10.1002/prot.20732
16. Soding J, Biegert A, Lupas AN (2005) The HHpred interactive server for protein homology detection and structure prediction. Nucleic Acids Res 33(Web Server issue):W244–W248. doi:10.1093/nar/gki408
17. Soding J (2005) Protein homology detection by HMM-HMM comparison. Bioinformatics 21(7):951–960. doi:10.1093/bioinformatics/bti125, bti125 [pii]
18. Margelevicius M, Laganeckas M, Venclovas C (2010) COMA server for protein distant homology search. Bioinformatics 26(15):1905–1906. doi:10.1093/bioinformatics/btq306
19. Margelevicius M, Venclovas C (2010) Detection of distant evolutionary relationships between protein families using theory of sequence profile-profile comparison. BMC Bioinformatics 11:89. doi:10.1186/1471-2105-11-89
20. Wu S, Zhang Y (2007) LOMETS: a local meta-threading-server for protein structure prediction. Nucleic Acids Res 35(10):3375–3382. doi:10.1093/nar/gkm251
21. McGuffin LJ, Roche DB (2010) Rapid model quality assessment for protein structure predictions using the comparison of multiple models without structural alignments. Bioinformatics 26(2):182–188. doi:10.1093/bioinformatics/btp629
22. Kryshtafovych A, Barbato A, Fidelis K, Monastyrskyy B, Schwede T, Tramontano A (2014) Assessment of the assessment: evaluation of the model quality estimates in CASP10. Proteins 82(Suppl 2):112–126. doi:10.1002/prot.24347
23. McGuffin LJ (2008) Intrinsic disorder prediction from the analysis of multiple protein fold recognition models. Bioinformatics 24(16):1798–1804. doi:10.1093/bioinformatics/btn326
24. Roche DB, Buenavista MT, McGuffin LJ (2013) The FunFOLD2 server for the prediction of

protein-ligand interactions. Nucleic Acids Res 41(Web Server issue):W303–W307. doi:10.1093/nar/gkt498

25. Bordoli L, Schwede T (2012) Automated protein structure modeling with SWISS-MODEL Workspace and the Protein Model Portal. Methods Mol Biol 857:107–136. doi:10.1007/978-1-61779-588-6_5
26. Berman HM, Burley SK, Chiu W, Sali A, Adzhubei A, Bourne PE, Bryant SH, Dunbrack RL Jr, Fidelis K, Frank J, Godzik A, Henrick K, Joachimiak A, Heymann B, Jones D, Markley JL, Moult J, Montelione GT, Orengo C, Rossmann MG, Rost B, Saibil H, Schwede T, Standley DM, Westbrook JD (2006) Outcome of a workshop on archiving structural models of biological macromolecules. Structure 14(8):1211–1217
27. Berman HM, Westbrook JD, Gabanyi MJ, Tao W, Shah R, Kouranov A, Schwede T, Arnold K, Kiefer F, Bordoli L, Kopp J, Podvinec M, Adams PD, Carter LG, Minor W, Nair R, La Baer J (2009) The protein structure initiative structural genomics knowledgebase. Nucleic Acids Res 37(Database issue):D365–D368. doi:10.1093/nar/gkn790
28. Bindschedler LV, McGuffin LJ, Burgis TA, Spanu PD, Cramer R (2011) Proteogenomics and in silico structural and functional annotation of the barley powdery mildew Blumeria graminis f. sp. hordei. Methods 54(4):432–441. doi:10.1016/j.ymeth.2011.03.006
29. Pedersen C, Loren V, van Themaat E, McGuffin LJ, Abbott JC, Burgis TA, Barton G, Bindschedler LV, Lu X, Maekawa T, Wessling R, Cramer R, Thordal-Christensen H, Panstruga R, Spanu PD (2012) Structure and evolution of barley powdery mildew effector candidates. BMC Genomics 13:694. doi:10.1186/1471-2164-13-694
30. Altschul SF, Madden TL, Schaffer AA, Zhang J, Zhang Z, Miller W, Lipman DJ (1997) Gapped BLAST and PSI-BLAST: a new generation of protein database search programs. Nucleic Acids Res 25(17):3389–3402, doi:gka562 [pii]
31. Finn RD, Bateman A, Clements J, Coggill P, Eberhardt RY, Eddy SR, Heger A, Hetherington K, Holm L, Mistry J, Sonnhammer EL, Tate J, Punta M (2014) Pfam: the protein families database. Nucleic Acids Res 42(Database issue):D222–D230. doi:10.1093/nar/gkt1223
32. Letunic I, Doerks T, Bork P (2014) SMART: recent updates, new developments and status in 2015. Nucleic Acids Res. doi:10.1093/nar/gku949
33. Roche DB, Tetchner SJ, McGuffin LJ (2011) FunFOLD: an improved automated method for the prediction of ligand binding residues using 3D models of proteins. BMC Bioinformatics 12:160. doi:10.1186/1471-2105-12-160
34. Jmol: an open-source Java viewer for chemical structures in 3D. http://www.jmol.org/
35. Zhang Y, Skolnick J (2005) TM-align: a protein structure alignment algorithm based on the TM-score. Nucleic Acids Res 33(7):2302–2309. doi:10.1093/nar/gki524, 33/7/2302 [pii]
36. Tucci V, Kleefstra T, Hardy A, Heise I, Maggi S, Willemsen MH, Hilton H, Esapa C, Simon M, Buenavista MT, McGuffin LJ, Vizor L, Dodero L, Tsaftaris S, Romero R, Nillesen WN, Vissers LE, Kempers MJ, Vulto-van Silfhout AT, Iqbal Z, Orlando M, Maccione A, Lassi G, Farisello P, Contestabile A, Tinarelli F, Nieus T, Raimondi A, Greco B, Cantatore D, Gasparini L, Berdondini L, Bifone A, Gozzi A, Wells S, Nolan PM (2014) Dominant beta-catenin mutations cause intellectual disability with recognizable syndromic features. J Clin Invest 124(4):1468–1482. doi:10.1172/JCI70372
37. Fuller SJ, McGuffin LJ, Marshall AK, Giraldo A, Pikkarainen S, Clerk A, Sugden PH (2012) A novel non-canonical mechanism of regulation of MST3 (mammalian Sterile20-related kinase 3). Biochem J 442(3):595–610. doi:10.1042/BJ20112000
38. Sugden PH, McGuffin LJ, Clerk A (2013) SOcK, MiSTs, MASK and STicKs: the GCKIII (germinal centre kinase III) kinases and their heterologous protein-protein interactions. Biochem J 454(1):13–30. doi:10.1042/BJ20130219
39. Dunwell TL, McGuffin LJ, Dunwell JM, Pfeifer GP (2013) The mysterious presence of a 5-methylcytosine oxidase in the Drosophila genome: possible explanations. Cell Cycle 12(21):3357–3365. doi:10.4161/cc.26540
40. Eswar N, Webb B, Marti-Renom MA, Madhusudhan MS, Eramian D, Shen MY, Pieper U, Sali A (2006) Comparative protein structure modeling using Modeller. Curr Protoc Bioinformatics Chapter 5:Unit 5 6. doi:10.1002/0471250953.bi0506s15
41. Matthews BW (1975) Comparison of the predicted and observed secondary structure of T4 phage lysozyme. Biochim Biophys Acta 405(2):442–451
42. Roche DB, Tetchner SJ, McGuffin LJ (2010) The binding site distance test score: a robust method for the assessment of predicted protein binding sites. Bioinformatics 26(22):2920–2921. doi:10.1093/bioinformatics/btq543

Chapter 24

From Structure to Function: A Comprehensive Compendium of Tools to Unveil Protein Domains and Understand Their Role in Cytokinesis

Sergio A. Rincon and Anne Paoletti

Abstract

Unveiling the function of a novel protein is a challenging task that requires careful experimental design. Yeast cytokinesis is a conserved process that involves modular structural and regulatory proteins. For such proteins, an important step is to identify their domains and structural organization. Here we briefly discuss a collection of methods commonly used for sequence alignment and prediction of protein structure that represent powerful tools for the identification homologous domains and design of structure-function approaches to test experimentally the function of multi-domain proteins such as those implicated in yeast cytokinesis.

Key words Cell division, Cytokinesis, Structure-function analysis, Sad kinase, Anillin

1 Introduction

Cytokinesis is the process by which a single-cell partitions its cytoplasm to give rise to physically separated daughter cells. It is required for cell proliferation and development of multicellular organisms, and well conserved from yeast to mammalian cells. A major feature of cytokinesis is the need for assembly and constriction of a cytokinetic ring that guides and/or provides force to drag the plasma membrane of the ingressing furrow. This requires the establishment of strong and dynamic connections between the cytokinetic ring and the membrane. Accurate coordination of cytokinetic ring assembly and constriction with chromosome segregation is also necessary to maintain genome integrity along generations. These events involve a large number of modular multi-domain proteins that can assemble into large complexes to build the cytokinetic ring and connect it to the membrane, or form signaling modules dictating the spatiotemporal regulation of cytokinesis. These proteins typically combine several domains of

Alberto Sanchez-Diaz and Pilar Perez (eds.), *Yeast Cytokinesis: Methods and Protocols*, Methods in Molecular Biology, vol. 1369, DOI 10.1007/978-1-4939-3145-3_24, © Springer Science+Business Media New York 2016

distinct functions such as protein-protein interaction domains, membrane binding motifs and catalytic domains [1–3]. In this chapter, we discuss the methods and tools available to identify protein domains within large modular cytokinetic proteins and get insights on their function. We use the example of the SAD kinase Cdr2 and the anillin-like protein Mid1 involved in the definition of the division plane in fission yeast to illustrate how these tools can be used to design experimental approaches.

2 Sequence Alignment

The function of a novel protein or protein domain can often be inferred from homologous proteins or domains in other biological systems. Genome sequencing projects have gathered an immense amount of data from an increasing number of organisms. Gene databases contain not only sequence data, but also gene expression patterns, alternative protein isoforms, protein localization data, protein family, and putative functions based on data established in other systems. Therefore, a very useful starting point is to align the sequence of the protein of interest with sequences from protein databases. This step may allow the identification of orthologs in other organisms whose function may be already established in the literature, which is possibly the first and best source to understand the function of your protein of interest. It can also establish the degree of similarity and evolutionary relationships between proteins of the same family within a given species. Protein sequences indeed accumulate changes in their sequences along evolution. The most frequent are simple amino acid substitutions, but insertions and deletions can also happen. Alignment of farther-related proteins shows stronger changes, such as protein fragment reorganization (e.g., N-terminal or C-terminal motor domain kinesins), or protein fusion events (multi-domain proteins).

2.1 Identifying Homologous Proteins by Sequence Alignment

Specific softwares like BLAST (*B*asic *L*ocal *A*lignment *S*earch *T*ool) [4] have been developed to efficiently compare primary biological sequences with those of databases. This program is based on the heuristic method in local alignments (i.e., finding suboptimal matches). BLAST first identifies and removes low complexity regions which may interfere with further analysis. It creates a collection of fragments whose initial length can be defined at will. With the help of a substitution matrix like BLOSUM62, BLAST chooses the highest score pairs (HSPs) from the query and organizes them so that they can be used efficiently for further analysis. Subsequently, it increases the length of the fragments, tests their new scores and organizes them for further analysis. BLAST then scans the exact matches from the database and, if possible, extends the fragments without gaps. BLAST tries to extend the matches by lowering the resultant score threshold enough to maintain the level of sensitivity for sequence detection. HSPs whose score is below the minimum threshold set to discriminate possible hits from random similarities are eliminated. The statistical significance of each HSP score is tested by analysis of the "Gumbel extreme value distribution." BLAST combines nearby HSPs and analyses the similarity significance by comparing the score of the newly formed HSP to individual ones. Gap introduction, taken into account by "open gap" and "gap length" penalties, impacts on the score of each alignment. Finally, BLAST reports matches with low enough expected values (E) to be confident of the significance of the alignment. It finally provides all pairwise alignments generated in the search and reports the percentage of the query sequence aligned and percentage of identity in the alignment (*see* Fig. 1a).

Several query options are available: blastn, (DNA against a DNA database); blastp, (polypeptide against a protein database); blastx (six frames of DNA against a protein database); tblastn, (protein against the six frames of DNA database); and tblatx (six frames of DNA against six frames of DNA database). BLAST offers the possibility to restrict the search to specific species. This is very useful to quickly find orthologs of yeast genes in specific animal species or vice versa. Figure 1a shows some of the sequences retrieved with such a BLAST search performed with the SAD-like kinase Cdr2 that organizes cytokinetic precursors on the medial cortex of fission yeast cells [5–9] selecting *Schizosaccharomyces pombe*, *Saccharomyces cerevisiae*, *Caenorhabditis elegans*, *Drosophila melanogaster*, and *Homo sapiens* databases. This search identified two homologs in *Schizosaccharomyces pombe*, the second Sad kinase Cdr1 and the more distantly related AMPK Ssp2; the Sad/Septin kinases Hsl1, Gin 4, and Kcc4 in *Saccharomyces cerevisiae*; Sad-1 in *Caenorhabditis elegans*; sugar-free frosting in Drosophila; and the two human Sad kinases Brsk1 and Brsk2. Interestingly, while all hits contained a serine/threonine kinase domain from the AMPK superfamily, only Cdr2, Gin4 and Kcc4, appeared to contain a lipid-binding KA-1 domain due to strong sequence divergence within this domain [10, 11].

BLAST offers additional algorithms useful to find distant relatives of a protein. PSI-BLAST (*P*osition *S*pecific *I*terative BLAST) first performs a regular BLAST with the query. It then assembles a multiple alignment to detect amino acid positions and performs an analysis of the frequency of amino acids for each position to generate a profile for the group of proteins analysed. A new search on the database is performed with the generated profile to find new proteins. This process is repeated until no new successful hit is retrieved, yielding an optimized profile for all proteins. PHI-BLAST (*P*attern *H*it *I*nitiated BLAST) limits the search to alignments that match a pattern in the query. DELTA-BLAST (*D*omain *E*nhanced *L*ookup *T*ime *A*ccelerated BLAST) uses a conserved

a

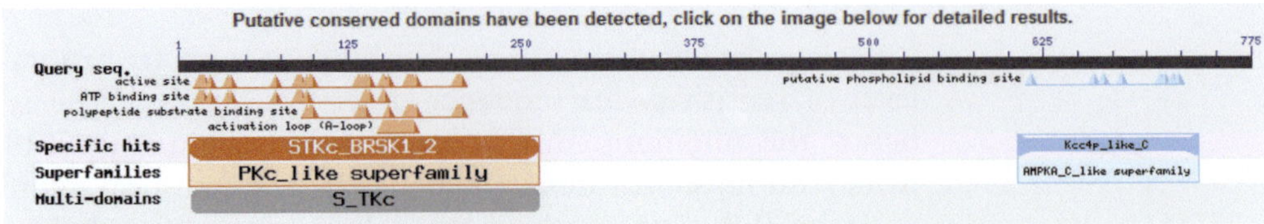

Sequences producing significant alignments	Query Cover	E value	% Identity
Serine/threonine protein kinase Cdr2 [Schizosaccharomyces pombe]	100%	0.0	100%
Hsl1p [Saccharomyces cerevisiae]	33%	1e-82	46%
Gin4p [Saccharomyces cerevisiae]	55%	3e-82	45%
Serine/threonine protein kinase BRSK1 [Homo sapiens]	40%	5e-80	44%
Serine/threonine protein kinase BRSK2 isoform 2 [Homo sapiens]	40%	3e-79	45%
Protein SAD-1, isoform a [Caenorhabditis elegans]	40	4e-79	42%
Sugar-free frosting, isofrom A [Drosophila melanogaster]	40%	1e-76	41%
Kcc4p [Saccharomyces cerevisiae]	56%	6e-76	43%
NIM1 family serine/threonine protein kinase Cdr1/Nim1 [Schizosaccharomyces pombe]	41%	3e-68	38%
Serine/threonine protein kinase Ssp2 [Schizosaccharomyces pombe]	40%	1e-66	39%

b

```
Brsk2  --EMSNLTPESSPELAKKSWFGNFISLEKEEQIFVVIKDKPLSSI-KADIVHAFLSIPSL  560
Brsk1  --EMSSLTPESSPELAKRSWFGNFISLDKEEQIFLVLKDKPLSSI-KADIVHAFLSIPSL  634
Cdr2   QPKRSFLRRLFSSEPSCKCVYASLVASELEHEILEVLRRWQLLGIGIADIIYDSVSAS-I  679
Kcc4   --NSVLLKKFSKGKILELEIHAKIPEKRLYEGLHKLLEGWKQYGL--KNLVFN-ITNMII  975

Brsk2  SHSVISQTSFRAEYKATGGPAVFQKPVKFQVDITYTEGGE--AQKEN----GIYSVTFTL  614
Brsk1  SHSVLSQTSFRAEYKASGGPSVFQKPVRFQVDISSSEGPEPSPRRDGSGGGGIYSVTFTL  694
Cdr2   SARIKRQNSLN-------------LKPVRFRISVLAEFFGS-------------QAVFVL  713
Kcc4   TGKLVNDSILF-------------LRSTLFEIMVLPNGDGR------------SLIKFNK  1010

Brsk2  LSGPSRRFKRVVETIQAQLLSTHD--  638
Brsk1  ISGPSRRFKRVVETIQAQLLSTHD--  708
Cdr2   ESGSSTTFDHLATEFQLIFEDKGFLD  739
Kcc4   KTGSTKTLTKLATEIOIILOKEGVLD  1036
```

Fig. 1 Sequence alignment of Cdr2 and related kinases. (**a**) Standard BLAST of full-length Cdr2 against *Schizosaccharomyces pombe*, *Saccharomyces cerevisiae*, *Caenorhabditis elegans*, *Drosophila melanogaster*, and *Homo sapiens* databases retrieves its predicted domain organization with an N-terminal kinase domain, a long central domain with no domain prediction and a C-terminal phospholipid-binding domain similar to Kcc4 KA-1 domain. A list of some of the sequences producing significant alignments with Cdr2 is shown below with BLAST reported query cover, *E* value, and percentage identity with Cdr2 in the aligned fragments. (**b**) Multi-alignment of the C-terminal regions of *S. pombe* Cdr2, *S. cerevisiae* Kcc4, and *H. sapiens* BRSK1 and BRSK2 performed by ClustalW2. Hydrophobic residues are shown in *red*, negatively charged residues in *blue*, positively residues in *pink*, and polar residues in *green*. Note the large conservation of the hydrophobic residues critical for proper protein folding

domain database to make a position-specific scoring matrix and searches a sequence database. DELTA-BLAST maybe the most powerful option to uncover distantly related domains. For example, the KA-1 domain of the human Sad kinase Brsk1 could be detected with DELTA-BLAST but not with PSI-BLAST nor PHI-BLAST using Cdr2 KA-1 domain as a query.

Finally, despite the fact that these algorithms are based on local alignments, it is sometimes useful to restrict the query to a specific domain of the protein to identify its closest relatives. On the contrary, removing a highly conserved domain from the query may help identifying regions with lower levels of conservation.

2.2 Multiple Sequence Alignments

A conserved region usually corresponds to a folded domain carrying a specific function. Once homologous domains have been identified using BLAST, multi-alignment algorithms can be used to refine alignments and pinpoint highly conserved residues that may be critical for the folding or for the specific function of the domain (e.g., residues of the catalytic loop of kinases). They can also reveal divergent residues that may explain functional variations (e.g., pseudokinases, like BubR1). From an evolutionary point of view, multiple sequence alignments allow the establishment of phylogenetic relationships. In that respect, a region with increased similarity does not necessarily involve a relation of homology: by evolutionary convergence, two regions may have evolved to a very similar one; physical limitations imposed by specific cellular structures also create bias in amino acids frequency (e.g., transmembrane domains, which are enriched in hydrophobic amino acids).

Early alignments programs compared the amino acids sequence globally, along all the length of the protein. This worked well for globular proteins but was a tremendous concern for multidomain proteins. For this reason, new local alignment methods were designed to find regions of similarity independently of nearby regions. These programs try to get the best fit that accommodates all query sequences. They try to position the most conserved amino acids, which may derive from a common ancestor and introduce gaps that represent inserts or deletions that happened along evolution. This is in fact a plausible hypothesis to explain the occurrence of mutations. Most algorithms reward an identical hit and penalize a substituted amino acid to give a final score to the alignment. Conservative substitutions receive a smaller penalty than non-conservative ones. Gaps are penalized according to their length.

Three different methods are used:

– Progressive Global Multiple Alignment Method used by Clustal, Multalign, or T-Coffee. These programs analyse the sequences pairwise to define the more closely related ones. In the next rounds, they sequentially add and adjust new sequences to the alignment generated previously. The user must check afterwards that the alignment obtained is the best option.

– Iterative Global Multiple Alignment Method used by DIALIGN2 or PRRN/PRRP. In this case, the consensus sequence is recalculated after the addition of a new sequence.
– Motif-based Global Multiple Alignment Method used by MEME, MATCH-BOX or PIMA. This method allows the alignment of sequences that share motifs that are differently organized or repeated along the sequence.

Figure 1b shows the example of such an alignment obtained with ClustalW2 [12] for Cdr2, Kcc4, Brsk1, and Brsk2 KA-1 domains.

3 Protein Structure Prediction

The structure that a protein acquires after folding conditions its function. Solving the structure of a protein is therefore highly valuable to interpret its function. Interestingly, although there are billions of protein sequences in nature, only over a thousand of folds have been described so far. As a matter of fact, phylogenetic relationships can sometimes be deciphered more accurately by analysis of protein folding than by analysis of the amino acid sequence. One example is the membrane-binding domain of Kcc4 whose identification as a KA-1 domain required solving its crystal structure, which revealed a similar fold than KA-1 domains of MARKs [10]. Unfortunately 3D structure prediction is lagging way behind the fast next generation sequencing methods. We briefly describe here bioinformatics methods to predict protein structure and how they can be used for functional studies.

3.1 Secondary Structure Prediction

Amino acids chains tend to fold in specific ways to adopt the lowest free energy state. The secondary structure represents the first level of folding of the polypeptide sequence. The basic types of secondary structures are α helices, β sheets, and loops. An α helix is a right-hand coil in which a hydrogen bond is established every four amino acids to stabilize the structure. Protein fragments rich in amino acids such as Ala, Met, Leu, or Glu and poor in Pro, Gly, or Tyr tend to adopt an α helix configuration. This bias is used by algorithms that predict secondary structure. β sheets are obtained by the parallel or antiparallel alignment of several amino acid chains. They are stabilized by hydrogen bonds, but contrary to α helices, the connections are established between non-adjacent amino acids. In general, β sheets are more difficult to predict than α helices and tend to be underrepresented. Loops are the regions of the polypeptide chain that connect α helices and β sheets with one another or between them. Since these regions are not under strong physical constrains, they tend to be much more variable between different members of a protein family, and point mutations, deletions or insertions can be indicative of the presence of a loop. Proteins also contain disordered regions. These fragments may contain low complexity sequences, with over representation of several amino acids. They usually lack hydrophobic amino acids since they cannot fold to create a pocket to hide them. These regions are often very flexible and tend to connect globular regions in the polypeptide chain. It has also been proposed that these unstructured regions may acquire a stable conformation upon binding to their targets. Finally, secondary structure elements can associate with each other to give rise to higher ordered elements such as helix-loop-helix, beta-hairpin, helical bundle, or coiled coil helices.

A number of algorithms to predict secondary structures have been developed. The GOR Method takes into account not only the probability of a single amino acid to be part of a specific secondary structure, according to databases of solved 3D structures of proteins, but also the contribution of adjacent amino acids to trigger a specific folding on each other. Machine learning methods use computational models such as artificial neural networks in which pattern recognition is achieved by association of specific sequence motifs to a database of solved secondary structures.

Several programs for secondary structure prediction are available. Since they are prediction software, we recommend validating the results obtained by performing predictions with several algorithms. Jpred is an algorithm based on sequential artificial neural networks. PSIPRED uses first a PSIBLAST to get a sequence profile; the first secondary structure prediction is obtained by an artificial neuronal network; a second neuronal network is used to filter the first predicted structure. PredictProtein uses a very stringent PSIBLAST to avoid false positives by using a filtered database to get the protein profile. It then searches for functional motifs in PROSITE database. It also includes the use of several other programs, such as PHD and PROF for solvent accessibility and membrane helix determination, COILS for localization of coiled-coil regions, or SEG for identification of low complexity regions.

3.2 Tertiary Structure Prediction

The folding of a protein is achieved by the spatial arrangement of the elements of its secondary structure defined above into protein domains that form spatially independent structural subunits that perform specific tasks, such as serving as localization signals or being responsible for a catalytic activity. Protein motifs usually refer to short regions with a specific amino acid sequence that is critical for the function of the domain. To determine the presence of a domain within a sequence, algorithms usually require the assembly of protein profiles, which are obtained by alignment of multiple sequences. Protein profiles are matrices that describe the probability for a specific amino acid to be found in a certain position. A more accurate method used for domain identification is the HMM (hidden Markov model), which calculates probabilities with a very rigorous statistical method, scoring both amino acid modifications and insertion/deletion events from a multiple sequence alignment.

A number of protein databases are available for the identification of domains.

- PROSITE is the major database for protein motifs and profiles. It is mainly manually curated, so an accurate validation of the data is expected.
- Pfam is a database of protein families with a vast collection of multiple alignments and HMMs for many domains. It contains two parts: PfamA, which is manually curated; and PfamB, which is automatically generated from data from SwissProt.
- SMART (Simple Modular Architecture Research Tool) is a similar database hosted by EMBL, which also contains annotation about the tertiary structure of the protein or critically important amino acids in a specific domain.

Information on the 3D structure of proteins is gathered at PDB (Protein Data Bank). Each newly solved protein structure is annotated. Here we can find data such as the experimental method used to generate the structure, its resolution or the authorship. Each structure contains a link to visualize it or download the corresponding PDB file that compiles the 3D coordinates of atoms solved by structural analysis.

There are several programs for the visualization of protein 3D structures. One of the first developed programs is RasMol. Chime or JMol were built on RasMol basis to make more accessible software. PyMol contains a powerful tool for creating 3D images of the polypeptide of interest with good resolution.

One of the greater challenges of protein structure prediction is to understand how an entire polypeptide chain can achieve its global folding. Although much progress is yet required to get accurate results, "ab initio" modeling is beginning develop, taking different methodologies.

- Energy-based methods try to predict the polypeptide folding in which the protein sits at a global minimum of free energy.
- Evolutionary covariation approaches require the analysis of a vast number of homologous sequences to determine the amino acids that co-evolved along evolution (e.g.: EVfold).

Since little progress has been obtained so far with "ab initio" approaches, comparative protein modelling is the most frequently used method for 3D structure prediction. It uses already solved structures as templates. It is generally considered as a good strategy due to the limited number of tertiary structure motifs found in nature. Two groups of modelling methods are available:

- Homology modelling; it assumes that two homologous proteins will share a similar 3D structure. Indeed structures are more conserved than amino acid sequences. If the 3D homology can be derived from the sequence analysis, a reasonable accuracy of the model is expected (e.g.: Modeller).
- Protein threading; it assesses the compatibility of a given amino acid sequence to fold like already known 3D structures (e.g.: Phyre).

As an example we used Phyre2 [13] to get a structure prediction for the KA-1 domain of Cdr2. Figure 2a shows the predicted secondary structure compared to Kcc4 KA-1 predicted and actual secondary structure defined in the crystal structure [10]. Phyre2 also generated a model for Cdr2 KA-1 shown in Fig. 2b, which is very similar to the model of Cdr2 KA-1 obtained with Modeller based on the solved structure of Kcc4 KA-1 [11].

Interestingly, comparison of Kcc4 KA-1 structure and Cdr2 KA-1 model allowed us to identify a series of basic residues crucial for acidic lipid binding on the surface of Cdr2 KA-1 that would have been difficult to identify on simple sequence alignments since they are not at the same position on the primary sequence than the basic residues with a similar function in Kcc4 [11]. Moreover, comparing Cdr2 KA-1 model to Kcc4 KA-1 structure allowed us to identify striking non conservative changes in the β4-β5 loop from charged to hydrophobic residues (Fig. 2b). Mutating these residues (see next paragraph for a method) revealed that they control Cdr2 KA-1 clustering, a specific property of Cdr2 KA-1 not shared by Kcc4 KA-1 [11].

4 From the Computer to the Bench: Designing Strategies to Unveil the Function of Your Protein

The tools presented so far may provide you with hypotheses for the function of your protein or protein domain that need to be validated experimentally. We will expose here a series of simple molecular biology techniques that may allow you to do so.

4.1 Isolating and Exchanging Protein Domains

An easy approach to test globally the function of a given domain is to isolate it or to exchange it for the domain of a well characterized orthologous protein whose function is already established, producing a chimeric protein. This can be easily achieved by simple PCR to isolate a domain or by using a fusion PCR method to produce chimeric proteins. To avoid folding issues, care should be taken to

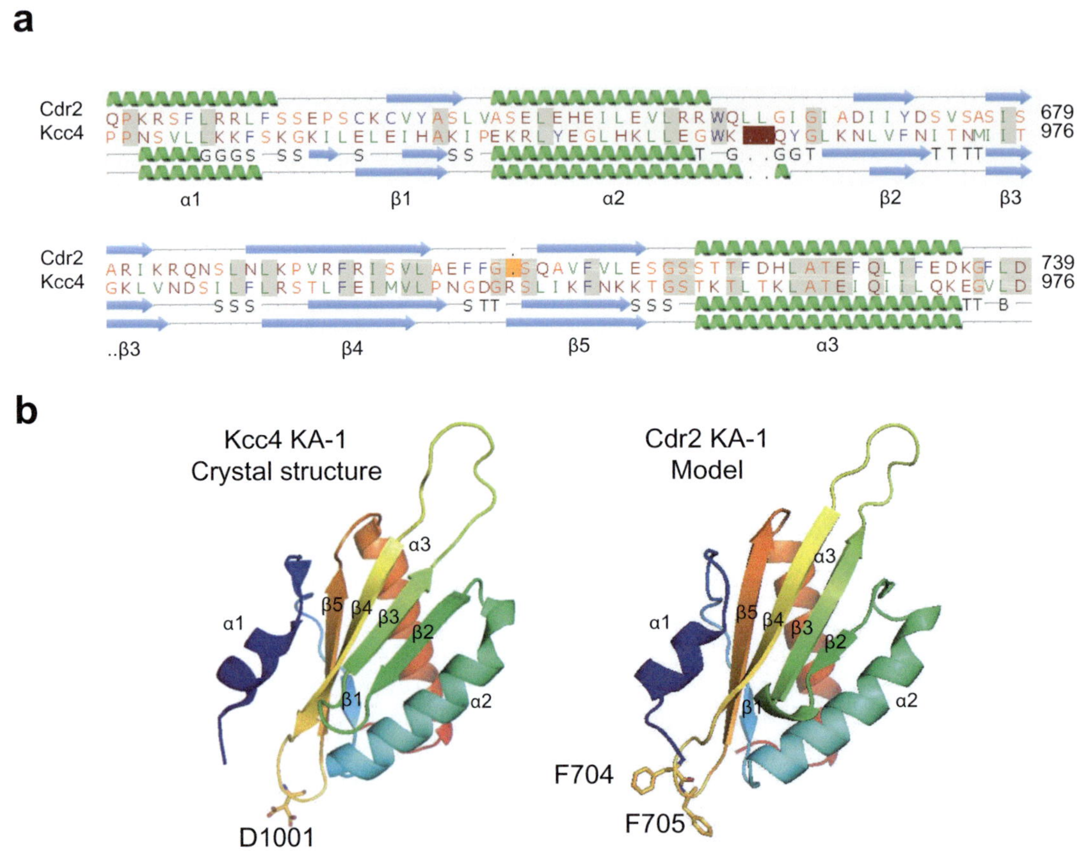

Fig. 2 Cdr2 KA-1 structure prediction and comparison to Kcc4. (**a**) Alignment of Cdr2 KA-1 domain with Kcc4 KA-1 using Phyre2. The predicted secondary structures are shown *above* and *below* the respective sequences and are compared to the secondary structure of Kcc4 KA-1 solved by crystallography (*bottom lane*, 3OST). *Green helix*: α helix; *blue arrows*: β sheet. (**b**) 3D model of Cdr2 KA-1 domain produced by Phyre2. Note that two hydrophobic F residues in β4-β5 loop of Cdr2 KA-1 are not conserved on the homologous loop of Kcc4 KA-1 which contains a negatively charged D residue

include complete domains in the fusion, based on alignments data, secondary structure predictions, or 3D models if available. Technically, one PCR is designed to produce a fragment encoding a region of the first protein ending by a ~20 nucleotides sequence identical to the beginning of the fragment encoding the domain of the second protein to be included in the chimera. A second PCR product just encodes the domain of the second protein. Then, in a second round of PCR, these two PCR products anneal in their overlapping region to prime a third PCR product corresponding to the chimeric gene that combines domains of the two proteins.

For example, we produced constructs encoding the isolated C-terminus of Cdr2 and Kcc4 including the KA-1 domains. This revealed that while both domains exhibit lipid binding properties

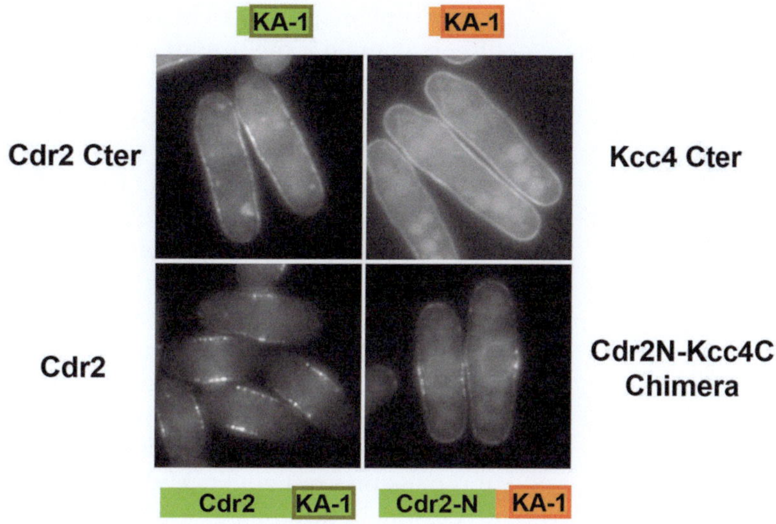

Fig. 3 Localization of Cdr2 and Kcc4 constructs. Localization of Cdr2 and Kcc4 C-terminal regions containing KA-1 domains, of full length Cdr2 and of a chimera containing Cdr2 N-terminus and Kcc4 C-terminal region. All constructs are fused to GFP at the C-terminus. Note that while both C-termini bind to the cell cortex, only Cdr2 C-terminus shows clustering properties. The Cdr2N-Kcc4C chimera reveals the clustering properties of Cdr2 N-terminus

only Cdr2 KA-1 possesses clustering properties (Fig. 3). A chimeric protein between Cdr2 N-terminal regions and Kcc4 KA-1 region further showed that Cdr2 N-terminus carried additional clustering properties that could only be revealed in the absence of Cdr2 KA-1 (Fig. 3) [11].

4.2 Creation of Point Mutations

As exemplified above with Cdr2 KA-1, multi-sequence alignments or domain modelling can pinpoint specific residues that may play a conserved or divergent role in the function of a specific domain compared to homologous domains. Mutagenesis of these residues is the best way to assess their actual function. Point mutagenesis by PCR amplification of a vector containing the gene of interest with oligonucleotides that include the desired mutations is traditionally used although this method may sometimes have a low efficiency. An alternative is to use the fusion PCR method mentioned above. In this case, two independent PCRs are performed to amplify the two halves of the gene to be mutated using overlapping internal oligonucleotides in the region of the desired mutations that include the actual mutation that needs to be introduced.

For example, mutagenesis of basic (Arg or Lys) residues from Cdr2 KA-1 surface and of a nearby non characterized basic domain to non-charged residues (Asn or Gln) allowed us to show that these residues participate in lipid binding and that the two domains cooperate to ensure a high avidity for membranes [11].

4.3 Performing Systematic Deletions

Some orphan proteins do not contain any conserved domain. Others contain a mixture of conserved domains and large regions with no domain prediction, or large disordered regions whose functions are difficult to assess. This is an indication for creating series of internal deletions that may help deciphering crucial regions of the protein carrying its localization or function, if functional tests are available. These internal deletion constructs may act as separation of function alleles of genes encoding proteins that combine several domains of independent function. Secondary structure prediction should be taken into account in the design of the deletion to avoid disrupting helices and β sheets that may fold in domains. These deletions can be produced using fusion PCR in a similar way as described above for chimeras.

We used this method to dissect the function of the N-terminus of the fission yeast anillin Mid1 that associates with the cytokinetic precursors in interphase, where it recruits in turn cytokinetic ring components to initiate contractile ring assembly at mitotic entry [3]. We created a series of 10 sequential 50 amino acid long deletions. Two deletions allowed us to identify interaction sites for the medial cortical node components Cdr2 (Mid1Δ400-450), and Gef2 (Mid1Δ300-350) [7, 14]; one identified a region targeted by the Polo-like kinase at mitotic entry to create an efficient interaction site for Rng2 that in turn allows Myosin II recruitment (Mid1Δ50-100), [15]; a fourth one (Mid1Δ450-506) identified a region promoting Mid1 nuclear import in addition to the classical C-terminal NLS sequence [7].

4.4 Expressing Mutants in Yeast in Replacement of the Endogenous Gene

The construction of an efficient integration module inserted in a plasmid can expedite the production of strains expressing the mutants of interest in replacement of the endogenous protein. This integration module should include 5′ and 3′ UTR of the gene of interest that will act as efficient recombination sites for the integration of the mutant gene at the endogenous locus. The mutant gene or gene fragment is inserted between the 5′ and 3′ UTR, placing it under the control of the endogenous promoter of the gene. It is followed by a terminator of transcription and a selection marker like the widely used KanMX6 cassettes for resistance to G418 antibiotic. This integration module is released from the plasmid by restriction enzyme cleavage prior to transformation using a classical yeast transformation method [16]. Its insertion at the original locus ensures that the mutant gene is present in single copy in the genome.

5 Concluding Remarks

This chapter exemplifies how the thorough analysis of the sequence and structure prediction of a protein of interest can help designing experiments to test its function experimentally. Cytokinesis is a process that requires modular structural, scaffolding, and signaling proteins that can greatly benefit from this type of previous-to-the-bench work. Since modular proteins control many other cellular processes, this guide may also be useful outside the cytokinesis field.

Acknowledgements

We would like to thank Carlos Vazquez de Aldana for very helpful discussions. A.P. is supported by Agence Nationale de la Recherche, Ligue contre le cancer Comité de Paris and Fondation ARC pour la Recherche. A.P. is a member of Labex CelTisPhyBio, which is part of Idex PSL*. S.A.R. received postdoctoral fellowships from Fundacion Ramon Areces and Marie Curie FP7 program.

References

1. Goyal A, Takaine M, Simanis V, Nakano K (2012) Dividing the spoils of growth and the cell cycle: the fission yeast as a model for the study of cytokinesis. Cytoskeleton (Hoboken) 68(2):69–88. doi:10.1002/cm.20500
2. Lee IJ, Coffman VC, Wu JQ (2012) Contractile-ring assembly in fission yeast cytokinesis: recent advances and new perspectives. Cytoskeleton (Hoboken) 69(10):751–763. doi:10.1002/cm.21052
3. Rincon SA, Paoletti A (2012) Mid1/anillin and the spatial regulation of cytokinesis in fission yeast. Cytoskeleton (Hoboken) 69(10):764–777. doi:10.1002/cm.21056
4. Altschul SF, Gish W, Miller W, Myers EW, Lipman DJ (1990) Basic local alignment search tool. J Mol Biol 215(3):403–410. doi:10.1016/S0022-2836(05)80360-2, S0022-2836(05)80360-2 [pii]
5. Young PG, Fantes PA (1987) *Schizosaccharomyces pombe* mutants affected in their division response to starvation. J Cell Sci 88(Pt 3):295–304
6. Morrell JL, Nichols CB, Gould KL (2004) The GIN4 family kinase, Cdr2p, acts independently of septins in fission yeast. J Cell Sci 117(Pt 22):5293–5302. doi:10.1242/jcs.01409, jcs.01409 [pii]
7. Almonacid M, Moseley JB, Janvore J, Mayeux A, Fraisier V, Nurse P, Paoletti A (2009) Spatial control of cytokinesis by Cdr2 kinase and Mid1/anillin nuclear export. Curr Biol 19(11):961–966. doi:10.1016/j.cub.2009.04.024, S0960-9822(09)00985-3 [pii]
8. Moseley JB, Mayeux A, Paoletti A, Nurse P (2009) A spatial gradient coordinates cell size and mitotic entry in fission yeast. Nature 459(7248):857–860. doi:10.1038/nature08074, nature08074 [pii]
9. Martin SG, Berthelot-Grosjean M (2009) Polar gradients of the DYRK-family kinase Pom1 couple cell length with the cell cycle. Nature 459(7248):852–856. doi:10.1038/nature08054, nature08054 [pii]
10. Moravcevic K, Mendrola JM, Schmitz KR, Wang YH, Slochower D, Janmey PA, Lemmon MA (2010) Kinase associated-1 domains drive MARK/PAR1 kinases to membrane targets by binding acidic phospholipids. Cell 143(6):966–977. doi:10.1016/j.cell.2010.11.028, S0092-8674(10)01308-5 [pii]
11. Rincon SA, Bhatia P, Bicho C, Guzman-Vendrell M, Fraisier V, Borek WE, Alves Fde L, Dingli F, Loew D, Rappsilber J, Sawin KE, Martin SG, Paoletti A (2014) Pom1 regulates the assembly of Cdr2-Mid1 cortical nodes for robust spatial control of cytokinesis. J Cell Biol 206(1):61–77. doi:10.1083/jcb.201311097, jcb.201311097 [pii]
12. McWilliam H, Li W, Uludag M, Squizzato S, Park YM, Buso N, Cowley AP, Lopez R (2013) Analysis Tool Web Services from the EMBL-EBI. Nucleic Acids Res 41(Web Server issue):W597–W600. doi:10.1093/nar/gkt376, gkt376 [pii]
13. Kelley LA, Sternberg MJ (2009) Protein structure prediction on the Web: a case study using the Phyre server. Nat Protoc 4(3):363–371. doi:10.1038/nprot.2009.2, nprot.2009.2 [pii]
14. Guzman-Vendrell M, Baldissard S, Almonacid M, Mayeux A, Paoletti A, Moseley JB (2013) Blt1 and Mid1 provide overlapping membrane anchors to position the division plane in fission yeast. Mol Cell Biol 33(2):418–428.

doi:10.1128/MCB.01286-12, MCB.01286-12 [pii]

15. Almonacid M, Celton-Morizur S, Jakubowski JL, Dingli F, Loew D, Mayeux A, Chen JS, Gould KL, Clifford DM, Paoletti A (2011) Temporal control of contractile ring assembly by Plo1 regulation of myosin II recruitment by Mid1/anillin. Curr Biol 21(6):473–479. doi:10.1016/j.cub.2011.02.003, S0960-9822(11)00166-7 [pii]

16. Bahler J, Wu JQ, Longtine MS, Shah NG, McKenzie A III, Steever AB, Wach A, Philippsen P, Pringle JR (1998) Heterologous modules for efficient and versatile PCR-based gene targeting in *Schizosaccharomyces pombe*. Yeast 14(10):943–951. doi:10.1002/(SICI)1097-0061(199807)14:10<943::AID-YEA292>3.0.CO;2-Y, [pii]10.1002/(SICI)1097-0061(199807)14:10<943::AID-YEA292>3.0.CO;2-Y

Index

Alberto Sanchez-Diaz and Pilar Perez (eds.), *Yeast Cytokinesis: Methods and Protocols*, Methods in Molecular Biology, vol. 1369, DOI 10.1007/978-1-4939-3145-3, © Springer Science+Business Media New York 2016

G

H

I

K

L

M

N

P

Q

R

S

MIX
Papier aus verantwortungsvollen Quellen
Paper from responsible sources
FSC® C105338

If you have any concerns about our products,
you can contact us on
ProductSafety@springernature.com

In case Publisher is established outside the EU,
the EU authorized representative is:
Springer Nature Customer Service Center GmbH
Europaplatz 3, 69115 Heidelberg, Germany

Printed by Libri Plureos GmbH
in Hamburg, Germany